DIGITAL ENTERPRISE CHALLENGES

IFIP - The International Federation for Information Processing

IFIP was founded in 1960 under the auspices of UNESCO, following the First World Computer Congress held in Paris the previous year. An umbrella organization for societies working in information processing, IFIP's aim is two-fold: to support information processing within its member countries and to encourage technology transfer to developing nations. As its mission statement clearly states,

IFIP's mission is to be the leading, truly international, apolitical organization which encourages and assists in the development, exploitation and application of information technology for the benefit of all people.

IFIP is a non-profitmaking organization, run almost solely by 2500 volunteers. It operates through a number of technical committees, which organize events and publications. IFIP's events range from an international congress to local seminars, but the most important are:

- The IFIP World Computer Congress, held every second year;
- open conferences;
- working conferences.

The flagship event is the IFIP World Computer Congress, at which both invited and contributed papers are presented. Contributed papers are rigorously refereed and the rejection rate is high.

As with the Congress, participation in the open conferences is open to all and papers may be invited or submitted. Again, submitted papers are stringently refereed.

The working conferences are structured differently. They are usually run by a working group and attendance is small and by invitation only. Their purpose is to create an atmosphere conducive to innovation and development. Refereeing is less rigorous and papers are subjected to extensive group discussion.

Publications arising from IFIP events vary. The papers presented at the IFIP World Computer Congress and at open conferences are published as conference proceedings, while the results of the working conferences are often published as collections of selected and edited papers.

Any national society whose primary activity is in information may apply to become a full member of IFIP, although full membership is restricted to one society per country. Full members are entitled to vote at the annual General Assembly, National societies preferring a less committed involvement may apply for associate or corresponding membership. Associate members enjoy the same benefits as full members, but without voting rights. Corresponding members are not represented in IFIP bodies. Affiliated membership is open to non-national societies, and individual and honorary membership schemes are also offered.

DIGITAL ENTERPRISE CHALLENGES

Life-Cycle Approach to Management and Production

IFIP TC5 / WG5.2 & WG5.3 Eleventh International PROLAMAT Conference on Digital Enterprise – New Challenges November 7–10, 2001, Budapest, Hungary

Edited by

George L. Kovács
Computer and Automation Institute
Hungary

Peter Bertók
School of Computer Science and Information Technology
RMIT University
Australia

Géza Haidegger
Computer and Automation Institute
Hungary

KLUWER ACADEMIC PUBLISHERS
BOSTON / DORDRECHT / LONDON

Distributors for North, Central and South America:
Kluwer Academic Publishers
101 Philip Drive
Assinippi Park
Norwell, Massachusetts 02061 USA
Telephone (781) 871-6600
Fax (781) 871-6528
E-Mail <kluwer@wkap.com>

Distributors for all other countries:
Kluwer Academic Publishers Group
Distribution Centre
Post Office Box 322
3300 AH Dordrecht, THE NETHERLANDS
Telephone 31 78 6392 392
Fax 31 78 6546 474
E-Mail <services@wkap.nl>

 Electronic Services <http://www.wkap.nl>

Library of Congress Cataloging-in-Publication Data

A C.I.P. Catalogue record for this book is available from the Library of Congress.

Printed on acid-free paper.

Printed in the United States of America.

CONTENTS

Life-Cycle Engineering

Process and Production Development

Energy Sector, General Planning

Digital Factory

Virtual Enterprises and Holonic Manufacturing

Decision Support in Engineering + Genetic Algorithms

Supply Chain Management - Manufacturing Management

Multi-Agent Systems, Manufacturing Control

Robotics

Index of contributors

PREFACE

The triennial PROLAMAT conferences focus on computer applications in manufacturing, and have already a respectable history. The first conference in Rome, Italy in 1969 and the next in Budapest, Hungary, addressed the issue of Programming Languages for Machine Tools, and created the acronym PROLAMAT. While the acronym has not changed, the topic has shifted considerably. The conferences have addressed Advances in CAM, Software for Discrete Manufacturing, Human Aspects of CIM and Globalization and Virtual Enterprises; the aim of PROLAMAT 2001 is to demonstrate activities and results that will characterize manufacturing in the 21st century. The conferences have covered a wide geographical area, and have travelled around the globe. They left the European continent for Glasgow, Scotland U.K. in 1976, then went on to Ann Arbor, Michigan, U.S.A. in 1979, followed by Leningrad, U.S.S.R. in 1982, and after visiting some European cities it arrived at Tokyo, Japan in 1992. The last few conferences went back to Europe.

The information technology revolution of the 1980s and 1990s made a strong impact on production technology as well as on its management, and many of today's enterprises can rightly be called digital enterprises. The pervasiveness of computers and networks has enabled the full computerization of production, and further on the whole supply chain. Decision support systems and other management support tools have also become more and more widespread.

As the web of computer networks gradually expanded, it opened up the possibility of global, computer assisted production, management, etc.. Different stages of the design and production process can now be conducted at different parts of the globe, and still be overseen by real-time control and management. Networked activities are not restricted to one enterprise: virtual enterprises allying different companies for the lifetime of a project are heavily relying on computer-enabled communication, cooperation and coordination, and business-to-business (B2B) commercial activities via electronic communication are becoming everyday practice.

The computing power of today's equipment has reduced the calculation time of modelling and simulation to such an extent that new approaches using computation intensive methods, techniques and algorithms have become feasible to employ. Using modelling and simulation tools a wide range of alternatives can be examined quickly to provide support for decisions. Design and manufacturing time has also been reduced, which allows faster response to market needs.

The 2001 PROLAMAT focuses on one of the greatest new challenges facing these digital enterprises: Life Cycle Approach in Management and Production. In an increasingly environment conscious world manufacturing and production is regarded as part of a larger picture: the product life cycle (production – use – disposal). This approach is characterized by three aspects: technology, economy and ecology (environmental impact). The PROLAMAT conference focuses on technology but also includes papers on the other two aspects; various solutions for the different activities are described in the papers. While elements of integrating several phases of the product life cycle have been apparent in earlier approaches, such as Design for Manufacturing, Design for Maintenance etc., the life cycle approach looks at the different phases from a holistic aspect.

Modelling can help in calculating the cost of a product, both in the manufacturing stage and in later stages of the product's life. As cost can be a decisive factor in a product's success, support tools can be used to facilitate the examination of economy aspects. Other uses of modelling include analysis and simplification of complexities in the supply chain that includes life cycle dependencies. Using the life-cycle approach a product model can be reverse engineered from the intended use and the given set of constraints, and the optimum design is derived from the desired outcome.

The principal message of the conference is that engineering has to embrace the whole life-cycle of the product, which includes environmental, social and economic sustainability of the production – supply chain customer use and disposition phases. Life cycle thinking integrates many different processes to give a complete view; from product design to decommissioning and disposal.

The importance of the topic was evidenced by the large number of papers submitted to the conference. Due to space limitations only some of them were selected for publication. The papers included in this volume were considered by the program committee to provide the best insight into these issues, and describe solutions for the problems. There were a number of people assisting in the editing process and organising the conference, and we wish to thank everyone who contributed to the success of this conference for their efforts. We hope that this conference will lead to further IFIP TC5 activities in the area, and which will also bring people from industry, research and education closely together.

The Editors

Integration of Human Intent Model Descriptions in Product Models

László Horváth
Budapest Polytechnic, John von Neumann Faculty of Informatics
e-mail: lhorvath@zeus.banki.hu

Imre J. Rudas
Budapest Polytechnic, John von Neumann Faculty of Informatics
e-mail: rudas@zeus.banki.hu

Carlos Couto
University of Minho
ccouto@dei.uminho.pt

Abstract: Advanced product model based style of engineering design together with internet based globalized project work relies upon advanced communication of model data. Model entities and their parameters represent decisions of engineers. Present day models of mechanical systems do not contain data about the background of human decisions. This situation motivated the authors at their investigations on exchange design intent information between engineers. Their concept was extending of product models to be capable to describe design intent information. Several human-computer and human-human communication issues were also considered. The complex communication problem has been divided into four sub-problems, namely communication of human intent source with the computer system, representation of human intent, exchange of intent data between modeling procedures and communication of the represented intent with humans. The paper is structured as follows. Firstly, main objectives of the research, earlier results and an approach to describe design intent in product models are outlined. Following this, utilization of human-computer procedures in modeling of design intent is characterized. Then main structure of model description of design intent is detailed. Finally design intent description in product modeling environment, is discussed and illustrated by examples.

Key words: Product modeling, Modeling of design intent, Model representation, Human-computer interaction.

1. INTRODUCTION

Engineering workplaces within a work group are situated in different geographical sites. Advanced, customer oriented product design involves frequent revision of decisions. Modification of a decision also requires modification of related decisions. A prerequisite of this activity is availability of information on the background of the original decision. This is why modification of an original decision requires contribution of the original decision maker or an other authorized engineer. The authors investigate application of computer description of design intent to replace original decision maker in product modeling. They proposed a computer method to assists the modification of decisions by modeling of design intent.

In advanced CAD/CAM systems successful efforts resulted excellent software tools for description of relationships between entities within the product model [3]. Logical continuation of these initial steps is mapping intent description entities to product model entities. This is the main content of the work presented in this paper. The authors investigated computer representation of design intent in advanced modeling systems during their earlier works [1]. They revealed characteristics and description of design intent [2]. Instead of individually created experimental environment, they decided to utilize commercial modeling systems that contain open surfaces for development tasks similar to intent modeling.

The paper is structured as follows. Firstly, main objectives of the research, earlier results and an approach to describe design intent in product models are outlined. Following this, utilization of human-computer procedures in modeling of design intent is characterized. Then main structure of model description of design intent is detailed. Finally design intent description in product modeling environment, is discussed and illustrated by examples.

2. AN APPROACH TO DESCRIPTION OF DESIGN INTENT IN PRODUCT MODELS

Design intent modeling is suitable for description and handling of all information and knowledge in the background of product related decisions.

A modeling system with engineering workplaces situated in different geographical sites was considered. A process parallel with product modeling has been established for modeling of design intent.

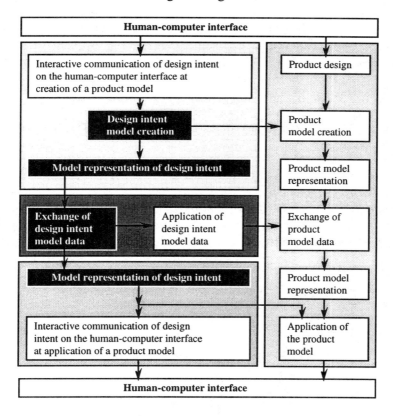

Fig. 1 Creating and processing of design intent

Creating and processing of design intent are outlined and related to normal product modeling activities in Fig. 1. Intent entities are created during product modeling in close connection with engineering decisions for product design. These entities are represented and related to results of the product design process then used by other product modeling and product model application procedures as well as humans. Modeling systems in the practice have various model representation capabilities and sets of entities. In the course of model data communication between different modeling systems, model data often should be translated and converted. This conversion is assisted by design intent information. Reference modeling and standard model description languages as EXPRESS make it possible to achieve integration of intent models into product model.

Four sub-problems have been identified in modeling of design intent. The first is gathering intent information in the course of interactive graphic communication between human and computer at creation of a model. The second is representing design intent information in a model and its mapping to other product model entities. The third is communication of intent model entities between concurrent engineering modeling systems of different representation capabilities. The fourth is how intent description can be communicated with an engineer on graphic user interface at an application of a model.

3. HUMAN-COMPUTER PROCEDURES AND DESIGN INTENT

Procedures for human-computer interaction (HCI) are being adapted and enhanced for effective communication based design intent modeling. Modeling of design intent utilizes knowledge acquisition, data access and authority control, functions for human interactions as well as analyses in human behavior, human-human communication and human error (Fig. 2).

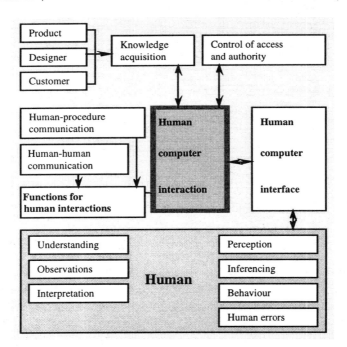

Fig. 2 Human related issues

Design intent often can be defined in some form of knowledge. The related knowledge is not only domain but also product, designer and customer related. Consequently, it can not be involved in the modeling system in one of the conventional ways. Members of a group of engineers who are working on a product design can describe or point to different knowledge sources. Sometimes knowledge is of personal nature. In other cases, access to given knowledge sources is allowed for several engineers in a work group.

Quality of an engineering decision depends on the performance of the human decision maker. This performance can be increased effectively by using of computer based decision assistance. At the beginning of her or his career an engineer is not well trained. Computer description of intents of skilled engineers can support beginners at their decision making. Computer can learn decisions of skilled and experienced engineers in order to support decisions of less skilled engineers. Moreover a threshold knowledge can be described and then used for the purpose of excluding untrained or careless engineers from decision making and avoiding fatal errors. This approach is cited as mutual adaptive human-computer interfacing [4].

A design intent often is based on observations. Design or planning tasks are solved during cognitive processes where results of decisions are found out by designers. In similar cases it is very difficult to describe the human thinking process. Sometimes variants for types, parameter value ranges or discrete parameter values can be defined in intent descriptions. The authors considered modeling of design intent for model variants besides the a single decided variant. This makes it easier to change the decision to a more appropriate one during the application of the original product model. Understanding a given situation can be improved by appropriate intent descriptions from other engineers. This is the case of human-human communication using intent description.

Equally important is the role of design intent description at prevention, detection, diagnose and correction of human and processing errors. Error related behaviors of humans seriously affect their decisions [6]. Creating and handling of design intent are determined by cognitive behavior of humans. Perception and inference functions are equally essential at reasoning based decision making of a human. Improper design of a human-machine system is often resulted in erroneous human activities [5]. A correct model of functions of humans who are in an interaction with computer procedures helps in understanding human-computer procedures. On the other hand role of computer generated suggestions in product design is important in avoiding errors. Engineers communicate computer systems through advanced graphic interface systems [7].

4. MODEL DESCRIPTION OF DESIGN INTENT

Entity parameters affected by an engineer decision can be defined as constraints and it can be related by constraints to other parameters of other entities. The proposed design intent modeling method offers different ways for definition of an intent at creating of a product model entity. Besides simple recording of intent information in intent model entities an intent creation procedure can work in close connection with the model creating procedure (Fig. 3). As an active way of intent creation, designer can be asked for intent information concerning a current decision. Similarly to the case of product model entity creation, authors suggest the application of the feature based concept for creating, relating and mapping intent model features. Intent feature libraries can be attached to product model feature libraries.

Easy to access background information packages can be linked to product model entities rather than involving them in model data structures by the application of Internet and related technologies. Large intent description data sets can be replaced by simple links in this manner. Remote sources of intents are maintained and actualized by engineers who created them.

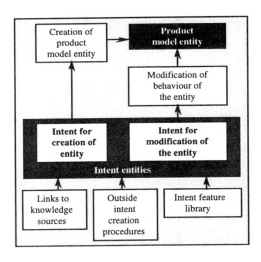

Fig. 3 Product model and intent entities

Remote access to background information is of outstanding importance when the applier of an intent is not authorized to use the knowledge source. Security measures are needed when running of a remote program is initiated at the application of the model. Industrial engineering practice produces very complex situations. As an example, members of a group of engineers can work on several design tasks for different companies. This style of work of engineers as well as the legislation background are to be taken into account at handling of design intent. An intent can be a simple identification of the decision maker such as the management of a company or an authority without any explanation.

Intent description is often created in the form of history of the background of a decision by using of a chain of intent entities. Fig. 4 shows a typical chain of design intent entities and several referred entities from the outside world. List of referred entities acts in the intent description as information source needed at processing of the intent description. Referred entities are accessed using links included in the intent description. History is considered as a chain of explanations for stages of a decision. Generic product models can involve generic intent models.

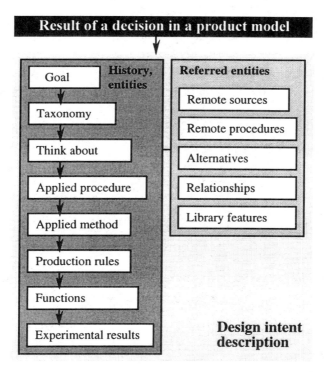

Fig. 4 Structure of a design intent

In the case of the example of Fig. 4 firstly a goal is defined for the decision then the related taxonomy is revealed. This is followed by a consideration of the applied procedure as a thinking process of the engineer.

In other words she or he does not apply any idea from handbooks, etc. Next, the applied method is selected taking into consideration of the choice that is offered by the selected procedure. Alternative procedures and methods can be involved or referred if necessary. The procedure needs input data that were defined using production rules, functions and experimental results. The origin of the experimental results is an important element of the intent description. Fig. 4 shows only one of the possible styles of history.

5. INTENT DESCRIPTION IN PRODUCT MODELING ENVIRONMENTS

The authors assumed utilization of inherent intent description capabilities of open architecture object oriented concurrent engineering systems as a starting point for their work. Concurrent engineering systems with professional software modeling tools serve human-computer communication, creating, storage and multiple access of models, and application of models for further development, analysis, production planning and demonstration. Open surfaces of object oriented concurrent engineering systems are used to access modeling procedures, model data and graphic user interface through application programming interfaces (API).

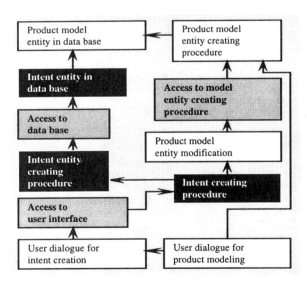

Fig. 5 Intent handling in an open architecture modeling system

A sketch of implementation of the proposed intent modeling is given in Fig. 5. User dialogue for product modeling is completed with dialogue for intent creation. Intent creation can be initiated automatically on the basis of the just defined model data. For example a computer procedure may ask the human that: "You defined an entity or its parameter. Describe why is it your decision and what to do if this decision should be changed by other engineers". In other cases intent creation is initiated by an engineer who describes: "This is why I decided as I did. If somebody should make some changes on this decision do the following..."

Access to user interface open system functions is necessary to establish communication between engineers and the intent creation procedures. The just defined intent entity can be used for modification of work of product model creation procedures. This is an active application of intent description. Active role of an intent can also be programmed for applications of product models. On the other hand engineers utilize intent descriptions at applications of product models. This is a passive application of intent model. Moreover, intent description can serve as a measure or even a command. Product model creating procedures and data bases are accessed from intent model handling procedures by open access functions of industrial modeling systems.

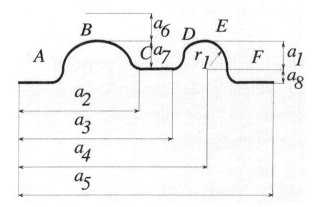

Fig. 6 Design intent at the creation of a section for creating a complex surface

Several typical considerations at modeling of design intent in the practice of modeling of mechanical parts are illustrated and discussed below through examples.

Complex surfaces are often created using complex curves that govern their shape. Fig. 6 shows a section of a complex surface that consists of six component surfaces (*A-F*). Structural requirements were considered then surface types and dimensions were decided. Relationships of dimensions were defined then values of some dimensions were calculated. Fixed values of dimensions were defined as constraints. Other dimensions are allowed to be modified within well-defined ranges during applications of the model. These ranges were not described in the part model. Instead, one of the possible values was defined as a constraint. To assist later modification of the related decisions and to prevent later changes of the dimensions to an illegal range, the allowed ranges of dimensions were recorded as design intent entities. Component surfaces A and F should be flat and component surface E should be cylindrical. All component surfaces are described using rational B-spline functions that offer free modification of their shape. Consequently, type of A, E and F surfaces should be constrained to avoid their modification as free form surfaces. There is a relationship between dimensions $A4$ and $R1$. Elements of the intent description for the above discussed example are:

- *Decision on relationship A_4-R_1*. This decision is based on a method by an expert. Several alternative solutions were defined. They were involved in the intent model.
- *Shape of component surfaces B and D*. Modifications of these free form surfaces are allowed. Although dimensions $A_{1\ min}$ and $A6$ limit the modification of these surface components.
- *Shape of flat component surface C*. This surface can be modified as a free form surface if mating surfaces of other parts are modified accordingly. Cost consequences can be estimated by a procedure that is related to the intent description and can be accessed through Internet.
- *Fillet surfaces* connecting component surfaces can be modified. Continuity between the connected surfaces must be maintained.

Description of design intent using the method proposed by the authors assists at the task of styling of the above detailed surface complex without changes that are not allowed by mechanical and manufacturing engineering. Styling engineer may create additional intent entities for additional styling and manufacturing planning procedures. A styled surface often produces problems or excess cost at manufacturing. A minor modification of the surface may result in cutting of the manufacturing cost dramatically. For this purpose design intent may involve description of allowable ranges of control point positions.

Fig. 7 illustrates basic relationships within and between parts of a mechanical assembly. The parameter model in the model of the *part 1* includes the relationship $A2=3,1\ A1$. The intent model includes a method that was used for creation of this formula. Sometimes parameters from the application environment determine parameters of some entities in the product design stage. At the same time quality of a design may be deteriorated by modifications that seem logical but contradict to the purpose of the product. Design intent description is essential in these cases and can be utilized at effective modification of the design. Engineers considered dimensions *A3* and *A4* as limited only by the contact constraints defined at the faces *F1* and F2. An other limiting factor can be the minimum allowed value of the gap between *part 1* and *part 2*.

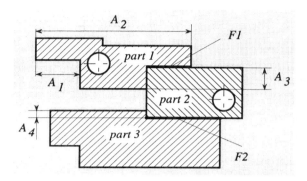

Fig. 7 Mechanical assembly

Is an intent count for much? Similar questions affect application of intent information. The answer is often subjective. Fortunately, decisions can be assisted by the description of origin, limits, strength and consequence of omitting of an intent. On the other side, engineers sometimes are not able to produce a good explanation for their decision. This is because intuition and anticipation are still important factors at decisions of skilled and experienced engineers. An intent description acts on behalf of the engineer who created it.

6. CONCLUSIONS

The authors proposed a method for modeling of the background of decisions using design intent descriptions. They proposed a method for mapping design intent model entities to product model entities. Intent modeling describes human thinking process and considerations behind human decisions. Intent description can be used by both model application processes and humans who handle them. Implementation of the proposed method assumes object oriented industrial modeling systems with open surface. Intent entities referred from the outside world are applied to link intent information to intent model that are not economical or not allowed to involve in the intent models. The proposed method can be considered as an extension of collaborative engineering. Outstanding importance of its application is where the original decision makers are not accessible and the only assistance at evaluation or modification of an earlier decision is the application of design intent description. Intent modeling is not modeling of what should be done by an engineer but what did an engineer and what are consequences of possible future work of other engineers. It can be considered as a well-prepared substitution of engineers by computer procedures.

ACKNOWLEDGMENTS

The authors gratefully acknowledge the grant provided by the OTKA Fund for Research of the Hungarian Government. Project number is T 026090. Authors also would like thank Hungarian-Portugal Intergovernmental Science & Technology Cooperation Programme for the financial support.

REFERENCES

[1] László Horváth, Imre J. Rudas: Attaching Knowledge to Product Model for Representation of Human Intent, *Proceedings of the 1997 IEEE International Conference on Systems, Man and Cybernetics*, Computational Cybernetics and Simulation, Volume 2, Orlando, Florida, USA, 1997, pp. 1580-1585.

[2] László Horváth, Imre J. Rudas: Representation of Human Intent in Computer Aided Engineering Design, *Proceedings of the Second IEEE International Conference on Intelligent Processing Systems*, Gold Coast, Australia, 1998 , pp 424-428.

[3] Kimura, F. - Suzuki, H.: A CAD System for Efficient Product Design Based on Design Intent, *Annals of the CIRP*, Vol. 38/1, 1989, pp:149-152.

[4] Yoshikawa, H. - Takahashi, M.: Conceptual Design of Mutual Adaptive Interface. *Preprint of Integrated Systems Engineering Conference*, Baden-Baden, 1994, pp. 221-226.

[5] Jose M. Nieves, Andrew P. Sage: Human and Organizational Error as a Basis for Process Reenginering: With Applications to Systems Integration Planning and Marketing, *IEEE transactions on Systems, man and Cybernetics Part A: Systems and Humans*, Volume 28, No. 6, 1998, pp 742-762.

[6] Pietro Carlo Cacciabue: A Methodology of Human Factors Analysis for Systems Engineering: Theory and Applications, *IEEE transactions on Systems, man and Cybernetics Part A: Systems and Humans*, Volume 27, No. 3, 1997, pp 325-339.

[7] L. J. Haasbroek: Advanced Human-Computer Interfaces ant Intent Support: A Survey and Perspective, 1993, IEEE International Conference on Systems, Man and Cybernetics, Lille, 1993, pp. 350-355.

Product Data Management as a Key Component of Integrated Enterprise Information Systems

Dr. J. X. Gao* and Mr. G. Bursell **
*School of Industrial & Manufacturing Science, Cranfield University, Bedford, MK43 0AL, UK
** Mabey & Johnson Ltd., Lydney Industrial Estate, Lydney, Glos. GL15 4EJ, UK.

Abstract: Faced with increasingly competitive markets, the need to reduce lead times and the need to react to customer demands more effectively, manufacturing resource planning (MRPII) systems and computer-aided design (CAD) have become commonplace in manufacturing companies.   These are, however, implemented in isolated and autonomous locations and without thought to their integration.   Where integration is achieved technically, operational, philosophical and organisational issues have often presented an inaccurate and uncoordinated solution resulting in sustained bottlenecks between design, manufacturing planning and production processes. This document presents a case study of a manufacturing company facing these issues in a technology driven manufacturing strategy and proposes a framework to resolve them, based around matured Product Data Management (PDM) technologies. Learning the lessons from earlier direct application integration, the notion that integration of these systems is not only an issue of data but also business processes, as manufacturing companies embrace collaboration with their suppliers, customers and suppliers is also presented.   This requires control, visibility and flexibility throughout enterprise information systems.

Key words: product data management, PDM, computer-aided design, CAD, manufacturing resource planning, MRPII, integration

1.   INTRODUCTION

The increasing use of CAD and design automation solutions generates vast amounts of data and documents.   The maturing of Product Data Management (PDM) as a means of managing this is becoming increasingly prevelant along with philosophies like Concurrent Engineering in manufacturing companies. However, bottlenecks are still being maintained between design and manufacturing with high degrees of redundancy, manual data entry and re-entry, duplication and errors. The role of PDM has changed, today playing a strategic role in removing these bottlenecks and promoting a single integrated solution with manufacturing resource planning (MRPII) systems.

2.   INFORMATION SYSTEMS' INTEGRATION

2.1   Design and Manufacturing Information Systems

The use of CAD at almost all stages of a typical design process is today widespread, from initial concept and specification through analysis and manufacture. Product Data Management (PDM) has evolved to handle the increasingly large amounts of data and complex product and process relationships generated by engineering design automation with CAD and remove the weakness in manufacturing planning systems of handling master and structural data during early cycles of the design process and even through product development where there are inherent uncertainties. A PDM system is a software tool used to solve manufacturers problems in managing and controlling their engineering data, especially related to new product introduction and engineering processes and, more recently, throughout a product's entire lifecycle [1]. The system comprises three functional categories: the Vault, User functions and Utility functions [2].

Computerised manufacturing planning systems have been in use since the early 1960s, initially as a method of determining and planning material requirements for production. As new tools were added, including production planning, master scheduling, capacity requirements planning and the ability to execute materials and capacity plans, Materials Requirements Planning (MRPII) systems evolved, highly integrated systems optimising materials, procurement and manufacturing processes and providing financial reports. The final evolution, into today's Enterprise Resource Planning (ERP) systems, includes finance, human resources, distribution and traditional accounting functions.

The cornerstone of both of these systems is the bill of material (BOM) and item master records. The BOM, or product structure, documents both part relationships and quantities as well as establishing the logical sequence of events in the manufacturing domain, whist, driven from CAD, it defines the finished physical hierarchical product structure used throughout an enterprise in engineering [3]. The product structure is reliant on the underlying part specification, or item master. From CAD the item master identifies and catalogues drawings whilst in the manufacturing domain it describes and documents what is constantly true about a part. Together, these elements form the basis around which integration revolves, effective only within operational and business processes [4].

2.2   Models of Integration

The bill of material or product structures are created and maintained separately in both engineering and manufacturing domains. This often involves a high degree of redundancy, manual data entry, duplication and errors. The scenario of maintaining separate engineering and manufacturing product structure representations has largely been maintained throughout interfacing and integration efforts [5], even though widely adopted philosophies such as Concurrent Engineering promote collaboration between design and manufacturing departments, processes and people in delivering tangible benefits. The co-existence of PDM and MRPII systems is a widely held key requirement in enterprise-wide requirements for manufacturing companies [6] however their integration provides an extensive range of benefits.

Whilst CAD integration with PDM systems is today classed as a commodity, there are a number of considerations in integrating PDM with MRPII systems [7]. Issues such as two master databases and data ownership contention are commonplace. Three typical models of integration can be identified [8] [9], either system becoming a single master data source or, as is more commonly found, item master and product structure information simultaneously exists in both systems, each mutually updating the other through formal engineering change mechanisms. Manufacturing systems are not well suited to the latter as access control mechanisms are required to progagate changes real-time. Where this has been achieved, any interruption in communication with the PDM system has the potential for BOM updates not being fed into manufacturing. The favoured approach is to create and modify BOMs in the PDM domain and feed them into manufacturing through change orders [10], with process workflow particularly well suited to this approach. More often, batch updates propagate changes over night, yet this does little for companies who require regular MRP runs and who would have inaccurate data. Rather, integration bespoke to each company's requirements. The methods of connectivity between PDM and MRPII systems are shown in Figure 1, showing both traditional dependency and abstraction, which have been greatly effected by recent technology developments. In bridging legacy MRPII systems, a gateway application is often favoured.

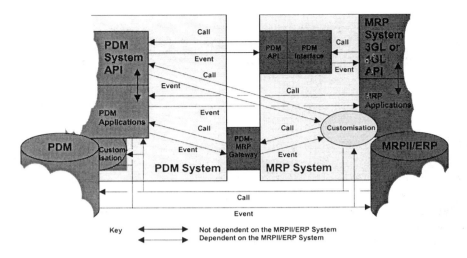

Figure 1. Integration connectivity between PDM and MRPII.

2.3   Contemporary Developments

There are encouraging signs that advances in information technology and related standards are finally delivering integration solutions. Both PDM and MRPII systems are embracing object-orientation, on top of operating system platform with object-based application programming interfaces, such as Microsoft's Component Object Model (COM) and ActiveX interfaces through to object integration frameworks like CORBA and distributed objects such as DCOM. The traditional restrictions of rebuilding interfaces when system configurations change or upgrades are installed are being replaced by integration techniques of providing a layer of abstraction outside of the core systems with integration specific APIs which provide high flexibility with low overheads [11].

However, many manufacturing companies maintain legacy second-generation MRPII systems, with closed 3GL APIs. Database standards such as SQL are removing the restrictions of application integration with these systems, where supported, yet the real developments in recent times have come with platform and system independent standards like Standard for the Exchange of Product Data (STEP) and the explosion of e-commerce with hypertext markup language (HTTP) and Extensible Markup Language (XML) [12] being exploited for integration. The XML specification has elicited support from many MRPII, ERP and PDM vendors with many such systems providing license free and platform independent 'connectors' [13] to major integration partner systems like SAP, Matrix, Baan and Sherpa. XML heads the re-emergence of neutral file formats for integration, applying a messaging protocol, which encapsulates structured data and metadata as well as content based on data model definitions, or schemas, held in Microsoft's Biztalk or RossettaNet frameworks. In addition, Biztalk Server provides messaging, and workflow capabilities to expand existing applications and information systems beyond their existing boundaries.

Further developments of note are the issues of scaleability in implementing PDM systems, often to world-wide locations. The development of standards and communications protocols today makes issues of platform dependence redundant.

3. CASE STUDY

3.1 Company Background

Mabey & Johnson Ltd is a privately owned company specialising in the design and manufacture of modular prefabricated steel bridging systems, which are supplied primarily from their facilities in Lydney, Gloucestershire to customers worldwide. Modular bridging systems are used in a variety of applications including temporary, permanent, floating and emergency bridging requirements. They are easy to transport and can be erected with unskilled labour with minimal engineering plant. Some 90% of the company's production is exported to over 130 countries, either ex-stock or to the client's own specification in a rapidly growing world market. This is especially prevalent in developing countries and those emerging from armed conflict that are expanding or repairing their transportation infrastructures.

The company has invested heavily into highly automated manufacturing technologies from which production throughput is predominantly in the form of standard products, fabricated by robots. These make up 80% of all production; the remainder being engineered to order, which is, processed through manual welding facilities and the use of sub contract manufacturers. The manual welding and fabrication facilities are maintained to accommodate non-standard requirements and to supplement the automated fabrication processes. As well as sales of bridging systems, a certain stock level for hire is maintained enabling the company to quickly react to requirements. Products with long manufacturing lead times are held as finished stock and demand is often volatile and highly uncertain.

Identifying rapid customer respfonse, high quality and flexibility as critical business success factors, the company commenced a rolling investment program in 1995 in the development of an enterprise-wide Management Information System (MIS), which is now nearing completion

[14].   This combines state-of-the-art manufacturing and information technology to reduce design and manufacturing lead times, including a reliance on 3D solid modelling CAD and Computer-Aided Engineering (CAE) tools. In a technology driven manufacturing strategy, the provision of an Engineering Support System (ESS) framework to centralise and coordinate activities and information between the various departments, processes and computerised engineering and manufacturing systems has been identified as the key facilitator to realise the full benefits of already highly effective islands of automation.

3.2   Proposed Engineering Support System framework

The proposed Engineering Support System (ESS) framework described is the result of an investigation into the availability of suitable information technology solutions, integration standards and technologies and operational business requirements. The purpose of this framework is to directly bridge support applications to the manufacturing planning and production domains and provide the foundation of computer-aided manufacture (CAM) integration, areas that were not considered in the development of the specialised hybrid planning framework implemented [14].   During this period a vast array of technologies, standards, suppliers and vendors were systematically reduced to a shortlist, however no integral solution existed and, as a consequence, a high degree of customisation and application development exists in all scenarios to present a solution. As a consequence, the ESS framework is constructed in four parts, supporting the operational focus of the company highlighted in the planning framework, material management and manufacturing execution by providing core level data and document control, process control, product structure management and engineering and manufacturing data and process integration. The foundation of the ESS framework is a PDM system, positioned as a layer around the manufacturing planning system to control the flow of information both in and out of this domain. An overview of the ESS framework is shown in Figure 2 and is described in subsequent sections.

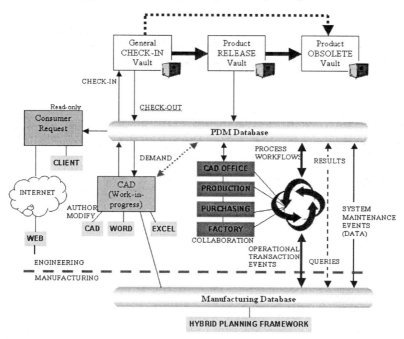

Figure 2. Proposed Engineering Support System framework.

3.3   Data and Document Control

The ESS framework will provide a coordinated engineering package of associated documents and data generated throughout a products entire life cycle.   Document and data classification will provide two categories, identifying documents and data used directly in production, such as fabrication drawings, and associated supporting documents and data, such as analysis results from produced early in the product design process.   All documents are revision controlled, with all changes tracked and the complex relationships and dependencies that exist managed to enable the engineering data packaging. This package would then be released to production.

Document storage is within a centralised vault, with security-controlled access.   This is an important issue, as considerable reworking of CAD models has been found due to a lack of control. A multiple vault topology will be utilised, separating released documents and data from checked-in work in progress and obsolete storage vaults.   This will also provide the archiving requirements for ISO 9000 quality standards compliance.   The engineering package is moved to the released vault upon approval, however, should any element of the package require revision, the entire package contents are moved back to the checked-in vault as associations and dependencies will be maintained across multiple vaults for control. This also means an impact analysis can be performed prior to changes being effected. The accessibility of the engineering packages is dependent on the roles of system users within the ESS environment. For instance, shop floor operators will have print (read-only) capabilities whereas CAD operators will have creation, modify and obsolescence capabilities.

Extensive search and retrieval capabilities, impacting throughout the ESS framework will vastly reduce the time taken to find information and documents in a centralised information repository as well as reducing duplication and redundancy in part creation. A plethora of uncoordinated hardcopy, computerised databases and manual storage mechanisms are currently in use. Automated coordinated part creation will be implemented within the ESS framework.

3.4   Process Control

The engine of the ESS framework is PDM-based workflow. This will be used to initiate, manage and track business processes; primarily approved release to manufacture, distribution and engineering change control. Processes will be modelled graphically within the PDM system in a flow chart format. The workflow definition is held within the PDM data model, allowing process flows to be adapted on the fly to changing requirements without the need to recompile a process before it can be used. Workflow augments business processes by packaging and routing associated data and documents between various stages, via approval or rejection, through departments and people in the enterprise. Traditional paper documents, such as engineering change request forms, will be amalgamated as data within an engineering change process workflow, collected or input through on-screen forms, matching company documentation and procedures. Workflow is not a linear process; indeed iterative, concurrent and serial activities will be found in almost all business processes. Adopting workflow as the core engine of the ESS framework will remove superfluous steps and inefficient activities in existing processes, as this will negate many of the benefits of this integration.

Visibility will be raised within business and operational processes, allowing workflows to be tracked, audited and archived and provide complete traceability in engineering as well as in manufacturing. Complex processes will also be handled as process workflows can also be nested and contain rule-based event triggers (decisions) to improve routing speed in a controlled manner. Manual processes can also be recorded and escalation policies will be enforced to ensure processes such as drawing approval, a problematic bottleneck, are completed. Combining drawing and manufacturing job information, such as planned start date, within drawing approval workflows, for instance, will enable problematic processes to be prioritised and, if necessary, escalated, to ensure all engineering information is released by the time production commences. User notification, found at various stages of workflows, is integrated with electronic mail messages and provides feedback and action requirements to users involved in workflows.

The role of workflow and process control in the ESS framework environment will ensure that if something needs to happen, it happens by the time it is needed. Conversely, if a process is not completed on time, visibility highlights who or what the bottleneck is. Engineering data, often dynamic in nature, is of no use to manufacturing if it does not arrive on time, yet in contrast; manufacturing data is relatively static and defined in detail. This fact underlines why processes must be managed to be effective. Errors, duplication, redundancy and lead times can all be reduced, yet a coordinated and centralised engineering package available to manufacturing will facilitate the high levels of flexibility required in manufacturing through adopted workflow processes.

3.5   Product Structure Management

The Configuration Management (CM) aspect of PDM in the ESS will manage two revision controlled product structure (BOM) views, or as-designed and as-planned configurations. The creation of product structures is exclusively performed in the PDM domain, utilising synchronised item masters from the manufacturing system. This is to enable the traditional push of data from engineering into manufacturing and is enforced as a series of business rules within operational process workflows. The development of the manufacturing product structure (MBOM), incorporating process routes, evolves from the intrinsically captured CAD-based engineering product structure (EBOM) far earlier in the design process, coupled with CM to manage changes that occur over time. Collaboration at an early stage will allow purchasing and planning functions to happen more as a consequence than as a demand and will allow impact analyses of engineering change requests to be preformed.

Product structures are built graphically, providing greater adaptive flexibility and visibility and are consolidated to provide the enterprise-wide product structure definition, although multiple configurations will inevitably exist in product development. The consolidation of engineering and manufacturing product structures is shown in Figure 3, providing coordinated effectivity controls for engineering change requests and orders.

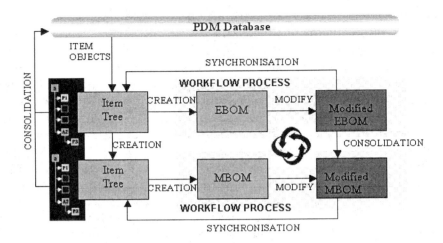

Figure 3. Engineering and Manufacturing BOM creation and control.

Manufacturing planning functions such as rough-cut capacity planning and procurement are significantly improved as a completed product structure is generated through the workflow transport in a distributed collaboration.

3.6   Data and Process Integration

Engineering and manufacturing data and process integration will take place at three levels between the PDM and manufacturing systems, however a feature of the ESS framework is that no direct integration or tethering of the two core systems takes place, rather a bi-directional messaging mechanism is implemented. Both systems are regarded as mutually controlling masters in the integration, driven through workflow processes. Both domains will work independently of each other, maintaining their own local data sources, utilising workflow processes to drive interaction.

While no workflow or CM capabilities exist in the manufacturing system domain, interaction between the two systems is implemented through XML messaging, encapsulating not only messaging content, but also metadata describing the content, structure, definition and context. Using Microsoft's Biztalk Server platform, integrated with PDM workflow, request messages are mapped to actions in one system, redirected through the Biztalk framework and received back as response messages in the second. Examples of this include a request for a completed and approved drawing from the manufacturing system following changes to the master production schedule to begin production and effect changes or create new component and assembly items. This would activate a process workflow in the ESS to progress the request and return the response with associated data. In order for this to be an effective approach, business data models and processes will require a complete definition but only need to be implemented once.

The three levels of messaging integration identified are:

1. *Process Integration.* Also described as operational transactional events, moves multi-part data flow processes between each system. Examples include engineering change requests and part identification. Message information is extracted and mapped to workflow nodes. This will be implemented by API integration between Biztalk Server and PDM workflow to transparently extend process workflows outside the PDM system and into the enterprise. The data requirements of these processes are held in schemas. Changes to a business process will therefore only require a change to the schema definition and routing of the process, as opposed to additional API integration. Irrespective of the computing platforms, Biztalk hands off messages to both systems but this is reliant on business processes being explicitly defined in both systems.

   Specific processes, such as engineering change, require external coordination. This will be performed by a user with detailed product and process knowledge as well as who utilises both systems and will involve dynamically re-routing workflow processes, reviewing and tracking process status and making strategic decisions to maintain optimum efficiency of the integrated systems' functions. Process integration will also allow engineering demand to be presented and scheduled outside of the manufacturing system to meet demand and production commencement lead times, especially following changes to the master production schedule in the manufacturing planning system.

2. *Data Integration.* In addition to triggered process flow events, the positioning of two perceived mutual master data sources requires maintenance transactions for synchronisation with transaction logging.

3. *Query Integration.* Principle information consumption in all areas will be through a web browser interface, where manufacturing job-related information, such as works instructions and process plans, is associated with the engineering package. This level of presenting ad-hoc read-only information from the manufacturing domain will be implemented through a COM OLE-DB service provider, implemented through visual basic API scripting on demand for fast data retrieval.

Non value-added functions are removed to reduced server-side processing, plying business processes and data to perform integrated functions as a consequence of operational processes, not in spite of them as has traditionally been the effect of integration.

4. CONCLUSIONS

In order to achieve the required levels of integration between PDM and MRPII systems, both core data and documents as well as business processes must be effectively managed together. The use of PDM systems and independent standards now affords companies with legacy MRPII systems the same benefits today's leading edge ERP systems provide.

REFERENCES

1. KIRBY, H. PDM – What do you need to know? *Works Management*, Vol. 52(5), 1999, 25-28.
2. CIMDATA INC. Product Data Management: The Definition. An Introduction to the Concepts, Benefits and Terminology. CIMdata Inc., 1996.
3. FRANK, D. N. The Business Case for Interoperability. *APICS*, Vol. 12(5), 1999, http://www.apics.org/magazine/May99/interoperability.htm
4. RUSK, P. S. The role of bill of material in manufacturing systems. *Engineering Costs and Production Economics*, Vol. 19(1-3), May 1990, 205-211.
5. HASIN, M. A. A. AND PANDEY, P.C. Analysis of present scheme of MRPII-CAD integration: problems and prospects. *International Journal of Computer Applications in Technology*, Vol. 9(2-3), 1996, 126-130.
6. BARTULI, G. AND BOURKE, R. W. The Best of Both Worlds: Planning For Effective Coexistence of Product Data Management and MRP II Systems. February 1995, http://www.pdmic.com/articles/bestbth1.html, http://www.pdmic.com/articles/bestbth2.html
7. GREGORY, A. Linking PDM systems to manufacturing. *Manufacturing Computer Solutions*, Vol. 4(4), April 1998, 34-36.
8. Miller, E. PDM/ERP REPORT: Integrating PDM and ERP. *Computer-Aided Engineering*, March 1999, http://www.caenet.com/res/archives/9903pdmerp.html
9. CANAL, B. The PDM – MRP Link. Proceedings of the Technical Program. NEPCON East '95. On Target with Today's Technology and Tomorrow's Challenges, 12-15 June 1995, Boston, USA, 84-92.
10. HALL, G. PDM: Integration challenge. *CADCAM*, Vol. 17(7), July 1998, 27-28.
11. FERMAN, J. E. Strategies for successful ERP connections: Techniques and their technologies for integrating systems. *Manufacturing Engineering*, Vol. 124(4), 1999, 48-60.
12. GOULDE, M. A. Microsoft's Biztalk Framework adds messaging to XML. Proposed framework for XML schemas and exchange of data. *E-Business Strategies and Solutions*. September 1999, 10-14.
13. MAINWARING, J. EAI - sticking it all together. *Manufacturing Computer Solutions*, Vol. 6(2), 2000, 28-30.
14. BEACH, R, MUHLEMANN, A. P, PRICE, D. H. R, PATERSON, A. AND SHARP, J. A. Information systems as a key facilitator of manufacturing flexibility: a documented application. *Production Planning & Control*, Vol. 9(1), 1998, 96-105.

Organizational Transformation and Process Modeling in Modern Telecommunications Companies

D.M. Emiris, D.E. Koulouriotis, Z. Chourmouziadou, V. Moustakis, N. Bilalis
Department of Production Engineering and Management, Technical University of Crete
73100 Chania, Crete

Abstract: The telecommunications industry has presented a healthy growth over the recent years, and generated combined revenues of approximately 700 billion Euros in 1995 worldwide, while it is estimated that in 2005, this number will rise to 970 billion Euros in the European Union alone. This growth is directly associated, apart of the strong direct dependence on informatics and digital technology, to the international policies followed, as regards the transition from a monopolistic environment towards a liberalized one, through various deregulation stages. Despite the obvious benefits for the end-user, which are further emphasized as a deregulation process evolves, the telecommunications market providers must face a continuously changing environment, which necessitates the research, selection and adoption of new technologies, the flexibility to adapt to rapid market changes through focused and flexible marketing programs, the agility to provide customized services, and the ability to integrate services, solutions, and equipment for the end-users. A critical step in the attempt to meet the above requirements is business modeling. The present work examines organizational methodologies for an entrant small-size telecom company, which would attempt to meet the above criteria. The study introduces the concept of process modeling in a general business environment and analyses the way in which a telecom company should be organized and structured. A SWOT (Strengths-Weaknesses-Opportunities-Threats) analysis initially defines the operational environment of the corporation. Case models are created as interim stages for the support of the interactions between the business entities. The external influences are enumerated and their role is highlighted. The business processes are then identified, analyzed and evaluated, and typical process flowcharts are generated and documented.

Key words: telecommunications, process modeling

1. INTRODUCTION

The telecommunications industry has presented a healthy growth over the recent years, and generated combined revenues of approximately 700 billion Euros in 1995 worldwide, while it is estimated that in 2005, this number will rise to 970 billion Euros in the European Union alone. This growth is directly associated, apart of the strong direct dependence from informatics and digital technology, to the international policies followed, as regards the transition from a monopolistic environment towards a liberalized one, through various deregulation stages. The EU, in particular, maintains a balanced policy between deregulation and obedience to the competition rules, and is based on several axes of priority, such as, the generation of new employment positions, the encouragement of technological innovation, the provision of the ability to the citizens to profit of the information society benefits, and the creation of conditions which will enable European corporations to strengthen their position in the international market.

Despite the obvious benefits for the end-user, which are further emphasized as a deregulation process evolves, the telecommunications market is a most competitive one for the involved corporations, not only the equipment manufacturers and technology generators, but also for the service provision ones. The latter face a continuously changing environment, which necessitates the research, selection and adoption of new technologies, the flexibility to adapt to rapid market changes through focused and flexible marketing programs, the agility to provide customized services, and the ability to integrate services, solutions, and equipment for the end-users. These factors often require the change of the operating philosophy for the incumbent telecom corporations, and certainly require a modern and flexible development strategy for new ones.

A critical step in the attempt to meet the above requirements is *business modeling*. This term is used herein to describe a symbolic language or protocol, both for representation and communication purposes, which includes graphical means (e.g. diagrams, flowcharts, other visual references) to depict inter-relationships and supporting means (e.g. reports). The need for business modeling stems from the need for effective knowledge management, efficient information flow within a company, support of basic processes and operations, and understanding the current conditions so as to enable future improvements [1,3]. There occur, however, certain conflicting concerns as regards the business modeling, namely: (a) the fact that the more complex the structure of a corporation is, the most complex, and thus more inflexible to changes, will be the business models that describe it; and (b) the fact that business models need to be quite detailed in order to be efficient and accurate, thus making them inefficient for information exchange and knowledge management.

The present work examines organizational methodologies for an entrant small-size telecom company, which would attempt to meet the above criteria. The study introduces the concept of process modeling in a general business environment and analyses the way in which a telecom company should be organized and structured. The common denominator in this attempt is the need to increase the productivity within a *lean* enterprise. The term *lean* is used to describe the ability of a corporation to produce added value at a low cost, so that the end-user receives an advanced service at a proper price. In this study, it is shown that the process-oriented management of a corporation enables to achieve a lean operation. Typical management schemes are contrasted and compared. The cost for the transition from an existing structured to a process-oriented structure, is investigated and identified.

The study applies the theoretical approaches to an actual telecom entrant. A SWOT (Strengths-Weaknesses-Opportunities-Threats) analysis initially defines the operational environment of the corporation. Case models are created as interim stages for the support of the interactions between the business entities. The external influences are enumerated and their role is highlighted. The business processes are then identified, analyzed and evaluated, and typical process flowcharts are generated and documented.

Several interesting conclusions result from this work. First, it is shown that for the successful representation of a company's operation, several software tools for process modelling and documentation are needed. It is recognized that, especially in the telecom market, today's corporations need to adopt a more flexible, process-oriented business structure. The analysis and modelling of business processes facilitates the identification and assignment of roles in an organization chart and enables a company to provide customized, client-oriented services through efficient processes, thus remaining effective and profitable in a competitive, rapidly-changing, disintermediative, digital techno-economic environment.

2. THE NEED FOR BUSINESS MODELING

The need to understand thoroughly and manage effectively the information in a company, has led to methods and techniques for the systematic organization of knowledge. A most important organization tool for knowledge and information is *modeling*. Modeling permits the discretization of knowledge and its presentation in a commonly perceivable manner, focusing in the importance of the human factor, as a company is inherently defined by the interactions between humans. In an attempt to identify the need for modeling in a company, it is important to stress the definition for a company. The nature of a company may be easily understood under the prism of relationships. A company (or an operation) may thus be defined as a closed network of interrelated persons, and is characterized by a unique identity. These relationships are responsible for the creation of the unique company identity and provide a coherence and integrity sense [2].

The identity of an operation, however, does not depend from the eventual persons: as long as the persons involved satisfy the given inter-relationship system, they can be any. On the other hand, if these relationships change, even if the people remain the same, the operation will also change; it will create a new identity. The identity of the company is directly related to its goal and its field of operation, thus permitting several companies to have similar identities. However, what makes a company unique is its structure, that is the way that the inter-relationships, which are developed within the company's frame, are shaped. In other words, it is the way a company is organized so as to achieve its goals. The need to understand in depth a company is thus translated to the need to depict and understand the inter-relationship system, from different perspective angles and different degrees of depth. Such an approach may be achieved through the modeling of information and the production of respective description reports [4,5,6].

The main obstacle in the attempt to deepen in a company's operation is the analysis and understanding of the interactions within the company. The use of a verbal code of communication in such an approach introduces a significant degree of fuzziness and cannot be considered efficient. On the other hand, the use of a specific terminology, although it may facilitate a department or a special team, by no means can meet the needs of the entire company and thus to be used for knowledge transfer. As a result, a technique for structuring a written code of communication in a way as to avoid fuzziness and to result in perceivable depiction forms is needed. In this context, business modeling is a widely used, efficient method to deepen into and to transfer information in a business environment.

Several reasons may trigger the modeling of a company's systemic components. The need to organize and manage systematically the information and knowledge (either provided or acquired) within a company, apart of being the main target of the *knowledge management* field, is the first reason, as it uses broadly modeling tools and emphasizes on the human factor. Secondly, the development of models within an operation may be used as a means for documentation and for personnel training in complex systems. The use of extensive documents, where processes and functions of an operation are described, is not the most recommended tools of knowledge transfer, as the length of a document may be an inhibiting factor in gathering the personnel's attention, as required. In contrary, the use of a diagram with a short descriptive report is much more likely to draw the attention. Since each person categorizes its duties with respect to the time that these require, by prioritizing them in an increasing order of time required, it is more probable that someone deals with a model rather than with an extensive document. Apart of this reason, the timesavings themselves are an important motive for development of business models, in order to communicate knowledge and information within a company.

Business modeling may also be triggered by the need to document the basic functions and processes that a company executes, as a risk prevention measure to protect the company from key people's leaving, moving, or substitution. Such a need arises each time a company's organizational structure undergoes a significant change, as in the cases, for example, of new product development, expansion to new markets, mergers, acquisitions, or buy-outs. In any respective case, the documentation of the basic processes and functions may offer a standardization pattern for the mode of operation and to create a basis for adaptation to changes, modifications or integration of new processes. Even in the case that a change in the organizational structure does not take place the processes documentation may be activated by the need to standardize the processes of product development and service provision.

Finally, the need to perform changes for the improvement of performance in a company may also trigger attempts to model the organization structure and the developed inter-relationships. Business modeling serves as a useful technique to conceive the present situation within a company and the faster examination of alternative improvement propositions. Through a model, several problematic points may be revealed, along with deficiencies and insufficiencies of the existing system. At the same time, the cost for the simulation of a system and for the examination of potential consequences caused by changes is much smaller compared to the recovery cost of a company after an unsuccessful deployment of improvement plans. The development of business models is thus encouraged before any attempt for re-organization or review within a company.

The development of business models is a fairly complicate task and it requires dexterities that should be developed. At the same time, there result three basic dilemmas as to the graphical representation of information, the nature of which needs to be understood, so as to efficiently deal with them. These dilemmas are further related to the diversification of the models developed. These dilemmas are:

- **The complexity dilemma**: *The more complex is the structure of an operation, the more complicate become the models that depict it. As a result, their margins of flexibility in description become narrower.* The answer to this dilemma is that the models should be kept as simple as possible. Any information that they carry must serve a purpose otherwise it should be omitted along with self-explanatory elements.

- **The control models dilemma**: *The more complex a model is, the more difficult is for its inherent information to be communicated and perceived; however, control cannot be made efficient if simple models are used.* The answer to this dilemma is that, on the one hand models should be kept as simple as possible by integrating only absolutely necessary information, but on the other hand, there should be a development within a company of such conditions as to permit the people who deal with monitoring and control to understand more complicate models. This may be achieved through the use of a common modeling code and continuous training.

- **The communication dilemma**: *The transmission of knowledge and information within a company requires communication models rather than complex control models or models of composite information systems; otherwise, communication is not successful.* The answer to this dilemma is that each model should be oriented to its user; that is, that a different model should be created for every possible user. In any case, a model is created in order to serve a purpose therefore, its role is predetermined. This role defines the complexity that each model should have, as well as the balance it should maintain.

3. COMPANY DESCRIPTION

The present work has examined organizational methodologies for an entrant small-size telecom company in an attempt to increase the productivity within a *lean* enterprise. The term *lean* is used to describe the ability of a corporation to produce added value at a low cost, so that the end-user receives an advanced service at a proper price. The steps outlined below have been followed in order to fulfill this task:

- The analytical organization plan has been developed and scrutinized, and the basic interfaces between the company's departments have been identified.

- The models for each business unit have been shaped along with the interactions among them, and between them and the external environment.

- The feasibility analysis, which in essence moves a company, has been implemented and the required entities that interact with the operation have been identified, as well as the job descriptions and their inter-relationships.

- The SWOT (Strengths-Weaknesses-Opportunities-Threats) analysis has been performed (Figure 1).

Strengths	**Opportunities**
• High quality technical infrastructure that enables processing of a large volume of simultaneous calls • Experienced personnel in telecom networks and troubleshooting • Call routing alternatives • Ability to provide services at better rates than our competitors • Ability to provide value-added services	• Ability to provide data services • Telecommunication network expansion and new markets deployment • Search for competitive prices for call termination
Weaknesses	**Threats**
• Absence of integrated marketing policy • Human resource responsibilities happen to overlap or generate gaps	• Fail to provide integrated telecommunication services • Competitive pricing policy • Loss of market share due to competitor's aggressive marketing strategy

Figure 1: *SWOT Analysis.*

The analysis of interactions, in any company, follows the principle that, for every department and process, one needs to produce outputs (e.g. products, services, etc.) using appropriate resources. The production of outputs requires the consumption or modification of input material, following certain rules or policies. As a result, assuming that in every operation and every department there exist certain processes executed, one can identify the above characteristics by studying the role of each process. These characteristics are: (i) the *inputs* (material, or information consumed or modified through the process), (ii) the *outputs* (materials, products, or information produced by the process), (iii) *mechanisms and resources* (key people, equipment, machinery, systems, etc.), and, (iv) *controls* (rules, policies, processes, limitations, orders, standards, etc., that define the process. These are depicted in Figure 2. Their fundamental property is that each characteristic may be further decomposed to a lower level of analysis, and vice versa, that is, each arrow in a specific level of analysis may result from the synthesis of partial characteristics of lower levels of analysis.

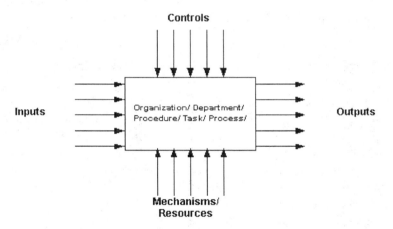

Figure 2: *Characteristics of Processes.*

The first step in the analysis of interactions includes the specification of the above characteristics for each process at every level, starting from the company as a whole and proceeding to the most detailed level. Such specification typically starts from the outputs, which are simpler to determine. Every process serves a specific purpose, which produces certain products. In the case where these products are not clearly specified or known, or they cannot be determined, then the process itself becomes questionable in terms of its need for existence. The second step is to determine the input data or material, which are consumed or transformed within the process context, so as to produce the outputs. Although rare, there exists the possibility that the input data and the output products are coincident; this would mean that the process does not produce an added value, and should be examined in terms of its need for existence. The difference between input data and output products may be specified by appropriate adjectives, which characterize the transformation that the process executes.

Once the input data are determined, the next step is to identify the mechanisms or resources, which are used for the process integration. The final step is the identification of the control under which a process is executed. All processes should be executed under at least one type of control. In cases where the identification of certain characteristics, such as inputs or controls, is fuzzy, then it is preferable for these to be characterized as controls as they constitute a form of input for every process.

The above type of analysis is depicted in Figures 3 and 4. In the first one, the case models (or main processes models) are shown along with their interrelationships for the company under study. *Case models* are interim models generated for the purpose of depicting cases and the way that business processes support certain interactions with external factors. In the second figure, one such main process, namely, the "New Customer Entry" process is further analyzed at a lower level, and the component processes are depicted along with their interactions and the interfaces with the outer environment.

Once the case models are created, they need to be decoded. Despite the fact that there is no established decoding method, in principle, inputs from the external environment are considered as data inputs or process activation commands. Similarly, outputs from the company to the external environment are considered as products or services generated from certain processes. Finally, the relationships between models signify interim products or services for inner customers or outer suppliers, or activation items for certain processes, such as the supply management when the supply control demonstrates insufficiency of the needs. For each case model, there may result one or more processes, which may be analyzed in turn, in subsequent levels of decomposition until simple tasks are determined.

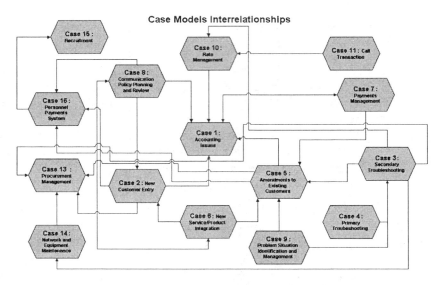

Figure 3: *Case Models Interrelationships.*

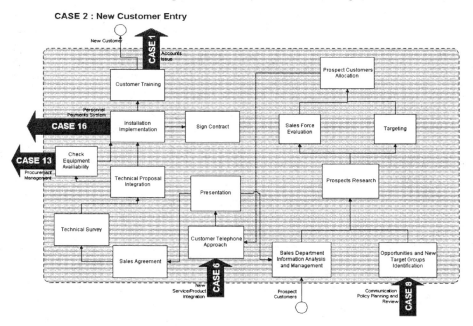

Figure 4: *Business Process Analysis Example.*

The depiction of processes is implemented using *Analytical flow charts*, in order to account for the temporal sequence of the processes. The temporal sequence is confined to the order of the particular processes and is defined by directed arrows. One such process, specifically, the *"Rerouting Decision Making Process"* is depicted in Figure 5.

10.1.1 Rerouting Decision Making Process

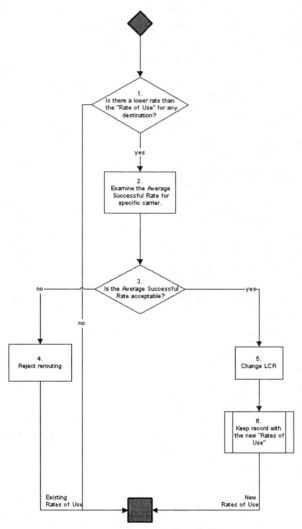

Figure 5: *Sample Process.*

4. CONCLUSIONS

The present work has examined organizational methodologies for an entrant small-size telecom company, which would attempt to meet the above criteria. The study introduced the concept of process modeling in a general business environment and analyzed the way in which a telecom company should be organized and structured. The study applied the theoretical approaches to an actual telecom entrant. A SWOT (Strengths-Weaknesses-Opportunities-Threats) analysis initially was used to define the operational environment of the corporation. Case models were created as interim stages for the support of the interactions between the business entities. The business processes were then identified, analyzed and evaluated, and typical process flowcharts were generated and documented.

Several interesting conclusions result from this work. First, it is shown that for the successful representation of a company's operation, several software tools for process modelling and documentation are needed. It is recognized that, especially in the telecom market, today's corporations need to adopt a more flexible, process-oriented business structure. The analysis and modelling of business processes facilitates the identification and assignment of roles in an organization chart and enables a company to provide customized, client-oriented services through efficient processes, thus remaining effective and profitable in a competitive, rapidly-changing, dis-intermediative, digital techno-economic environment.

5. BIBLIOGRAPHY

[1]   Armistead, C. and Rowland, P. *Managing Business Processes: BPR and Beyond.* Wiley, 1996.

[2]   Espejo, R. *et al. Organizational Transformation and Learning: A Cybernetic Approach to Management.* Wiley, 1996.

[3]   Johnson, Dr. and Edosomwan, A. *Organizational Transformation and Process Reengineering.* Kogan Page, 1996.

[4]   Scheer, A.-W. *Business Process Engineering, Reference Models for Industrial Enterprise.* Springer-Verlaag, 2nd Edition, 1994.

[5]   Scheer, A.-W. *ARIS-Business Process Frameworks.* Springer-Verlaag, 3rd Edition, 1999.

[6]   Scheer, A.-W. *ARIS-Business Process Modelling.* Springer-Verlaag, 3rd Edition, 1999.

Logical Communication Levels in an Intelligent Flexible Manufacturing System

János Nacsa
(nacsa@sztaki.hu)
Computer and Automation Research Institute (www.sztaki.hu)
Budapest, Hungary

Abstract: In most cases in intelligent manufacturing applications the communication functions depend on the capabilities of the intelligent tool (e.g. expert system). Three different types of working mode and different logical levels of the communication of an intelligent cell-controller in a CIM environment is shown in the paper. These levels are implemented - of course - within the same protocol. After this concept the paper explains the connection between these logical levels and the communication protocols and ontologies. Finally a simulation tool introduced in the second part of the paper was developed to examine the effects of the different communication messages and to analyse how the different type of messages can be measured in a real Flexible Manufacturing System (FMS) environment.

Key words: distributed artificial intelligence, knowledge communication, flexible manufacturing system

1.   INTRODUCTION

The practical problems of the communication of expert systems (ES) in CIM (Computer Integrated Manufacturing) applications can be divided into two parts. One is the hardware-software connection (physical) and the other one is the logical connection between the controller(s) and controlled devices. This decomposition was very useful both in the design and implementation phase during the last projects of our CIM Laboratory. If this decomposition is not so sharp many problems may occur during the development and specially in maintenance later on.

There are relatively easy programming interfaces (etc. C/C++) in most available ES shells. These interfaces provide data transfer with external tasks, stations, etc. They support clear and easy programming to reach objects, to call procedures, to set and get variables, etc. The interfaces are dedicated to specific software tools of the ES and they are general towards the external world without being able to take into account the requirements of the given application. So nearly each CIM implementation requires special software development to cover the gap between the external world and the ES.

2.   LOGICAL COMMUNICATION LEVELS

The communication functions depend on the capabilities of the expert system. The way of learning and knowledge handling determine the logical levels of the communication. Three different levels of the logical communication among intelligent controllers in a CIM environment were defined [6] as are shown in Fig. 1. Typically the messages of these three logical levels are coded in the same way, are using the same protocols and are mixed according to the communication features of a given manufacturing environment.

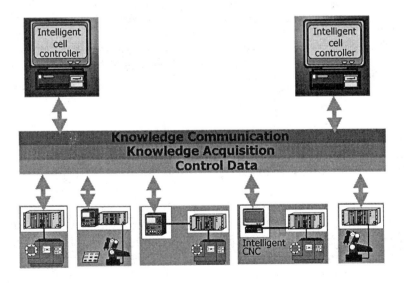

Figure 1. Three logical communication levels

- There is no intelligence in the CIM environment.
- All the AI solutions in the system are working individually, so there is no logical connection among them.
- The intelligent systems co-operate with each other in a hidden mode. Hidden intelligence means in this term that the knowledge based technology is applied only inside the cell controller and it has no specific actions via the communication channel. A typical example of the hidden case if a KB system is built up on the top of a traditional control system using its original communication.

On the higher logical levels the Knowledge Acquisition and the so-called Knowledge Communication levels were separated. These other two levels have messages if and only if the 'intelligence of the cell controller is not hidden'.

In the case of Knowledge Acquisition, the intelligent node/module initiates special data acquisition to update, validate or grow its knowledge. According to the general accepted terms: one can say about passive knowledge acquisition - where only existing data are collected remotely - and active knowledge acquisition - where also remote procedures are activated to get the necessary data.

When a KB system shares its knowledge (new or modified) it belongs to the Knowledge Communication level as Buta and Springer named it [1]. In this case the different intelligent nodes/modules ask/answer to each other as agents.

IT must be stated that the communication messages of most real and pilot KB based applications for FMS control belong to the lowest, possibly to the middle logical level.

3. KNOWLEDGE SHARING VERSUS LOGICAL LEVELS

In an environment where the different nodes want to share their knowledge, they have to "understand" each other. It is a big problem of the common understanding if the logical meaning of the different facts, objects and data are different in the different modules and subsystem. It is specially true in the manufacturing systems where the complexity of the individual elements are high and the terminology even among the different human field-experts is not the same. In the literature it is known as ontology problem [3].

Analysing the different knowledge sharing models and initiatives 4 different levels of ontologies can be distinguished:

1. There is no ontology - sub-symbolic or procedural representation is used in the module.
2. Closed ontology - the module can not share its knowledge with others.
3. Individual ontology - the module has a special - typically procedural - API to provide some achievement of its knowledge.
4. Junction ontology - the module can share its knowledge with others.

The Table 1. shows the correlation between the level of ontologies and the logical level of communication introduced here:

Table 1. Logical levels and ontologies

	closed	individual	junction
control data	+	unnecessary	unnecessary
knowledge acquisition	cumbrous	+	unnecessary
knowledge communication	not possible	cumbrous	+

It is clear that a nearly one-to-one mapping can be found, so the experimental levelling of the logical messages proves its usability.

4.   COMMUNICATION PROTOCOLS VERSUS LOGICAL LEVELS

In the world of FMSs there are many different types of protocols for different purposes, unfortunately there are more protocols than necessary. In this chapter the different protocols are examined in the communication point of view: how they determine the meaning of the messages.

The communication nodes can be active or passive; clients, servers or none of them (peer to peer). These categories are not related to the content of the communication and are handled typically in lower layer protocols. Three types of protocols can be distinguished:

- Traditionally CNCs and robots are closed from the network point of view. Their proprietary DNC channel allows only minimal information exchange.
- In the 90s many message based protocols came to the front that were based on the object oriented technology (e.g. CORBA) or similar (e.g. MMS). In these solutions in the network point of view the manufacturing devices seem to be virtual devices.
- The intelligent agent based experimental systems are using the speech act communication that has many features of human communication origin. The most important agent communication

language is the KQML (Knowledge Query and Manipulation Language [2]).

The main difference in the knowledge communication between the message based and speech act types of protocols is that the first one does not contain any information how to process the given message [4].

Table 2. shows again the correlation between the different types of protocols and the logical level of communication:

Table 2. Logical levels and types of protocols

	procedural	message based	speech act
control data	+	+	unnecessary
knowledge acquisition	+	+	+
knowledge communication	not possible	cumbrous	+

In this case the mapping is not one-to-one, but the levels approximately fit the types of protocols.

5. SIMULATION OF COMMUNICATION IN AN FMS

Introducing the logical communication levels they can be used to examine the distribution of the intelligence in a CIM environment. To start this evaluation a previously developed FMS simulation system [5] was utilised.

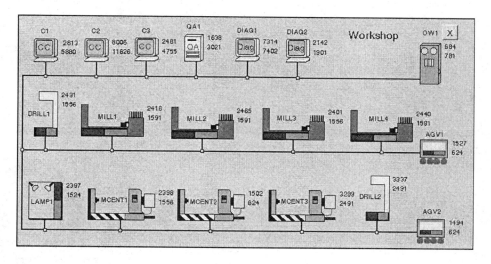

Figure 2. FMS used during the simulation tests

The experimental FMS (Fig. 2) contains 3 machining centers, 4 mills, 2 drills and 1 measuring machine served by 2 AGVs. This FMS is controlled via 3 cell controllers plus 1 quality assurance and 2 diagnostics PCs. The outside world is modelled by a single node. Some devices were set as "intelligent" either as knowledge acquisition or knowledge communication node or both. Different random sources filled the system with messages.

The effects of the different disturbances (machine down, quality problems, unexpected order etc.) could be analysed with this simulation system, and - among others - the flux of the different logical level messages can be visualised.

6. CONCLUSIONS

Three levels of experimental categorisation was introduced for the different manufacturing messages in a CIM environment. The correlation of these levels and the ontologies and protocols were analysed.

7. REFERENCES

1   Buta P, Springer S: Communicating the knowledge in knowledge-based systems, Expert systems with applications, 1992, 5, pp. 389-394
2   Finin T., Labrou Y., Mayfield J.: KQML as an Agent Communication Language. Software Agents, AAAI Press, 1997, pp. 2291-316
3   Gruber T.R.: The Role of Common Ontology in Achieving Sharable, Reusable Knowledge Bases, Proc. of 2nd Int. Conf.: Principles of Knowledge Representation and Reasoning, Cambridge, MA, pp: 601-602, Morgan Kaufmann, 1991
4   Kaula R.: Communication Model for Module-Based Knowledge Systems, in. Knowledge Based Systems, Academic Press, Vol. 1, 2000
5   Kovács G., I. Mezgár, S. Kopácsi, J. Nacsa, P. Groumpos: A hybrid simulation-scheduler-quality control system for FMS. In: Mechatronics. The basis for new industrial development. Proceedings of the joint Hungarian-British international mechatronics conference. Budapest, 1993. Southampton, Computational Mechanics Publ., 1994. pp. 655-662.
6   Nacsa J., Kovács G.L.: Communication problems of expert systems in manufacturing environment, in AIRTC'94 (ed. Crespo), pp. 377-381, Preprints of the Symposium on Artificial Intelligence in Real-time Control, Valencia, Spain, 1994, IFAC

8. BIOGRAPHY

János Nacsa is a research associate of the CIM Laboratory of the Computer and Automation Research Institute, Budapest, Hungary. He published more than 50 papers and led different research projects. His research topics are open systems and application of artificial intelligence in manufacturing.

NC Program Simulation with the Capability of Generating Alternative Process Plan for Flexible Manufacturing

Ferenc Erdélyi[1] and Olivér Hornyák[2]
[1]*PhD, Associate Professor*
[2]*Assistant Professor*
University of Miskolc, Faculty of Mechanical Engineering
Institute of Information Science, Department of Information Engineering

Abstract: This research effort focuses on improving the flexibility of production engineering for manufacturing industry by means of providing technological alternatives. A new generation of NC simulators required supporting the decision making of production engineering. In this paper a conception of NC simulator trying to fill this gap is given. This is a step forward realising Virtual Manufacturing.

Key words: CAPP, Robust NC Technology, Simulation, OOSE, Virtual Manufacturing

1. INTRODUCTION

In flexible manufacturing the applications of alternative and robust technology process plans play more and more important role in the practice of production management.

In a Computer Integrated Manufacturing (CIM) environment a Computer Aided Process Planning (CAPP) application is responsible for generating the process plans including NC part programs. Market demands requires these process plans to be optimised considering the actual business goals [1,3].

Optimised process plan is a selected plan of some alternatives that meets all the valid constraints and guarantees the extreme (minimum or maximum) of an objective function preferred by the production management. For example, time, cost, volume produced or their weighed portfolio could be one of these goals.

The optimal control theory suggests the use of robust control that assures less sensitivity for the changing of process parameters or constraints, and for the change of the goal function, in more general sense.

In analogy to robust control we define the robust technology process plan. It represents that kind of planned technology process, which is not sensitive to technology parameters or constraints, and forms a group of alternatives (population) from which the production management can easily select the appropriate one in accordance with the actual demand [4,6,7].

With CAPP (Computer Aided Process Planning) application tools, there are some new possibilities to make alternative process plans. These are as follows:

- Making alternative plans in the early period of product and process planning with concurrent engineering methods, in the pre-manufacturing phase. These serve for preparations to manage the changing environment during the life cycle of the manufacturing of the products.
- Making robust plans that exist in a single form but can easy be transformed accordingly to the different requirements of the actual production management goal. (This is an extended application of the processor – post processor principle proposed in APT (*Automatically Programmed Tools*) system at first 1955-58.
- Making plan-classes by means of integrated components that provide a possibility to generate plan object instances in the period of execution. (This is the principle of object oriented modelling methods.) The use of Group Technology (GT) and the intensity-based technology planning approach suggested by Hungarian researchers (*Tóth & Erdélyi*, 1997) can be regarded as this kind of methods [7].

2. GENERATING ALTERNATIVE NC PROGRAMS

In most factories, programming of NC machine tools is supported by computer applications (Computer Aided NC Programming) traditionally named CAD/CAM systems. These software applications support the geometric modelling of parts and tools as well as generation of tool paths resulting Cutter Location Data (CLD) files.

In case of conventional post-processing, the geometric model of the machining must remain unchanged; the generated NC program must fulfil the syntactical (formal) requirements of the available NC controller. The NC programming systems support this task by providing numerous postprocessor components. The up-to-date CAD/CAM systems offer the extension of technological capabilities. The volume and surface modelling, parametrical design, introduction of feature based geometrical modelling has become the base of creation of robust technological alternatives.

In the course of generating alternatives a certain number of Shop Floor Control (SFC) parameters can be used. The selection from these alternatives is an experience-based task. Some of these parameters are:
- accuracy parameters,
- tool parameters,
- tool path strategy,
- depth of cut in subprograms (in macros),
- technological intensity suggested by *Tóth*, 1988, (e.g.: in cutting operations: cutting rate or the rate of stock removal [8]),
- feed rate or cutting speed.

The comparison and the evaluation of the alternatives require some production performance indexes. Such indexes are the operation time and cost, the list of active tools, or the type of the machine (with the name of the postprocessor) etc.

There are two further significant techniques for making robust NC program alternatives that have possibility to create different performance indexes: These are as follows:
1. Creating alternatives in depth of cut structure or strategy of touring tool path.
2. Finding the optimal technological intensity in accordance with the different objective functions and constraints.

The first method has its advantage when using feature based programming macros. Fanuc, Siemens and other CNC controllers have had the capability of using this technique since the early '80s. This facility was introduced to support the Workshop Oriented Programming (WOP) and the user macro oriented manual programming. (e.g. the L95 macro family of Siemens SINUMERIC 810-840 which supports roughing cycle of turning operations). These macros have parameters that can be altered to create alternative part programs. The advantage of this method is that there is no need for re post-processing, and the alternatives are handled real-time by the executive tasks of the controller. However it has the disadvantage evaluating the objective function of the process posterior only.

The technological data (depth of cut, speed of cut, federate) can be involved to this task. New idea in this area is the optimisation of technological intensity that is applicable considering the objective functions and constrains coming from the production management in the early phase of process planning.

Generation of multiple NC programs is a time consuming task. When the depth of cut does not appear as a parameter then it can be calculate by actual geometrical states and actual motion co-ordinates. When a spindle speed is given then the cutting speed doesn't appear explicitly, it is a function of cutting diameter (or the tool diameter in case of milling and grinding).

The NC programming applications allows the user to define these parameters interactively. When the NC program is manually generated, the macros supported by the controller can be parameterised properly. The same situation is in case of WOP when a special descriptive language of a certain controller can be used.

3.   EXTENDING THE NC SIMULATION

The suitability of generated program variants must be examined. The most important criteria are the syntactical check and geometrical verification. Most of the NC programming systems includes simulator to verify the NC program with the use of graphical animation. However, these simulators do not offer information for MES (Manufacturing Execution System) except the operation time and some base data. It is noticeable that only the cutting time can be exactly calculated, the other times including tool changing times, rapid tool motion times, the set up, preparation and finishing times of a batch depends on the machine tool, the used tools, the alignment of tools etc.

The advanced flexible and adaptive Manufacturing Executions Systems requires the pre-estimation of further production parameters. Such as:

- operation cost,
- set up and operation times,
- cutting force, tool wear, tool cost,
- power of cutting process, consuming of energy, machine tool utilisation,
- prediction of dimensional accuracy, quality of surface, rate of waste product.

At present there are no simulators in the industrial practice, which can provide such services. The reason for this phenomenon can be found in the difficulties of modelling of machining operations.

Some of the indexes mentioned above are non-linear functions of the machining parameters. From Information Technology (IT) point of view these functions can be modelled using Artificial Intelligence (AI) techniques.

The research that has recently been in progress in the Department of Information Engineering at the University of Miskolc is targeted to the development of a new NC part program simulation tool for this advanced simulation. The main components of the simulator are as follows:

– NC part program code verification component
– Process simulator component, with
  • Process definition component
  • Process animation component
– Process analyser component
– Secondary Post Processor component

Figure 1. depicts the system architecture of the simulator with the most relevant interfaces to NC programming environment.

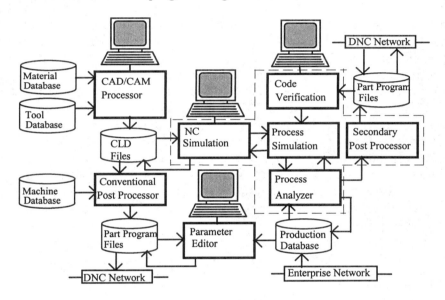

Figure 1. Components of extended NC simulator

4.   SOFTWARE ENGINEERING TOOLS FOR NC SIMULATION

Object-oriented (OO) programming has been revolutionizing software development and maintenance. When applied to simulation systems, OO programming also provides an opportunity for developing new ways of thinking and modelling.

In this section some software engineering aspects of the existing application and the future work is given.

During the Object Oriented Analysis and Design a set of classes has been worked-out for representing the geometric and technology entities of NC simulation. The use of OO technology makes it possible creating interfaces with these classes to other existing classes. It also supports software modules being under development or going to be developed in the future. The instances of these classes are the simulation objects involving all the information what an NC program can contain. Some aggregate management indexes of NC cutting operation (e.g. machining time, cost) can be evaluated as the time-integrated sum of certain attributes. Simulators in common use also provide graphic animation to check the NC program for semantic errors. Due to enclosing properties of objects each object is responsible for its representation during the animation. This also means that the attributes and member functions of an NC entity are suitable not only for modelling a real existing object but for supporting the software representation as well. This type of abstraction has a major effect on implementation and should be considered when mapping between objects and the process being modelled.

The process analyser of the NC simulator uses the management indexes coming from the process simulator. Most of them appear as non-linear functions of the cutting parameters. It is expedient to use AI methods for modelling those relationships. An Artificial Neural Network (ANN) module can be used to predict the cutting forces and tool life. Time related attributes could be directly predicted from the feed-rate and the tool path data of NC program. After computing these parameters, many other state variables can easy be estimated.

By means of the Secondary Post Processor component some new instances of the parameterised NC program class can be generated.

This introduces a new generation of NC simulation.

5.   CONCLUSIONS

The paper discussed the technological trends of Computer Aided Process Planning. One of these possibilities is the use of process simulation over geometric simulation of NC part programs. Process simulator can support the production management in decision making, when goals of business change frequently. This is a step for realising Virtual Manufacturing in the field of NC turning. This paper and the background activity contribute to these aspects.

6.   ACKNOWLEDGEMENTS

The research summarized in the paper has been continued within the framework of *Production Information Engineering Research Team (PIERT)* established in the Department of Information Engineering and supported by the Hungarian Academy of Sciences. The research has also been supported by FKFP project entitled *"Object Oriented Modelling of Manufacturing Processes"* (Id.no.: 0275, headed by *F. Erdélyi*). The financial support of the research by the aforementioned sources is gratefully acknowledged.

REFERENCES

1.   Chryssolouris, G. (1992): *Manufacturing Systems, Theory and Practice.* Springer, New York.
2.   Chryssolouris, G., Mavrikos, D., Fragos, D. & Karabatsan V. (2000): *A Virtual Reality Based Experimentation Environment for the Verification of Human Related Factors in Assembly Process.* Robotics and CIM. V.16. No.4. August: p. 267-277.
3.   ElMaraghy, H.A. (1993): *Evolution and Future Perspectives of CAPP.* Annals of the CIRP, Vol. 42/2: p. 739-751.
4.   Erdélyi, F & Hornyák, O. (2001): *Simulation tools for supporting robust process planning in the field of NC turning.* Proceedings 3rd Workshop on European Scientific and Industrial Collaboration, June: p. 27-29.
5.   Hornyák, O. (1999): *Object Oriented Software Engineering for Simulation of NC Machining Operations,* X Workshop on Supervising and Diagnostics of Machining Systems. Innovate and Integrated Manufacturing, p. 96-102.
6.   Phadke, M.S. (1989): *Quality Engineering Using Robust Design.* Prentice Hall, Englewood Cliffs.
7.   Tóth, T. & Erdélyi, F. (1997): *The Role of Optimisation and Robustness in Planning and Control of Discrete Manufacturing Process.* The 2nd CIRP World Congress in Intelligent Manufacturing, Budapest, Hungary, p. 205-210.
8.   Tóth, T. (1988): *Automatizált műszaki tervezés a gépgyártástechnológiában, Akadémiai doktori értekezés [Computer Aided Design and Planning in Production Engineering].* (in Hungarian), Budapest, DSc Dissertation, Hungarian Academy of Sciences.

A General Model of Auxiliary Processes in Manufacturing

Mezgár[1], I., Kovács[1], G., Gáti[1], G., Szabó[2], Cs.
[1] *CIM Research Laboratory, Computer and Automation Research Institute, Hungarian Academy of Sciences, E-mail: mezgar@sztaki.hu*
[2] *G. Tim-Co. Ltd.,1082 Budapest, Hungary, Kisfaludy u. 40.*

Abstract: In all sizes of manufacturing enterprises, daily shop-floor activities include a number of tasks not directly related to the main production flow. These complementary activities concern, among others, recycling of used material, productive resource maintenance. The goal of the research work was to capture the knowledge related to these activities and based on this knowledge to generate a general process model applicable for describing recycling-, maintenance- and strategic analysing processes. The paper introduces the general model of an air conditioner refurbishing process and some further applications of this model through a case study.

Key words: maintenance, modelling, non-manufacturing process, recycling, simulation.

1.   INTRODUCTION

Shop-floor processes are becoming more and more complex because of the auxiliary processes that are running parallel with the main production flow (e.g. maintenance, tooling), because of the organisation structures required by the enhanced flexibility, and because of the increasing requirements originating from the green manufacturing concept [4]. Current process models fail to capture and integrate at an appropriate semantic level, all the knowledge elements necessary to describe this complex scenario. Usually, application-dependent models are available at different levels, with different goals [3].

CAPP systems provide very detailed descriptions of single operations at design level, production planning and control packages support highly approximated product-oriented process models; production schedulers and simulators support resource-oriented models. In addition, all of these models consider only the main production flow, as activities that run in parallel with the main production process (e.g., maintenance and recycling) are not properly and fully represented in these models [1].

In order to solve the problems originating from this modelling inaccuracy, and parallel give an answer for the market demands, a project named EPSYLON (Enhanced Process Modelling System For Lean Operations Management, No. 25359) was organized. The main objectives of the project were:

1. define, implement and validate a new software for managing process information at the highest possible semantic level, with the possibility of translation it into the STEP standard;
2. integrate the new process model platform with existing scheduling and monitoring packages, in order to improve the current manufacturing operation management;
3. develop new DSS applications that fully exploit the new process model expressiveness, to support recycling, maintenance, simulation and strategic analysis.

The project consortium consisted of innovators, technological partners and end users from six different countries. The expertise of the consortium members cover all scientific, technological, functional and organisational needs, which have been identified as necessary to carry out the project. The paper gives a short overview on the theory of the new modelling approach, the main software modules, and then introduces the results through presenting the different models of a refurbishing process in a railway rolling stock and engineering service firm.

2. THE EPSYLON SYSTEM

2.1 The new modelling approach

The most important part of the project was the development of the proper representation form of processes that can be applied in different applications.

The process representation is realised through modelling primitives with the following advanced features:

(a) definition of operation types in abstract terms, capable to describe the different kinds of shop-floor activities (e.g., machining, assembly) independently of the context-dependent and resource-dependent details that characterise execution of single operation instances;

(b) representation of processes as sequences of operations, with parallel, alternative and cyclical paths, directed to a specific goal (e.g., disassembling a returned product, performing a given maintenance procedure);

(c) parametrisation of process and operation definitions, to capture engineering and manufacturing degrees of freedom (e.g., alternative routings for the same component), along with their mutual dependencies and constraints;

(d) description of resources in terms of their structure and variable features (e.g., machine tools configurations), capabilities and independent behaviour (e.g., operation timing, parallel operations execution);

(e) partial or complete specification of process features, in terms of included operations, involved resource instances, location inventory, to build different factory scenarios.

2.2   System Architecture and Modules

The enhanced process representation will support decision-making at all levels of factory operations management. With reference to the below architecture scheme, the following software modules have been implemented based on the proposed process model (see Figure 1.).

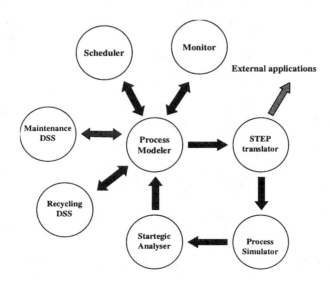

Figure 1. The EPSYLON system architecture.

Process Modeller. This module provides functions for creating, editing and consultation of process representation schemas based on the EPSYLON model primitives. Process modelling is solved on two levels; conceptual and instance levels. The conceptual level is that dealing with part, operation and resource types, classified according to their general features. The instance level describes the actual factory in terms of existing instances of the above entity types, their specific configurations, relationships, past, current and foreseen states.

STEP translator: translation of EPSYLON process representations into the corresponding STEP representations. The new process model must be capable of both linking to any factory information and control system existing at the user company and also be capable of importing and exporting data to other organisations.

Scheduler (integrated module): integration between the Process Modeller and an existing short-term production planner, extended and improved to take advantage of the new process representation in managing non-standard demand items.

Process Monitor (integrated module): integration between the Process Modeller and an existing process monitoring system, extended and improved to maintain an updated image of the factory in its current status.

Recycling DSS: supports to manage the processes originated from taken-back products, scraps and by-products to be recycled, taking into account all the variable and optional process features (e.g., type and status of the returned product, defects frequency and incidence on the recycling process, disassembly and recovery costs);

Maintenance DSS: supports the user in selecting the appropriate actions (e.g., defining the sequence of operations in reaction to a failure), taking into account general rules and data drawn from experience.

Simulator and Strategic Analyser: development, simulation and evaluation of alternative process scenarios, e.g.: new shop-floor configurations, "green" manufacturing options, introduction of alternative manufacturing, maintenance and inspection methods. Simulation proposes future scenarios in response to a distribution of external events either user-defined or generated by the system on a probabilistic basis.

2.3 Expected benefits

The main goal of the project is to produce useful, applicable results for the industry. The expected benefits for the end-users can be classified into three main groups:

1. *Improved management of process data.* The system allows enterprises to capture, to understand and to formalise knowledge of their shop-floor processes. The adoption of generalised, family-based primitives will allow a reduction of the representation size, in most cases of above 50% over traditional structures.
2. *Control over complementary operations.* The system will support documentation, planning and monitoring of non-manufacturing operations, like recycling and maintenance, as currently is given only for manufacturing ones.
3. *Improved decision-making functions.* The common process representation will provide a unified, detailed and updated reference for strategic decision-making.

The feasibility of the above improvements, along with the costs and time required for achieving them, was assessed through experimentation of the EPSYLON prototype on user firms in the Consortium.

3.   THE CASE STUDY

3.1   The firm

The firm where one of the pilot implementation took place was a railway rolling stock and engineering service SME. The main activity of the company is to carry out reparation and maintenance on the units of high speed wagons used in international and national inter-city traffic. The periodic maintenance, the reparation of several subsystems of the wagons, like the air-conditioner units, are done by the firm.

The main activities of the firm are usually like a work order, because the reparation and the maintenance jobs are not possible to plan in advance. Unfortunately the periodic maintenance also depends on the traffic, so it is often not done in time, only with deviations. The production is like a small batch production, because the customer needs small quantities. The products are not fixed in advance, the firm works for his customer in a very wide range according to their daily expectations.

3.2   Description of the air-conditioner refurbishing process

The main revisions of the different devices are the most precisely determined activity of the company. It has to be taken on all equipment in every four years. Although there are several types of these devices, the processes are very similar. For the process model validation the refurbishing of air conditioner has been selected, as this process is rather complex, but at the same time the tasks and their sequences can be described well, and represent the typical refurbishing process at the firm. The process-oriented description of air-conditioner refurbishing can be found in details in Deliverable D02 [2].

This chapter introduces a part of the refurbishing process of the air conditioner by means of the EPSYLON Process Model entities and their relationship. The whole documentation concerns production flow for the air conditioner refurbishing activity and provides a clear representation of what can be done with the primitives of the EPSYLON process model. The process model is suitable for representing all the possible operations (assembly, transformation, disassembly and test), alternative production paths and parts (manufactured, recycled, wasted, etc.) that are involved in a production process flow.

Three basic productive processes have been identified: manufacturing process, disassembly process and recycling process. The relations between these three types of process are the following: the results of a disassembly process are disposed, recycled or further disassembled, the results of recycling processes are used by manufacturing processes, the results of manufacturing processes can feed, after usage, proper disassembly processes.

The disassembly operational schema as it can be seen on Figure 2., is a tree that, starting from the root part, produces some leaf parts. A syntactic exception is given by the possible presence of disassembly alternatives, with or without disassembly joins, and of path alternatives, always with path splits. It has no secondary outputs since unsuccessful cases are explicitly modelled by alternatives. The only types of allowed operations are transformation and disassembly. Leaves can be recovered parts, returned parts (from lower level modules), and wasted materials. Root is always a returned part.

The recycling operational schema is a sequence of transformations (with possible secondary outputs) that produces a recycled part from a recovered part. A syntactic exception is given by the possible presence of path alternatives, always accompanied by path splits. The only allowed operations are transformations. Leaf is always a recovered part, root is always a recycled part.

The mixture of these two types of operational schemas can bee identified in the first step of the air conditioner (AC) refurbishing process. Figure 2. shows that AC is disassembled into six sub-units, while the gas in the AC starts its recycling process. The refurbishing of each part is described with the use of these symbols in the whole model.

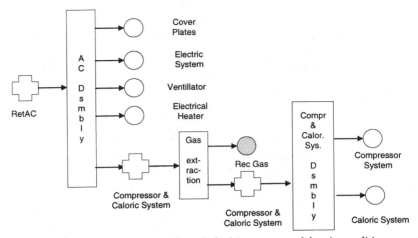

Figure 2. First steps of the refurbishing process of the air-conditioner.

3.3   Models of the refurbishing process

3.3.1   Actual EPSYLON configuration

As a first step an experimental configuration was installed at the firm. This configuration had more goals, as the validation of the software and the models itself, and besides these ones the user training, collecting more experience were the additional goals.

The configuration that was installed at the firm had to cover the practical needs of the firm, so the following configuration had been tested during the experiment phase:
-    Process Modeller (PM),
-    Recycling Decision Support System (RDSS),
-    Simulator.

The STEP translator was an active part of this configuration as the communication among the simulator and the other modules could be realised through this interface.

3.3.2   The models

The models introduced in this chapter all represent the different phases of the refurbishing process of the electric system, which is a subsystem of the air-conditioner.

The models describing the refurbishing processes are the following, according their sequence of application:

1. *operational schema* that is the base of modelling. (Fig. 2)
2. *general process model* (GPM) by using the PM software module (Fig. 3).
3. *RDSS model* conversion from the GPM.
4. Conversion the GPM into *STEP format.*
5. *Simulation model* generation from the STEP format.

The operational scheme is a user-friendly representation of the process, which contains all important process information and can be developed by a person who has little experience in modelling.

The Process Modeller module has the most important role in the applications, as this is the base for all applications. The PM is generated from the operational scheme. It is also investigated from the aspect, how it could be used to serve process data for other applications as the EPSYLON.

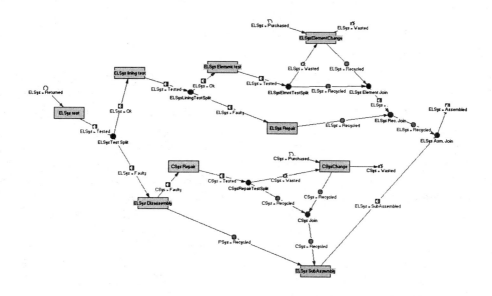

Figure 3. The ELSys process model

The connection between PM and RDSS is the following; the process model can describe the different phases of the life cycle. The RDSS is appropriate to follow and handle these phases and it is possible to predict (on a certainty level) some of the unscheduled corrective actions. By checking and processing (if it is needed) these failures and repair them with the planned maintenance action, can save a lot of time and can extend the MTBF (mean time between failure) of the actual unit.

```
#356=location(17,'Aux. 2','Forgogep javito',5,100,,100,6,,0)
#357=location(18,'Loc',",0,0,,1,8,,2)
#358=location(19,'Loc',",0,0,,1,8,,2)
#359=locationcategory(6,'Indoor',,0)
#360=locationcategory(7,'Outdoor',,0)
#361=locationcategory(8,'Shopfloor',,0)
#362=locationinstance(13,'Ind1-1',13,,0,",0,0,0)
#363=locationinstance(14,'Outd1',15,,1,",0,0,0)
#364=operation(77,0,6,'AC Disassembly',",,,,,0)
#365=operation(78,0,6,'ELSys test',",,,,,0)
#366=operation(79,0,6,'ELSys lining test',",,,,,0)
#367=operation(80,0,6,'ELSys Disassembly',",,,,,0)
#368=operation(81,0,6,'CSys Repair',",,,,,0)
#369=operation(83,0,6,'ELSys Repair',",,,,,0)
#370=operation(84,0,6,'ELSys Element test',",,,,,0)
#371=operation(85,0,6,'ELSys SubAssembly',",,,,,0)
#372=operation(86,0,6,'AC Assembly',",,,,,0)
```

Figure 4. The STEP representation of ELSys model

The STEP "module" represents the openness of the EPSYLON system. Through this module it is possible to export/import process description to/from other systems. In Figure 4 a part of the converted Process Model of the electronic system can be seen.

The simulation model is build from the STEP representation in the Taylor simulation system. The simulation was used in the pilot case for resource analyses, for definition of bottle - necks and initial schedule generation.

The process model can be modified easily according to the actual needs. The possible developments are; to create a more complex process by adding more subprocesses, develop a deeper model by giving more details. The increased usability of the model – closer to real-life application – results increasing motivation of the user.

An important aspect was the user's reaction during the project. Having seen the computer model and the possibilities (e.g. in use of RDSS) user started to show more enthusiasm, and in the final model building/user training phase they were nearly fully convinced on the usefullness of the software system.

4. CONCLUSIONS

The paper introduced a new modelling approach to represent non-manufacturing processes in a general way. This model can be applied by different application oriented programs increasing communication/data exchange speed among applications and decreasing the risk of multiple data handling, and the time of multiple data input. The set of decision systems based on this model offers help among others, in the management of maintenance, recycling and monitoring processes. The general model supports the analysis of these processes as well.

At user sites the practical results are that the system allows enterprises to capture, understand and formalise knowledge of their shop-floor processes, at a proper detail level and with the amplest possible scope. This radically improves all activities inside (and in many cases also outside) the firm. By using the information system based on the formalised process descriptions the work of both the managers and the workers on the shop floor will be more effective.

5. REFERENCES

[1] Bonfatti F., Monari P. D., Paganelli P. A rule-based manufacturing modelling system, *International Journal of Computer Applications in Technology* (IJCAT), **Vol 10**, 1-2, 1997.

[2] EPSYLON Consortium. Deliverable D02 – Process Model, June, 1998.

[3] Kurbel K., Rautenstrauch C. Integrated Planning of Production and Recycling Processes, MIM '96 Int. Conference, Leicester (UK). 1996.

[4] Tipnis, V.A. Evolving Issues in Product Life Cycle Design, *CIRP Annals,* **Vol. 42**, 1. 1993.

Modeling and Management of Production Networks

E. Ilie-Zudor, L. Monostori
Computer and Automation Research Institute, Hungarian Academy of Sciences, Budapest, Hungary
Tel: +361-4665644, Fax: +361-4667503, e-mail: {laszlo.monostori, ilie}@sztaki.hu

Abstract: The paper attempts to illustrate the concept behind a newly emerging area of research, the concept of production networks. The materialization of this paradigm requires clear definitions of the terms involved, setting up a reference architecture, design and development of supporting platforms and appropriate control methodologies with protocols and mechanisms. The main goal of the paper is to overview the different kinds of production networks, with special emphasis on architectures, advantages and open problems.

1. INTRODUCTION

One of the newly emerging areas of research, representing a real challenge for the planning and management of production systems, is related to the paradigm of production networks. Regarding the concept of production networks, different, interrelated concepts can be found in the literature: i.e., virtual enterprise, supply chain, extended enterprise. For the time being, there is not a unique, universally accepted definition yet for these terms. Usually, no clear distinction is made between virtual and extended enterprises, though everyone distinguishes the concepts of virtual factories (enterprises) and virtual manufacturing.

Extended enterprises, *virtual enterprises* and *supply chains* are considered in the paper as very similar concepts, each of them being a *production network (PN)* formed from independent companies collaborating by sharing information, skills, resources, and having the same goal of exploiting market opportunities. A kind of production network focusing on collaboration between companies, is a *virtual enterprise (VE)*. If the collaboration network incorporates a dominant company imposing the rules of the information exchange, we call it *extended enterprise (EE)*. The alliance that focuses on the chain aspects is then the *supply chain (SC)*.

2. DEFINITIONS AND CATEGORIZATION
2.1 Definitions

In the related literature, a quite large number of definitions is presented. In the following we specify some of them.

Virtual enterprise

In [3] a virtual factory is defined as a community of dozens of factories each focusing on what it does best, all linked by an electronic network that would enable them to operate as one, flexibly and inexpensively regardless of their location.

A virtual enterprise is a temporary consortium formed by real autonomous companies on the basis of strong collaboration to exploit fast changing worldwide opportunities quickly, which a single company is unlikely to realize [35]. Other definitions of the term "virtual enterprise" can be also found in [33], [17], [34], [18], [4], [13].

Supply chain

The supply chain is a set of activities, which span enterprise functions from the ordering and receipt of raw materials through the manufacturing of products through the distribution and delivery to the customer [10].

The supply chain of a manufacturing enterprise can be defined as a world-wide network of suppliers, factories, warehouses, distribution centers and retailers through which raw materials are acquired, transformed into products which are then delivered to customers [28].

Other definitions of the term "supply chain" can be also found in [2] and [5].

Extended enterprise

In [32] the virtual enterprise and the extended enterprise are considered as two different notions. The extended enterprise is seen as a network of closely collaborating independent partners, where the goal is to achieve competitive advantages by forming formal linkages (contracts) and to maintain distributed cooperation throughout the network. An enterprise usually belongs to more than one network.

Virtual enterprise- Supply chain

According to [28] the Supply Chain Management focuses on the chain aspects and is related to the life cycle of products, while the Virtual Enterprise focuses on the collaboration among the related manufacturing factories.

Virtual enterprise - Extended enterprise

The difference between the extended enterprise and the virtual enterprise is that the virtual enterprise lasts shorter period of time, and has less formal partnerships [32].

The concept of extended enterprise, the closest "rival" term, is better applied to an organization in which a dominant enterprise "extends" its boundaries to all or some of its suppliers. The VE can be seen as a more general concept including other types of organizations, namely a more democratic structure where the cooperation is peer to peer [7].

Virtual enterprise- Virtual manufacturing

A virtual manufacturing system is defined as a computer system, which can generate the same information about manufacturing system's structure, states, and behaviors as we can observe in a real manufacturing system [13].

The virtual manufacturing model can be used for creating a test field for conducting experiments on the extended enterprise models.

2.2 Classification

The different types of alliances between enterprises can be categorized not only as virtual, extended, supply chain, but also according to criteria presented in Figure 1.

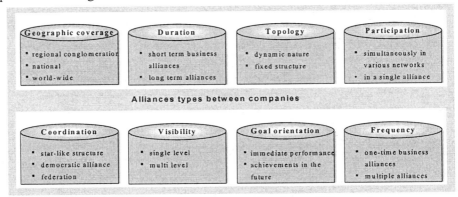

Figure 1. Categorization of alliances according to different criteria

In a dynamic type production network, enterprises can dynamically join or leave the alliance according to the phases of the business process or other market factors.

The structure of the *star-like alliance* has in the center a dominant company, defining "the rules of the game" and imposing its own standards to the other partners that form a relatively fixed network of suppliers. In a *democratic alliance*, all participants cooperate on an equal basis, keeping their autonomy, but joining their core competencies. The *federation* is a common coordination structure created between companies that previously have been partners in a successful alliance and have realized its benefits.

A participant in a *single level alliance* can only see its direct neighbors, while in a *multi level* one, has some visibility over other non-direct levels.

3. APPROACHES, ARCHITECTURES

The latest approaches to realize extended enterprises, which involve building virtual enterprise architectures, enterprise integration and supply chain management, are based on the agent theory [7],[8],[12],[15], the mobile agents technology [5], [19], [17] or the neural networks technology [13].

3.1 Virtual enterprise

Main functional requirements

In [8] and [7] the main functional requirements for the design of an architecture for industrial virtual enterprises are presented. The work is partly done in the framework of the Prodnet II (Production Planning and Management in an Extended Enterprise, (http://www.uninova.pt/~prodnet/) project. The aim is the development of reference architecture and supporting infrastructure for virtual enterprises particularly suited for small and medium sized enterprises in the area of metalomechanics.

The fundamental features of this architecture are incorporated in two main modules: Internal module and Cooperation module. The enterprises in the network are viewed as nodes that add some value to the process. The nodes can exchange information at the same time. Every node is extended with a cooperation module. The Cooperation module supports the information exchange among the nodes. The Internal module is connected to the Cooperation module via a "mapping interface" and the Cooperation modules of different enterprises are connected to the network via a "network infrastructure interface". The Internal Module comprises the complete structure of the company's information and all the internal decision making processes.

Within this architecture two kinds of nodes are defined: Network Coordinator node, as the regulatory component of the enterprise network, and Member Enterprise node, which will store the information about the enterprise itself, and provide the external connectivity.

To support several important information management requirements of the virtual enterprises the PEER [1] federated object-oriented information management system was used.

Agile scheduling

In [17] and [18] a prototype of a multi-agent system for agile scheduling and the extension for the operation in a virtual enterprise environment is described. The system is called MASSYVE (Multi-agent Agile manufacturing Scheduling Systems for Virtual Enterprises) and is part of the MASSYVE INCO-DC KIT Project. MASSYVE uses the HOLOS framework [20] as a baseline for advanced scheduling and the PEER information management framework for the information integration.

The HOLOS [22] scheduling system allows a system to be custom-tailored for each particular enterprise and, at the same time, to be reconfigured and adapted whenever new production methods, algorithms, production resources, etc., are introduced or changed.

In this framework an instance of a scheduling system is interactively and semi-automatically derived from the HOLOS Generic Architecture (HOLOS-GA) [21] reference model, supported by the HOLOS Methodology [22], based on Agent Oriented Programming [30] constructs.

The agent classes used in HOLOS are: the *Scheduling Supervisor (SS)* (agent that performs the global scheduling supervision and it is the unique system's "door" to other systems), the □*Enterprise Activity Agents (EAA)* (the real executors of manufacturing orders), the □*Local Distribution Centers (LDC)* (represent functional clusters of EAAs in order to avoid announcement broadcasting, also responsible to select the most suitable agent for a certain order after a negotiation process), and the □*Consortium (C)* (temporary instances created to supervise the execution of a given order).

HOLOS uses the Contract-Net Protocol coordination mechanism to support the task assignments to agents, and the Negotiation [9], [24] method to overcome conflicts taking place during the scheduling phases. The interaction between the agents is only vertical and agents cannot change the set of other agents that they can communicate with.

The HOLOS control hierarchy uses one *global manager* (the agent SS), some *functional managers* (the agents LDC), and assuming that a shop floor is usually composed of production resources (the agents EAA) with small production capacities. A HOLOS agent can establish communications with four kinds of external entities: other HOLOS agents, the end-user, a CIM Information System and the production resources.

MASSYVE proposes a three-layered federated database architecture to support the sharing and exchange of information both within each multi-site enterprise and among different enterprises uniformly, based on PEER: Intra-organization Federated Layer, Federation of HOLOS systems, Federation of Virtual enterprises.

Holonic based framework

A framework for virtual enterprises based on holonic principles is presented in [12]. The framework is presented through two views: the global and the local view. From the global point of view, authors consider two kinds of holons: Virtual Enterprise (VE) holon and Member Enterprise (ME) holon.

The VE holon is seen as the global coordinator of the virtual enterprise and it is located at the top level of ME holons. The VE holon is not imposing anything to the other holons, but acts as an assistant to them, and provides goals, constraints, warnings, suggestions, knowledge sharing services, etc. The ME holons, from the global view, are located on the same level.

Seen from the local view, holons are distributed on three levels: on the first level the ME holon can be found, on the second level the planning holon and the scheduling holon, and on the third level the task holons and resource holons.

The paper discusses also the VE control under the holonic framework. This includes planning, scheduling, resource sharing, dealing with organizational changes.

Partner selection

In [15] the problem of choosing the enterprises forming the Virtual Enterprise in a multi-agent environment is discussed. Besides the enterprises, which are seen as agents, the framework comprises a VE coordinator agent and information servers agents. The goal of enterprises is to form the most favorable group that would satisfy a certain need. The Coordinator Agent decomposes the goal in sub-goals and forms a VE goal hierarchy. This hierarchy can be dynamically changed during the process of partner selection.

The partner selection is treated as a distributed constraint satisfaction problem.

Subcontracting

In production networks different kind of subcontracting may take place [16] due to technological reasons, and referred as *technology-driven subcontracting*, due to insufficient capacity, referred as *capacity-driven subcontracting*, due to strategic reasons for keeping the cooperation with a certain supplier, subcontracting between factories of the same company, etc.

Subcontracting between suppliers and producers, or between producers and customers is called *classic subcontracting*.

To achieve a better reliability enterprises in the network should allow and assure some information sharing regarding machine loading, availability, orders progress, planned demands, stocks. In order to help increasing the redundancy in networks, a software tool called FAST/net was developed. The software provides information about the status of orders, resource availability, partners' stocks. FAST identifies bottleneck systems, marks orders suitable for subcontracting, and makes calculations on security stocks.

Reference model

A reference model for virtual enterprises can be found in [4]. Enterprises searching for business opportunities form an aggregation named *Virtual Industry Cluster (VIC)*. When a business opportunity can be exploited, a *Virtual Enterprise Broker (VEB)* will search through different Virtual Industry Clusters for enterprises with the appropriate competencies. The enterprises in question will form a *Virtual Enterprise (VE)*. When a VIC has a VEB of its own, the VIC becomes a *Virtual Organization (VO)*. The number of enterprises in a VIC is varying, enterprises may enter or exit the aggregation at any time.

The VEB obtains and offers business opportunities to the members of VIC and sets up the VE. The VIC provides information about the competencies of the enterprises involved. The VE will run the business opportunities and take part in its reconfiguration.

The VO has to manage alliances, and define and manage strategies.

3.2 Internet enabled virtual enterprises and supply chains

The Collaborative Agent System Architecture (CASA) and the Infrastructure for Collaborative Agent Systems (ICAS) have been developed for implementing Internet enabled virtual enterprises and for managing the Internet enabled complex supply chain for a large manufacturing enterprise, though, initially they were proposed as a general approach for Internet based collaborative agent systems [28], [27].

The approach comprises the following elements: Cooperation Domain Servers, Yellow Page Agents, Local Area Coordinators, Collaborative

Interface Agents, High-Level Collaboration Agents, Knowledge Management Agents, Ontology Server.

The cooperation domain server receives all the messages sent by the agents in the cooperation domain and forwards the message transparently to its destination and may also record all the transactions.

Yellow Page agents accept messages for registering services and record this information in a local database.

A local area coordinator acts both as a representative of the area to the outside world, and a manager for the local agents within the area. It also provides an interface service to the outside world.

Collaborative Interface Agents □are supposed to be communicative, semiautonomous, collaborative, reactive, pro-active, adaptive, self-aware and mobile. High-Level Collaboration Agents are introduced to increase the basic collaboration services provided by Yellow Page Agents and Local Area Coordinators agents. Agents do not communicate with each other directly, but those working together form cooperation domains. Each agent in a cooperation domain routes all its outgoing messages through the cooperation domain server. All Cooperation Domain Servers are connected with the Ontology Server for translating messages from different agents into a common format.

Five mechanisms developed in previous projects (MetaMorph, MetaMorph I, MetaMorph II) are used: Agent-Based Mediator-Centric Organization, Task Decomposition, Virtual Clustering, Partial Agent Cloning, Adaptation and Learning.

3.3 Supply chain

A hybrid agent-based architecture for manufacturing enterprise integration and supply chain management is proposed in the framework of MetaMorph II (http://imsg.enme.ucalgary.ca/research.htm#top) project [25], [29]. The objective of the project is to integrate the manufacturing enterprise's activities such as design, planning, scheduling, simulation, execution, and product distribution, with those of its suppliers, customers and partners into an open, distributed intelligent environment. The architecture was named Agent-Based Manufacturing Enterprise Infrastructure (ABMEI).

MetaMorph II architecture considers a main manufacturing enterprise central to the supply chain management. Each manufacturing enterprise has to have at least one Enterprise Mediator. Partners, suppliers and customers are dynamically connected with this main enterprise through other kind of mediators via the Internet/Intranet.

Mediators are agents, called mediator agents. They provide primarily message services and promote cooperation among intelligent agents, furthemore learning from the agents' behavior. All manufacturing resources (e.g.: machines, tools, workers, AGVs, etc.) are modeled as resource agents and are coordinated by layered mediator agents. Parts are also modeled as agents. Part agents do not communicate directly with resource agents. All manufacturing task requests are sent to the Resource Mediators.

MetaMorph II combines the mediation mechanism based on hierarchical mediators and the bidding mechanism based on the Contract Net protocol to solve the cooperative negotiation among resource agents. The authors propose two scheduling mechanisms: Machine-Centered Scheduling and Worker-Centered Scheduling. The present implementation of ABMEI consists of four mediators: Enterprise Mediator, Design Mediator, Resource Mediator and Marketing Mediator.

Static and mobile agents infrastructure

A supply chain infrastructure in a textile-manufacturing environment is presented in [5]. The framework consists of static and mobile cooperating agents, organized in three functional levels: the supply chain level, the factory level and the user level. Mobile agents have behavior, state and location. Agents share the same communication language and a common vocabulary. They use code mobility as well.

Generic reusable enterprise model

The architecture of the Integrated Supply Chain Management System presented in [10] is based on a set of cooperating agents. There are two types of agents: functional agents, for planning and controlling activities in the supply chain, and information agents, for information and communication services. The functional agents in this architecture are the Order Acquisition agent, the Scheduling agent, the Resource Management agent, the Dispatching agent, the Transportation Management agent, and the Logistics agent. Supply chain agents exist within an Enterprise Information Architecture (EIA), having a generic reusable enterprise model in the center, in order to support the integration of supply chain agents.

The EIA is responsible for finding the information that a certain agent asks for, and also for distributing the information an agent wants to share, to the agents that is interested.

4. ADVANTAGES

The involvement in production networks like these, has numerous advantages for the member enterprises, such as:
- increased reaction to changes in market condition and demands, because the participation in the network involves a better flexibility and agility;
- possibility to complement core competencies in order to be able to share some market opportunities;
- ability to specialize in few areas, representing the core competencies;
- new market opportunities, by increasing the geographic coverage;
- improved quality and responsiveness to market opportunities;
- elimination of capacity bottlenecks and reduction of capital investment through resource sharing;

- making global life-cycle orientation possible;
- increased reuse and recycling in waste management;
- increased redundancy, as more than one partner may produce the same service or product.
-

5. PROBLEMS, OPEN QUESTIONS

In every production network, one of the main problems is represented by the information interchange through the network. In addition to this, problems caused by differences in culture, trust issues are added. A scheme of the issues to be considered is presented in Figure 2.

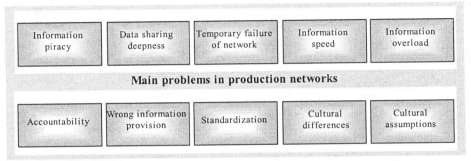

Figure2. Issues to be considered when independent companies join into alliances

5.1 Information interchange

Piracy. The security mechanisms and the access rights have to be very well implemented, in order to protect the network against unauthorized access to information, which could lead to secret information reveling, unauthorized utilization of resources, result in the alteration of information without any gain to the intruder.

Deepness. In some cases, partners agree to share data by making internal information visible for each others, such as the capacity of a particular shop floor, the orders progress, material flow status, etc., in order to adjust their own planning. An issue to be dealt with in this case is how deep a partner can allow to the others to see into his internal structure.

Temporary failures. The redundancy of information should be ensured, because temporary failures might happen in the network.

Speed. In order to work as a single integrated unit companies must be able to inter-operate, share and exchange information in real time. As the communication between enterprises may induce high network traffic, it is necessary to speed up the exchange of information.

Overloads. The high network traffic mentioned above can cause not only slowness in the network, but also overloads. Therefore, the data volume in a network has to be limited. This can be done through the reduction of the need for communication or the reduction of details.

Accountability. It is necessary to guarantee that the sender and/or receiver should not deny the transaction after it has taken place [17].

Incomplete or wrong information provision. Enterprises in the network do not consider the global optimization as their main purpose, but they act selfishly, often having antagonic goals, and therefore, sending wrong or incomplete information on purpose, even though a general willingness to cooperate can be assumed.

Standardization. Because of a different technical background, at the time they enter a production network, companies may use different software, different standards, different communication protocols. In order to exchange and share information with each others, enterprises must agree on the use of some common standards.

At the moment, most enterprise networking experiences are based on the implantation of the STEP standard for the exchange of technical product data and of the EDIFACT standard for commercial data exchange. In agent based system developments, standards as CORBA (Common Object Request Broker Architecture) (http://www.omg.org/corba/) have been used for inter-agent communication, and for developing agent based manufacturing systems, KQML (Knowledge Query and Manipulation Language, http://www.cs.umbc.edu/kqml/) as a common communication language for agents, with KIF as a common content format. The situation regarding quality management has also to be considered.

5.2 Cultural differences

Enterprises can enter a PN such as a virtual enterprise e.g. from different geographic areas. In this case, new issues that have to be dealt with will appear: Different legislation in different countries; Different time zones; Different currencies; Different technical and cultural background; Different educational level; Team-work experience, etc.

5.3 Cultural assumptions

Between the members of a PN must exist an ethic code for providing a base for trust and the warranty of cooperation between members. Companies are supposed to manifest solidarity, collective orientation, trust [10].

For enterprises (agents), in order to cooperate, there must be cultural assumptions such as [34]: agents are constraint-based problem solvers, agents have the ability to generate more than one solution, agents have the ability and authority to realize local goals and possibly relax a sub-set of constraints if the global solution is further optimized, etc.

6. CONCLUSIONS

Different concepts (virtual enterprise, supply chain, extended enterprise) behind the general term of production networks were discussed in the paper, with the aim of making some order in the closely related approaches. Special emphasis was given on the characteristic features of these architectures with their advantages and disadvantages. The elimination of the enumerated open problems requires further research efforts combining the approaches of different disciplines, such as information technology, artificial intelligence and operation research.

A list of "live" links related to the concept of production networks (Virtual/Extended Enterprises, Supply Chains) can be found, among other things, on the following web page: http://www.sztaki.hu/sztaki/ake/ai/manufacturing/.

7. ACKNOWLEDGEMENTS

This work was partially supported by the National Research Foundation, Hungary, Grant Nos. T026486 and T034632.

References

[1]   Afsarmanesh, H.; Wiedijk, M.; Hertzeberger, L.O.; Negreiros Gomes, F.; Provedel, A.; Martins, R.C.; Salles, E.O.T. 1995: **A Federated Cooperation Architecture for Expert Systems Involved in Balanced Automation Systems**, L.M. Camarinha-Matos and H. Afsarmanesh (Eds.), Chapman & Hall, Jul 1995.

[2]   Barbuceanu, M., Fox, M.S., 1997: **Dynamic Team Formation and Management in an Agent Structured Supply Chain**.

[3]   Bell Canada Company, 1996, Harvard Business Review, July/August 1996.

[4]   Bremer, C. F., 2000: **From An Opportunity Identification to its Manufacturing: A Reference Model for Virtual Enterprises**, Annals of the CIRP, vol. 49/1, 2000, pp. 325-329.

[5]   Brugali D., Menga G., Galarraga G., 1998: **Inter-Company Supply Chain Integration via Mobile Agents.**, In The Globalization of Manufacturing in the Digital Communications Era of the 21st Century: Innovation, Agility, and the Virtual Enterprise. Kluwer Academic Pub. 1998, http://www.polito.it/~brugali/.

[6]   Camarihna-Matos, L.M., Afsarmanesh, H., 2000: **Frameworks for Agile Virtual Enterprise in Manufacturing,** IFAC-MIM Symposium on Manufacturing, Modeling, Management and Control, Rio, Patras, Greece, 12-14 July, pp. 84-89.

[7]   Camarinha-Matos L.M., Afsarmanesh H., Garita C., Lima C., 1997: **Towards an Architecture for Virtual Enterprises**, Keynote paper, Proc. 2nd World Congress on Intelligent Manufacturing Processes and Systems, Springer, Budapest, Hungary, June 1997, pp. 531-541.

[8]   Camarinha-Matos, L.M., Afsarmanesh H., 1997: **Virtual Enterprises: Life Cycle Supporting Tools and Technologies**, Handbook of Life Cycle Engineering: Concepts, Tools and Techniques, A. Molina, J. Sanchez, A. Kusiak (Eds.), Chapman and Hall

[9]   Davis, R.; Smith, R., 1983: **Negotiation as a Metaphor for Distributed Problem Solving, Artificial Intelligence**, N 20, pp. 63-109.

[10]   Fox M.S., Chionglo J.F., and Barbuceanu M., 1993: **The Integrated Supply Chain Management System**, Internal Report, Dept. of Industrial Engineering, University of Toronto.The Integrated Supply Chain Management Project, http://www.eil.utoronto.ca/iscm-descr.html.

[11]   Hongmei G., Biqing H., Wenhuang L., Shoujo R., Yu L., 2000: **The Application of Holonic Manufacturing Paradigm to the Virtual Enterprise Control**, Proceedings of 16th World Computer Congress 2000, Information Technology for Business Management, Aug 2000, Beijing, China, pp. 270-276.

[12]   Kazuaki Iwata, Masahiko Onosato, Koji Teramoto, Suguru Osaki, 1997: **Virtual Manufacturing Systems as Advanced Information Infrastructure for Integrating manufacturing Resources and Activities**, Annals of the CIRP Vol. 46/1/1997, pp. 335-338.

[13]   Lau H.C.W., Chin K.S., Pun K.F., Ning A., 2000: **Decision supporting functionality in a virtual enterprise network**, Expert Systems with Applications, 19/ 2000, pp. 261-270.

[14]   Lutz, S., Wiendahl, H-P., 2000: **Concept for Production-Management in Networks**, Proceedings of 16th World Computer Congress 2000, Information Technology for Business Management, Aug 2000, Beijing, China, pp. 424-431.

[15]   Macgregor, R. S.; Aresi, A.; Siegert, A., 1996: **WWW.Security, How to Build a Secure World Wide Web Connection**, IBM, Prentice Hall PTR.

[16] Mezgár I., Kovács G.L., Paganelli P. 2000: **Co-operative Production Planning for Small and Medium-Sized Enterprises**, International Journal of Production Economics, No. 64, pp. 37-48.

[17] Monostori, L.; Szelke, E.; Kádár, B.: **Intelligent techniques for management of changes and disturbances in manufacturing**, Proceedings of the CIRP International Symposium: Advanced Design and Manufacture in the Global Manufacturing Era, August 21-22, 1997, Hong Kong, Vol. 1, pp. 67-75.

[18] NIIIP 1999: Vision of the National Industrial Information Infrastructure Protocols (NIIIP), http://www.niiip.org/vision.html.

[19] Rabelo R., Camarinha-Matos L.M., Afsarmanesh H., 1998: **Multiagent perspectives to agile scheduling**, Proc. Of BASYS'98 - 3rd IEEE/IFIP Int. Conf. On Balanced Automation Systems, Intelligent Systems for Manufacturing (Kluwer Academic), pp. 51-66, ISBN 0-412-84670-5, Prague, Czech Republic, Aug 98.

[20] Rabelo R., Camarinha-Matos L.M., Afsarmanesh H., 1999: **Multi-agent-based agile scheduling**, Journal of Robotics and Autonomous Systems (Elsevier), Vol. 27, N. 1-2, April 1999, ISSN 0921-8890, pp. 15-28.

[21] Rabelo R., Spinosa M., 1997: **Mobile-agent based supervision in supply chain management in the food industry**, Proceedings of Agrosoft'97 - Workshop on supply chain management in agribusiness, Belo Horizonte, Brazil, Sept. 97, pp. 451-459.

[22] Rabelo, R.J., 1997: **A Framework for the Development of Manufacturing Agile Scheduling Systems – A Multi-agent Approach**, Ph.D. Thesis, New University of Lisbon, Portugal.

[23] Rabelo, R.J.; Camarinha-Matos, L. M., 1995: **A Holistic Control Architecture Infrastructure for Dynamic Scheduling, in Artificial Intelligence in Reactive Scheduling**, Eds. Roger Kerr e Elizabeth Szelke, Chapman & Hall, pp.78-94.

[24] Rabelo, R.J.; Camarinha-Matos, L. M., 1996a: **Deriving Particular Agile Scheduling Systems using the HOLOS Methodology**, International Journal in Informatics and Control.

[25] Rabelo, R.J.; Camarinha-Matos, L. M., 1998b: **Generic framework for conflict resolution in negotiation-based agile scheduling systems, Proceedings IMS'98** – 5 th IFAC Workshop on Intelligent Manufacturing Systems, Gramado – Brazil, pp.187-192.

[26] Shen W., Norrie D.H., 1998: **An Agent-Based Approach for Manufacturing Enterprise Integration and Supply Chain Management,** In G. Jacucci, et al (eds.), Globalization of Manufacturing in the Digital Communications Era of the 21st Century: Innovation, Agility, and the Virtual Enterprise, Kluwer Academic Publishers, 1998, pp. 579-590.

[27] Shen, W., Norrie, D.H., 1999: **Implementing Internet Enabled Virtual Enterprises using Collaborative Agents,** In Camarinha-Matos, L.M. (ed.), Infrastructures for Virtual Enterprises, Kluwer Academic Publisher, pp. 343-352, (http://imsg.enme.ucalgary.ca/ publicate.htm#ISG publications in 1999).

[28] Shen, W., Ulieru, M., Norrie, D.H. and Kremer, R., 1999: **Implementing the Internet Enabled Supply Chain through a Collaborative Agent System**, Proceedings of Agents'99 Workshop on Agent Based Decision-Support for Managing the Internet-Enabled Supply-Chain, Seattle, WA, May 1-5, pp. 55-62.

[29] Shen, W., Xue, D., Norrie, D.H., 1998a: **An Agent-Based Manufacturing Enterprise Infrastructure for Distributed Integrated Intelligent Manufacturing Systems**, in Proceedings of PAAM'98, London, UK. (A hybrid agent-based approach for integrating manufacturing enterprise activities with its suppliers, partners and customers within an open and dynamic environment. Description of its functional architecture, main features and a prototype implementation.) (http://imsg.enme.ucalgary.ca/).

[30] Shoham, Y., 1993: **Agent-Oriented Programming, Artificial Intelligence**, N 60, Elsevier, pp.51-92.

[31] Szegheo, O., 1999: **Extended Enterprise-the Globeman 21 way**, Procedigs of the Second International Workshop on Intelligent Manufacturing Systems, sept. 1999, Leuven, Belgium, pp. 405-410.

[32] Walton, J.; Whicker, L., 1996: **Virtual Enterprise: Myth & Reality**, J. Control, Oct 1996.

[33] Wiendahl H.P., Helms K., Höbig M., 1998: **Management of Variable Production Networks-Visions, Management and Tools**, Annals of the CIRP, vol. 47/2, 1998, pp. 549-555.

[34] Yu L., Biqing H., Wenhuang L., Wu C., Hongmei G., 2000: **Multi-Agent System for Partner Selection of Virtual Enterprise**, Proceedings of 16th World Computer Congress 2000, Information Technology for Business Management, Aug 2000, Beijing, China, pp. 287-294.

[35] Zhou, Q., Ristic, M., Besant, C.B., 2000: **An Information Management Architecture for Production Planning in a Virtual Enterprise**, in The International Journal of Advanced Manufacturing Technology, (2000) 16: pp. 909-916.

A Framework for a CAD-Integrated Tolerance Optimisation System

Ross Eadie and James Gao
School of Industrial & Manufacturing Science
Cranfield University, Bedfordshire, U.K. MK43 0AL
ross.eadie@edwards.boc.com, james.gao@cranfield.ac.uk

Abstract: Current assembly modelling and tolerance analysis functionality offered by most CAD vendors extends to the evaluation of a design in terms of its form and fit. There is no evidence of these tolerance optimisation tools offering the designer an 'in-context' perspective of process capability: an assessment of whether the required tolerances are achievable from the available manufacturing process. There is also a lack of provision for the direct use of experimental design data, within the CAD system environment. The aim of this research is to enable these prototype and production test results to be used in conjunction with a manufacturing capability model to establish the best compromise between product performance and process capability. This research program is to establish a framework with which process capability data can be captured, interpreted and subsequently used in a CAD system, making it readily available for the designer to use in the pursuit of a more robust product. The framework will be demonstrable in the context of vacuum pump design and manufacture in BOCE (BOC Edwards), the company hosting the research activity.

The framework will have the potential to reduce the level of first time test failures that are attributed to products containing parts that are not made to specification. It will also eliminate the time and resource overhead incurred in redesigning parts, which cannot be manufactured due to the specification of unachievable tolerances. As a result, machining scrap will also be reduced. In addition, the designer will be more confident that a 'robust' product has been produced, with reduced sensitivity to normal variations in the manufacturing process. Where a 'close' tolerance is absolutely vital, and the current manufacturing process is not capable, the framework will highlight this early in the design phase and allow the purchase of additional (and possibly long lead time) tooling or fixturing, thereby providing effective support to the new product introduction process.

Keywords: Process capability, CAD, PDM, SPC, tolerances

CHALLENGES OF TOLERANCE SPECIFICATION

Manufactured parts are rarely made to exact specifications: variation exists in the size of parts and their constituent features because of variations in the chosen manufacturing process. The amount of variation is a function of the cost and the quantity of the finished product. The purchase price of the product will dictate its prime cost, which in turn, dictates the available manufacturing processes with which it can be economically made. Selection of an appropriate manufacturing process is therefore critical to the commercial success of the product in the market, and to the profitability of the manufacturer. Of equal importance is an appropriate tolerance specification to satisfy a product's functional requirements, and to gain an understanding of how the dimensional variation of critical components in an assembly will affect overall product performance. Once a functional envelope has been established, this can be compared with the available manufacturing technology and processes, which are economically suitable to the product quantities and price of the product.

Often, the performance envelope of a product is determined by knowledge of the manufacturing capacity, particularly where it is desirable to have tight dimensional control on critical components that directly influence product performance and reliability. It is vital to understand manufacturing capability here because over-tight tolerances are often relied upon to guarantee performance, but with the penalty of high manufacturing rejects and first-time test failures. Effectively, the tolerances are being used to control the process, rather the process being inherently capable of achieving the desired tolerances. The product designers therefore must know at the development stage the required part tolerances to satisfy its specification, and must also understand the *capability* of the available manufacturing technology to deliver these tolerances.

Most manufacturing companies, particularly with an established quality department, collect data on the dimensional accuracy of their manufacturing processes. This can be achieved during the pre-production sample approval stage where a batch of production representative parts is 100% inspected. For production parts, statistical process control (SPC) is used, where only critical features on each part are measured. These data are displayed as control charts showing the measurement trend over time and are used to detect whether the parts are within tolerance and whether the mean values of the features measured are centrally disposed about the mid-

value in the tolerance range.   Along with other data held on product
performance and reliability, the objectives of the framework are too:

- Provide a generic process capability model applicable to the available
  manufacturing resource.
- Produce a method to link this process capability model with the latest
  statistical process (SPC) data.
- Produce a method of comparing the process capability available with the
  tolerance specification sought within the CAD environment.
- Advise the designer as to whether the component tolerance specification
  will enable the product to deliver the desired performance, given that it
  is manufactured with a capable process.
- Introduce product data management (PDM) and other enabling
  technologies to dynamically link departmentally maintained design,
  manufacturing and inspection databases.

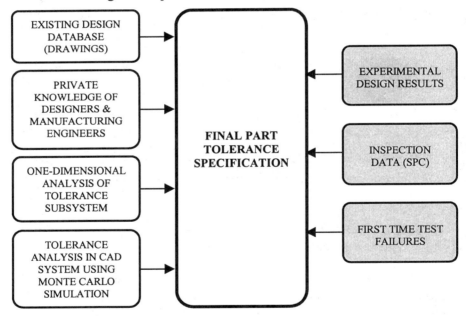

Figure 1: Process Inputs for Part Tolerance Definition

Figure 1 illustrates the key inputs in the tolerance specification
process. The left-hand side lists the inputs that the designer traditionally
establishes prior to deciding a final tolerance specification. Designers may
refer to examples of similar solutions (from a drawing database for example)
as a reference for tolerancing, using a solution already in production, or

from their own personal experience or that of others. The danger here is that a 'legacy' tolerance may not functionally suit the part as used in its *new* assembly. In addition, by referencing a drawing in isolation, there is little understanding of how capable the current manufacturing technology is at achieving this tolerance, and whether as a consequence, there would be a high scrap or rework rate. A potentially unsuitable tolerance would then be perpetuated into the new design. A more sophisticated approach is to use worst-case (arithmetic) tolerancing, performed as a one-dimensional tolerance stack-up either by manual calculation or in a spreadsheet package. The success of this approach relies on an understanding of the tolerance chain and the correct use of the component feature minimum or maximum condition [1]. However, this analysis is external to the CAD modelling environment. At a higher level of sophistication, there are CAD-based tolerance analyses tools available using Monte Carlo statistical simulations to evaluate clearance conditions in 3D digital assemblies (such those from Tecnomatix or VSA). Figure 1 also highlights three additional inputs (on the right-hand side) that, in contrast, are not formally used in the tolerance assignment process within the CAD system, probably since there are currently few ways to explicitly link current manufacturing capability or product performance with the final tolerance specification:

- Experimental design results: During the product development process, prototypes are built and tested, with critical components deliberately made at upper and lower tolerance limits to test the possible range in performance. If pre-production prototypes are manufactured, these may be built in a statistically significant batch number to accrue product performance data that could be used in a series of experimental designs to test the sensitivity of critical components to dimensional variation.
- First-time test failures: Manufactured products usually undergo some form of quality testing before shipment, usually expressed as a series of metrics. This information is reviewed in isolation as trend data, but is often not explicitly correlated on a continuous basis with other available data such as machine tool life and component inspection results.
- Component inspection data: statistical process control is often used to monitor the manufactured output at the component level, expressed in various indices such as C_p (potential capability), C_{pk} (process capability) and C_{pm} (a modified process capability, emphasising 'on target' performance) – with the proviso of a stable process [2]. But this process data is rarely fed back to the design stage of a new or modified product, where the selection of a particular tolerance may exceed the capability of the manufacturing facility.

Most of this information remains isolated, and is often never visible to designers responsible for specifying tolerances. At the very least, the data is not available in a timely, seamless fashion to expedite its use. Deciding the tolerances for a part is traditionally a consultation process between departments or organisations. It is a time consuming, iterative task, which is conducted by review and may require several drawing up-issues. Furthermore, the time-honoured activity of drawing production (now usually generated from a geometric CAD model) does not interactively call for the use of process capability data. Yet this is the stage at which tolerance decisions are made, and where SPC data should be transparently available to support tolerance assignment. The tolerance framework aims to address this requirement, and this paper introduces the fundamental concepts of its architecture, and the direction of the research programme. The research will establish how this data can be captured, stored and interactively used in a CAD system, in the pursuit of a more robust product. The framework will also support the *association* of data pertaining to the part, including tolerance analyses contained, for example on spreadsheets stored separately from the CAD model.

OVERVIEW OF THE PROPOSED FRAMEWORK

Architecture of the Framework

The bedrock of the entire framework is the integration of the PDM and CAD systems. Figure 2 illustrates the underlying software environment

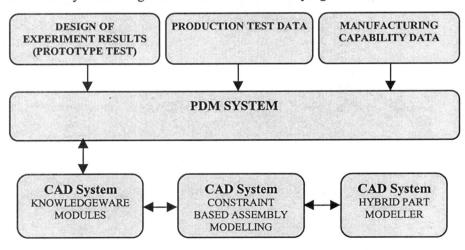

Figure 2: Software Environment of the Framework

of the framework. The PDM system manages, stores and shares the data pertaining to production and prototype testing, together with process capability. The data relating to these activities may not actually reside in the database shipped with the PDM system: they may actually exist on a remote part of an organisation's network possibly in a variety of database or application formats (for example, an Excel spreadsheet or an Oracle database).

The CAD system is used to create, modify and store constraint-based geometric models and assemblies. These part models should be built on a feature-based approach, enabling a logical decomposition of the part for subsequent analysis and for ease of modification [3]. The CAD system also exploits the growing knowledgeware functionality from vendors. The PDM system is essentially used as a medium to centralise and format the required data for the CAD system's knowledgeware front-end.

Definition of A General Process Capability Model

At the heart of the framework is a generalised model to represent the available manufacturing capability (Figure 3). This is the main logic 'engine' of the framework, which (with user interaction) will consider the function of the part, its feature composition and its position in the context its parent assembly within the CAD environment. The model is intended to

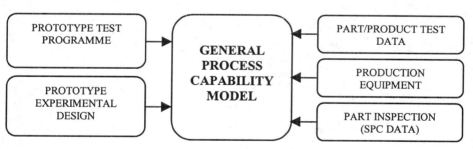

Figure 3: General Process Capability Model

address the commercial needs of a variety of manufacturing industries including fabrication, metal machining, plastics moulding and printed circuit board manufacture. It would therefore undertake to support the common denominators from these activities, which could include data from the manufacturing process tools (such as machining centres), component inspection data, and test data at a component or product level. In addition to data relating to production, it should also be able to consider prototype design and testing. Before actually using the framework, the functional

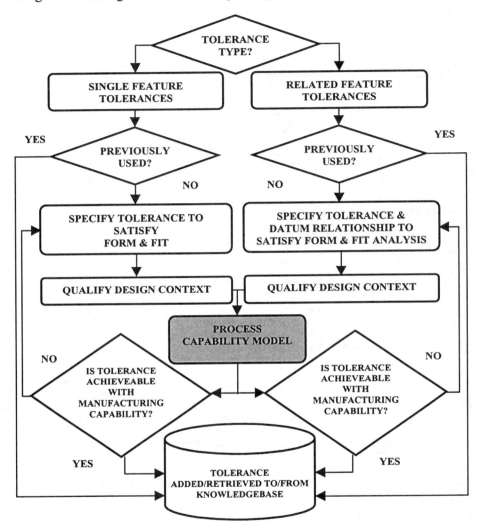

Figure 4: The Concept Framework

subassembly will be constructed using the logical constraints available in the CAD system. Typically, these might include adjacency, concentricity and co-axiality conditions, depending upon the richness of functionality in the CAD system. This step enables the designer to view the part in context, and helps to define the primary and any secondary functional datums on the part. Once the assembly definition is complete, the next step is to identify the geometric and linear tolerances for dimensional control. As previously discussed, there are tools available both internally (for example the Valysis product) and externally (such as an Excel spreadsheet) to the CAD environment to support functional tolerance allocation.

At this point the new framework is used, and is illustrated in Figure 4. The designer selects whether the tolerance is a single feature tolerance (for example flatness, straightness or cylindricity), or a related feature tolerance (for example parallelism, position or concentricity). The initial tolerance from the functional form and fit analysis is called up. The framework also considers the context in which the tolerance assignment is being made – if there is additional demand from other factors such as deep bore machining or limitations with the position of the fixture for example. The proposed values are then passed into the process capability model for evaluation, providing feedback to the designer as to whether the chosen tolerances are suitable for the available manufacturing process technology and resources. Where relevant, the process capability model can also be used to evaluate if the product is likely to meet performance requirements, which can be disseminated from Experimental Design (for example Taguchi results). If there are anticipated capability issues, the designer can re-specify a new tolerance value, or, if the value is acceptable, the feature and its tolerance may be stored via the CAD system's knowledgebase for future reuse. As mentioned earlier, the interface should be integral with the CAD system, which centralises both the decision-making process and the justification for the tolerance selection. The framework will implement Open Database Connectivity (ODBC) to enable the CAD/PDM system to link into the most commonly used relational database systems such as Access, SQL Server, Oracle and Visual FoxPro [4]. It is worth noting that it would be commercially desirable to enable the framework to use the Object Linking and Embedding Database (OLE DB) specification, which defines interfaces for gaining access and manipulating all types of data. In short, it enables the use of both relational *and* non-relational databases and will ensure the greatest flexibility of implementation [5].

An Industrial Case Study

BOC Edwards designs and manufactures vacuum pumps and accessories for the Scientific, Chemical and Semiconductor manufacturing industries. The company employs a cellular manufacturing strategy, and is currently installing shop floor co-ordinate measuring machines (CMMs) in each machining cell. The intention is to build a manufacturing database, primarily displayed in terms of control charts deriving C_p and C_{pk} indices. At present the structure of the database is driven by the requirements of the Quality Department - a somewhat limited audience. There are no plans at present to externalise this data to other departments within the company, and certainly no means of feeding the raw data back into the product design. The research program undertaken at BOC Edwards will review the range and depth of the currently available SPC data, and investigate data on pump first-time test failure. It will also review data on tool life, stored in the controller memory on each machining centre, and investigate characteristics of different machine tools: for example accuracies of position and circular interpolation. A Correlation of the data for machining centre tool life, SPC data and pump first time test failure can be made.

The first step is to develop a process capability model specifically for BOC Edwards. The task here is to define the constituent elements of the model to represent the particular processes of the company, encompassing manufacturing capability (manifested as SPC data), prototype performance test results (from experimental design), machining-centre characteristics, tool life data and production first-time test results (Figure 5).

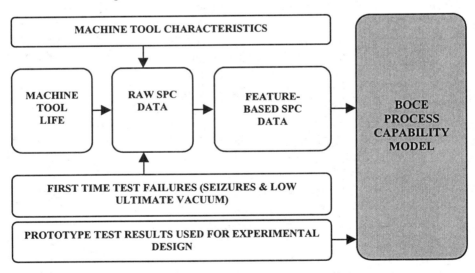

Figure 5: Relationship of Key Elements in the BOCE Process Capability Model

SPC data will be the main element of the capability model. In BOC Edwards, SPC data will be available for all critical part tolerance evaluations. However, this data has to be presented in a useful way to the designer, on a feature basis, which is how the design data is organised and stored in the CAD system. The process capability model should cater for the available manufacturing resource, (i.e., machining centres) since each available machine tool will have a set of physical machining characteristics (such as machining envelope and available accuracy), which can be stored as attributes. The tool life trend on these machines will also be considered, particularly in the context of machining critical component features. This will help establish the worst-case process capability, where worn tooling will allow the process mean to drift off the tolerance mid value. These SPC results can also be correlated with the first time test results database, illustrating the effect on product performance of critical parts, which are not made to nominal size. In addition, the process capability model is designed to allow the use of experimental design test data (to determine 'robust' design). The model should consider the relationship and weighting of these factors in the context of a feature-based tolerance definition. The model will be devised in a way consistent with its intended implementation as a CAD system integrated program.

Figure 6 illustrates the key components of the prototype system. The PDM system underpins the entire implementation, which will be used to manage, store and share information relating to this engineering research

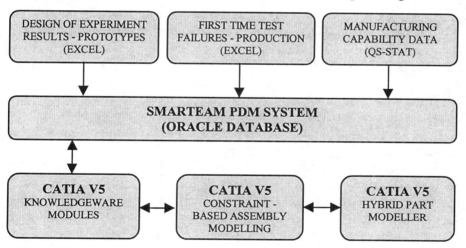

Figure 6: Components of the Prototype Tolerance System

project. The product selected is SmarTeam (from Smart Solutions Ltd), which when shipped with CATIA V5 from Dassault Systemes, is equipped with an Oracle database. Process capability data will be accessed from the company QS-STAT database. The PDM system will enable access to the experimental design results used for prototype testing, held in Excel spreadsheet format. CATIA V5 will be the primary part and assembly-modelling engine for the system. Dassault Systems have a suite of Knowledgeware modules for CATIA, which facilitate the creation of dedicated applications, completely integrated within the CAD environment.

INDUSTRIAL BENEFITS

The key benefits of the research will be to pioneer a system to reduce first time product test failures, reduce the scrap and rework rate, and to reduce the time and human resource absorbed in redesign due to unachievable manufacturing tolerances. The overall objective is a more robust product, insensitive to normal variations in the manufacturing process. The framework also supports the new product introduction process by identifying any shortfall in available manufacturing capability: the project management team can then plan to make new capital investment or identify a suitable subcontract source.

CONCLUSIONS

This CAD-integrated optimisation system addresses the shortfall of current tolerance analysis systems which do not provide an 'in-context' perspective of process capability or make provision for the direct use of experimental design data, within the CAD system environment. The aim of this research is to enable these prototype test results to be used in conjunction with the manufacturing capability model to establish the best compromise between product performance and process capability. The research therefore aims to endow the designer and the manufacturing resource planner with a complete vision of how the tolerance chosen for form and fit, compares with the capability available from the manufacturing process to achieve the target tolerance. It therefore complements existing functionality in CAD/CAM systems, in particular the ability to create constrained assemblies; to apply tolerances to geometric models; and the use of these tolerances in statistical analysis. In the research company, the overall impact of the project will be to reduce the level of first-time test failures that are attributed to pumps containing parts that are not made to specification. It will also eliminate the need to rework and redesign parts that cannot be manufactured due to originally specifying unachievable tolerances and reduce machining scrap. In addition, the designer will be confident that a 'robust' product has been produced, which is not subject to normal variations in the manufacturing process.

ACKNOWLEDGEMENTS

Many thanks to Ian Currington, David Wong, Sia Abbaszadeh, Alan Holbrook and Mo Kalta at BOC Edwards for their encouragement and financial support of the research.

REFERENCES

1. **Creveling, C.M.** (1997), 'Tolerance Design - A Handbook for Developing Optimal Specifications', published by Addison Wesley Longman, Inc.
2. **Berezowitz, William A., Chang, Tsong-How** (1995), 'Capability Indices – Somewhere the Point Got Lost', Technical Paper MS95-148, published by Society of Manufacturing Engineers, Dearborn, Michigan, United States.
3. **Eadie, R.G., Gao, J.X.** (1999), 'An Industrial Implementation of Manufacturing System Integration Through Optimisation of CAD Data', Proceedings of the 15th International Conference on Computer-Aided Production Engineering, published by the University of Durham, England.
4. **Caron, Rob, Larsen, Paul** (1999), 'Data Interoperability using Enterprise-Wide Data Sources', published on the Internet by Microsoft Corporation msdn.microsoft.com/library/backgrnd/html/dataint.htm.
5. **Vaughn, William R.** (1998), 'Hitchhiker's Guide to Visual Basic and SQL Server', published by Microsoft Press, United States.

Integration of Cables in the Virtual Product Development Process

Elke Hergenröther and Stefan Müller
Fraunhofer Institute für Graphische Datenverarbeitung e. V.

Abstract: *In car- and aircraft industry virtual environments became an important technology in the product development process. The goal is to replace the physical mock-ups step by step with digital mock-ups. Using digital mock-ups the development time and the mock-up production costs can be reduced. The assumption is that digital mock-ups should have the same functionality as the physical mock-ups. Today bendable elements like cables, pipes or rubber plugs are not integrated in the virtual prototype development process. In this paper we analyse the integration of virtual cables in the digital mock-up development process. We also present a prototypical application in a virtual environment (VE), which one can use for the installation of a wiring harness and for assembly.*

Key words: manufacturing, virtual prototyping, user interface for virtual environments

1. INTRODUCTION

In both the car- and the aircraft industry virtual environments have become an important technology in the product development process. The goal is to replace the physical mock-ups step by step with digital mock-ups. Using digital mock-ups reduces both the development time and the mock-up production costs. The assumption is that digital mock-ups should have the same functionality as the physical mock-ups. Therefore, the integration of cables, pipes and rubber plugs as bendable elements in the used application is necessary. Until today these types of elements are not sufficiently integrated in the virtual prototype development process. To compensate these disadvantages we developed a cable installation and manipulation application in a virtual environment. This application can be used for the installation of a cable harness and for the assembly and disassembly simulation.

2. ANALYSE OF THE ACTUAL SITUATION IN THE DEVELOPMENT PROCESS

The construction phase of a car can be divided in two different parts. At first the mock-up is designed in the design phase. Here the shape of the car or the aircraft is fixed and it is planed how much space the single elements like motor, air filter and so forth get. Afterwards the different departments in the industry develop the single elements. The digital result of this phase is a CAD-model. In the assembly phase, which follows the design phase, it will be controlled if the CAD-models of the single elements are suitable. Until today the assembling is tested on physical mock-ups the most time. In this article only the digital assembling is of interest. The errors detected during the assembly phase have to be corrected in the following design phase. The design phase and assembly phase are interrelated to each other. Therefore it is important that an application for the installation and manipulation of wiring harness can be used in both the design phase and the assembly phase.

2.1 The Design Phase

During the design phase the two-dimensional plans for the cable installation are transformed by hand in a three dimensional model. Therefore CAD-Systems like CATIA [CATIA] or Adams [Adams] are used. In CATIA the cable shapes are calculated as splines. The cable length can be fixed and the maximal flexibility can be chosen. Also the start and the end orientation of the cables are adjustable. The disadvantage of the cable simulation module in CATIA is that not every combination of parameters (cables start and end-point, material components, etc.) produces a cable. If no cable could be created the user gets no hints, why the creation failed. Also the physically behaviour of the cable in the field of gravitation could not be simulated in CATIA. This missing physically behaviour, the non-intuitive usability and the sometime failed cable creations are the main disadvantages of the cable simulation in CATIA and perhaps in other CAD-systems, too. Another other kind of problem is that CATIA does not notice which cables are included in a harness. This missing information makes the manipulation of a harness very difficult.

During different interviews with persons, who are working in different departments of car construction, we had to learn that the exchange of datasets between different systems is most important. The best situation is that one can do all tasks in one system, because during every transformation of datasets information gets lost. Now the real advantage of the common CAD-systems is obvious. In the most CAD-systems one can do all tasks, which are needed during the construction phase in the same system. In most cases a constant dataset is more useful than a more efficient application.

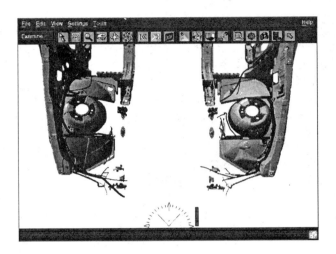

Figure 1. Cable installation with CATIA

2.2   The Virtual Assembly Phase

After the construction phase the virtual prototype has to be assembled. In this phase, the assembly phase, virtual reality applications are used. The CAD-dataset has to be converted in a dataset for a virtual reality (VR) application. In the VR application errors can be detected interactively. The most common kind of failure is the penetration, which is not visible in the CAD-system. Since a few years VR is used to detect if the designed parts match together. For example to test if a screw and a thread is suitable. After an error was found the CAD-model has to be modified. Of course here the data exchange could be also necessary, but until today the advantages of virtual reality applications are more useful than the advantage of a constant dataset. Actually none of the known VR-systems includes deformable cables, pipes and rubbers for the assembly test.

3. SYSTEM FOR THE INSTALLATION AND MANIPULATION OF WIRING HARNESS

The advantage of the immersive 3D-environment (like CAVE or Powerwall) is that one can see the virtual product in its real size. In such an environment the virtual assembly tests can be performed in a very realistic manner. During the virtual assembly it should be tested if the designed elements match together and the best way for assemble different elements should be find out. Therefore the flexible elements like, cables, pipes or rubber plugs have to be pushed aside in a natural way.

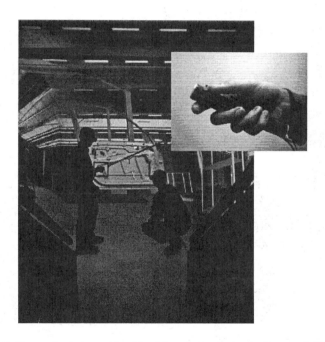

Figure 2. Cable manipulation with a Flystick as input device in the cave [IGD-CAVE].

To realize this prototypical application some developments were necessary. In this chapter the single developments are descript. We had to develop a cable simulation algorithm, a harness management, function selection and some interaction techniques and metaphors. After the chapter study you will see our goal was the implementation of an effective an easy usable.

3.1   The Cable Simulation

The most important demands are visually correct results, real-time calculation and stable calculation behaviour. Another significant demand for a cable simulation used in an assembly application is the constant cable length during the simulation. If the cable is fixed on both ends then it should not get longer while the user pushes it aside. To fulfil these demands we developed a kinematics based cable simulation algorithm presented in [Hergen'00]. The bending property is also important in this context. The kinematical chain is complemented with mass points and rotation springs. Thereby the simulation of different materials is possible and consequentially a different bending property can be calculated.   So the algorithm can simulate deformable, but inelastic and relatively stiff objects very well. It is a hierarchical method to ensure a real-time calculation on different processors. During the user interaction, a coarser resolution of the cable is simulated. If the user stops the cable movement, the resolution of the cable becomes refined, and so the surface becomes smoother.

3.2   The Wiring Harness

In reality, a vehicle needs a lot of cables and most of them are handled in a cable harness. So the next challenge of this application is the handling of the high complexity of the harness. To reduce the complexity of the cable harness for the simulation, we divide it into different cable sections.

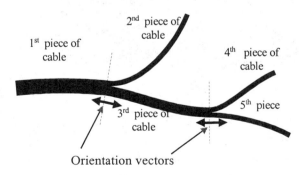

Figure 3. A wiring harness assembled from five sections

In Figure 3 one can see an example of a divided harness. After the simulation of the individual harness pieces, the different cables should result in a smoothly coherent cable harness. We solved this problem using orientation vectors (Figure 3). They are placed at the start point and at the end point of every piece of cable and describe its orientation. It is also necessary to notice the cable, which build a harness. Our system memorizes the cable of a harness.

3.3   Function Selection

Also in our application we need the possibility to switch between various functions in a very simple and fast manner. In tests with different menu layouts it turns out that the pie menu [Callah'88] is the most suitable for our application. A pie menu is defined as a circle with 4 or 8 evenly distributed menu items (Figure 4). When opening the menu, the cursor is in the centre of the circle. The advantage of that type of menu is that all items are at the same distance from the initial position of the cursor.

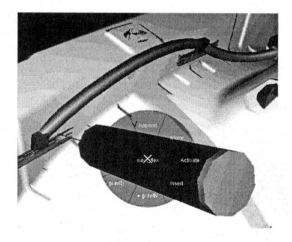

Figure 4. The pie menu

3.4   Create a Harness With the Laser Beam Technique

In this area of application the interaction with the single cables and with the whole cable harness is very important. To realize an effective input device, we adopt a well-known method for placing and selecting the cable's attachment points in the scene: the virtual laser beam [Bowman'97]. It is a very intuitive technique to select objects, which are not far away. The virtual beam is emitted from the hand in the direction the user points. An attachment point is selected if the beam penetrates it and the user triggers an event, like pressing a button on the input device.

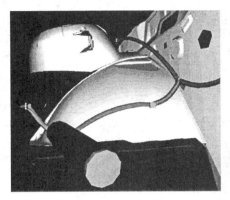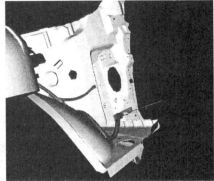

Figure 5. Cable installation in Virtual Environments from different points of view

3.5   Virtual Reality Input Devices

For immersive environments we use the Flystick [Knoepf'00] as input device. It is a tracked pen, equipped with three buttons. The position selection with the Flystick is intuitive and rather accurate. Desktop applications are more common than immersive environments are.

Figure 6. Using the Phantom as input device

However, the problem of interaction in the 3D space remains. CAD applications offer the use of a spacemouse, a 2D mouse and a keyboard. Using a spacemouse is difficult for an untrained person. Therefore we search a more intuitive input device. We decided to test the application with the Phantom [Phantom]. The Phantom is an absolute input device for 3D desktop applications. It is a small robot arm with up to 6 degrees of freedom. The direct user input device is a pen with one button. In our tests we use the phantom without force feedback. They show that that the position selection is fast and quite precise.

3.6   Interaction With Cables During the Assembly

During the assembly phase the cables have to be pushed aside very often. Thus we had to realize this functionality in our application at first. In reality the user does this task mechanically, because it is so easy. But in VR-systems that task is not realized till this day. At first we use the laser beam method. Original, we used this method to position of the harness attachment points (chapter 3.4). Now we use the laser beam technique to push the cable aside. At first we have to select one attachment point and than the cable is pushed to the point where the beam penetrates an object. Using the beam, the cable could only move along another object. That is suitable during the installation phase, but during the assembly phase the cable should be moved freely. So the laser beam technique is unsuitable for the assembly phase.

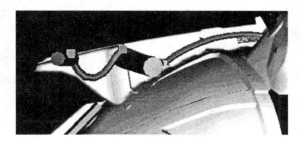

Figure 7. Moving the cable with the virtual copy of the Flystick

It makes more sense to push the cable away with the virtual copy of the input device. Figure 7 shows this operation. Pushing the cable aside by using the Flystick is very similar to pushing the cable aside by hand. Unhappily this method is unsuitable for a virtual assembly application. The reasons for this will be understandable after the explanation of the method's realization. If a collision between cable and virtual Flystick is recognized then the cable and the Flystick model are connected with each other on the collision point. If the user moves the Flystick then the cable moves, too. The problem is the connection of Flystick and cable. The cable *glues* on the Flystick, because the connection point cannot be changed during the interaction except the user triggers an event, which separates cable and Flystick. The steady connection has two disadvantages. The Flystick occludes the cable most of the time. The second disadvantage is the not existing visual feedback if the cable cannot be pushed or bowed any more. Without a visual feedback the user could not work with the application. The lack of user support is the reason why this method is not usable.

The next idea was to use of a second copy of the Flystick, the ghost. During a successful cable pushing the ghost is not visible. The ghost becomes only visible if the cable's movement stops. In this case the ghost represents the movement of the Flystick. This method is more effective than the former methods, but also with the ghost the user does not know why the cable does not move. To give the user a really informative feedback, we developed the cable ghost.

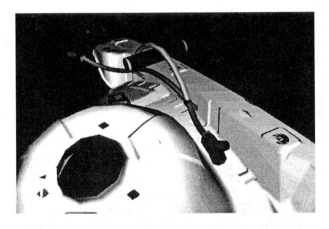

Figure 8. The cable ghost (green)

The principle is similar to the Flystick ghost. If the cable movement stops, because it cannot be stretched or bowed anymore, then the cable ghost becomes visible. The cable ghost can be realized as a spline or as an elastodynamical simulation. We chose a mass-spring simulation. On one hand the ghost cable can follow every movement of the Flystick, because it is highly deformable. On the other hand the kinematically simulated cable shows the realistic behaviour of the cable. Together both cable representations build a well usable and effective interface for the interaction with not elastically deformable objects.

3.7 Discussion of the Interaction Paradigms

Our fundamental idea was the development of a user-friendly application. During the interaction the user should be supported as much as possible. This principle is reflected in all parts of the application also in the cable simulation itself. We made tests to choose the most accepted input device. It is the Flystick for immersive environments and the Phantom for desktop applications. The used interaction features like pie-menu or laser beam are accepted techniques. The use of a deformable ghost object is new in this type of application. The ghost object is highly deformable and can represent all user movements. The deformation of the ghost is the pattern for the deformation of the virtual cable model. This combination of the two cable representations provides the almost mechanical user interaction and the information about the real cable shape.

4. CONCLUSION AND FUTURE WORK

The digital development process in the car- and aircraft industry can be divided in two parts: the design phase, where the construction of the product is done, and the assembly phase, where the correction of the mock-up is done. The presented application has many functions, which are needed during the design phase [Hergen'01a][Hergen'01b]. The integration of haptical feedback has to be done in the future. This will make the cable installation faster and more accurate. Also the connection to a CAD-system is needed. At the moment we work on the expansion of the functionality for the assembly phase. The development of the cable ghost was a very important development step, because this paradigm enables an almost mechanical user interaction.

5. REFERENCES

[Adams] http://www.adams.com
[Bowman'97] D. A. Bowman, An Evaluation of Techniques for Grabbing and Manipulating Remote Objects in Immersive Virtual Environments: *Symposium on Interactive 3D Graphics*, Providence, USA, 1997, 35-38.
[Callah'88] J.Callahan, D. Hopkins, M. Weiser, B. Schneiderman, An empirical comparison of pie vs. linear menus, *Proceedings of SIGCHI 1988*, New York, USA, 1988, 95-100.
[CATIA] http://www.catia.com

[Hergen'00] Hergenröther, E., Dähne P.: Real-Time Virtual Cables Based on Kinematics Simulation, *The 8th International Conference in Central Europe on Computer Graphics, Visualization an Interactive Digital Media'2000*, Plzen, Czech Republic, 2000, 402-409.

[Hergen'01a] Hergenröther, E., Knöpfle, C.: Cable Installation in Virtual Environments, *Proceeding of the IASTED International Conference on Modelling and Simulation* , Pittsburgh, USA, 2001, 276-280.

[Hergen'01b] Hergenröther, E., Knöpfle, C.: Installation and Manipulation of a Cable Harness in Virtual Environments, *Proceeding of the IASTED International Conference on Robotics and Manufacturing*, Cancun, Mexico, 2001, 240-244.

[IGD-CAVE] www.igd.fhg.de/igd-a4/index.html

[Knoepf'00] C. Knöpfle, G. Voß, An intuitive VR user interface for Design Review, *Proceedings of Advanced Visual Interfaces 2000*, Palermo, Italy, 2000, 98-103.

[Phantom] Developed by MIT Touch Lab: http://touchlab.mit.edu/

CONSTRAINED FITTING — A KEY ISSUE IN REVERSE ENGINEERING CONVENTIONAL PARTS

Pál Benkő
Tamás Várady
Computer and Automation Research Institute, Budapest
benko@sztaki.hu

Abstract   Constrained fitting is the process of approximating segmented point sets simultaneously by multiple surfaces while certain geometric constraints, such as tangency, perpendicularity, parallelism, concentricity are satisfied. This technique is particularly important in the context of reverse engineering, where geometrically and topologically consistent, near to perfect CAD models must be created. In this paper various representational and numerical problems are discussed in order to make constrained fitting computationally efficient even for large, multiple point clouds. Special emphasis is taken to resolve contradicting constraints. Simple examples of handling smooth profile curves and tangentially connected face sets are also presented.

Keywords: reverse engineering, analytic surface fitting, constraints, minimisation

Introduction

Reverse engineering is the process of converting 3D, multiple view measured data into an evaluated boundary representation model [9]. There are many publications concerning the reconstruction of geometrically demanding free-form objects [4], but in this paper we concentrate on topologically complex solid models bounded by simple analytic surfaces, such as planes, cylinders, cones, spheres, tori and small blending surfaces [5]. While in many graphical and computer vision applications approximate models are sufficient, our main interest is to create such boundary representation models, which can directly be used in commercial CAD/CAM systems through standard data exchange interfaces, such as IGES or STEP. In this way, the benefits of using the existing CAD/CAM technology can be directly applied for the reconstructed objects as well, including (re)design, analysis, simulation and NC manufacture.

The reverse engineering procedure can be decomposed into several phases, including data acquisition (typically by laser scanners), merging multiple point clouds, triangulating and decimating data sets, segmenting a point cloud, fitting surfaces for individual regions, building B-rep models and reconstructing blends. As it has been analysed in [1], an ideal reconstructed model must satisfy several requirements in order to be usable in an engineering environment. The bounding surfaces must be accurate, their surface type and the related parameters must be reliably estimated. Complex linear extrusion surfaces and surfaces of revolution must be recognised, and represented as special, smoothly connected surface elements.

For B-rep model building *smooth edges* must be explicitly detected and reconstructed, otherwise not only imperfect models are generated, but the reconstruction process may crash at the intersection of nearly tangential surfaces. In various cases, such as vertices with more than three valencies, special *topological constraints* need to be satisfied. Difficulties emerge partly due to noise, which is always present in the measured data, partly because the estimation of the various geometric entities, which also carry certain errors. The final model must be consistent from both topological and geometric points of view; this is why a higher level intelligence is necessary to incorporate certain 'likely' constraints into the reconstructed model. A final group of requirements concerns various — local or global — engineering constraints [3, 6], such as enforcing parallelism, perpendicularity, tangency, concentricity, or symmetry amongst various geometric entities.

In this paper the concept of constrained fitting is introduced, which is a technique to overcome the above mentioned difficulties. While there is a vast literature on constraints in solid modeling and on individual surface fitting, there is only one project known to the authors, where these two areas have been coupled for reverse engineering applications — see [10]. The novel method presented here is computationally efficient and capable of *resolving conflicts* between constraints according to a given priority. Computational considerations of applying so-called *auxiliary elements*, and speeding up the computation by separating the terms holding the point data and the surface parameters are also discussed. Finally, several application examples are given, such as, the reconstruction of smooth, constrained profile curves, the decomposition of smooth, multiple surface regions and enforcing various types of general constraints.

1. Mathematics of constrained fitting

Given a set S of curves and surfaces, and for each $s \in S$ a point set P_s to be approximated. Each entity is characterised by a hypothetical curve or surface type and unknown parameters a_s, which are concatenated in a vector a. There is a given set of constraints in the form of $c_i(a) = 0$ what we want to satisfy. The constraints are specified by the user or derived by the reverse engineering system itself. Our goal is to approximate simultaneously several surface elements while the constraints are also satisfied.

The problem can be stated as minimising a function $g : \mathbb{R}^n \to \mathbb{R}$ on the zero set of another function $c : \mathbb{R}^n \to \mathbb{R}^m$ (the concatenation of c_i's). In our case g is formulated as

$$g(a) = \sum_{s \in S} \alpha_s \sum_{p \in P_s} f(p, a_s)^2,$$

where $\alpha_s \in \mathbb{R}$ is a weighting term and $f(p, a_s)$ approximates the distance of point p from surface s.

The unconstrained minimisation by Newton-Raphson iteration would go by picking a starting point, and computing a step from the second order Taylor-approximation at the current point. Since g is a sum of squares, this approximation is positive semi-definite. The usual method for marching on a given surface is stepping on the tangent plane and projecting the result to the surface. Projection to the implicit surface $c^{-1}(0)$ can be solved again by Newton-Raphson iteration using the first order Taylor-approximation of c at the current point. To avoid double iteration, these methods are combined in our constrained minimisation scheme (as described in details in [2]). In each iteration step of the minimisation just the first step of the projection iteration is made; an approximation of the tangent plane is computed, and the minimisation step is performed on it. Computing the tangent plane approximation is done sequentially, by traversing all constraints. In this phase it is recognised whether a constraint follows from the previous ones or contradicts them. In these cases the constraint is ignored: in the former case nothing needs to be done, in the latter case nothing can be done.

At the end of the iteration the result is on the constraint surface, and the gradient of g is orthogonal to it. Initial values for the iteration are generally obtained from independent unconstrained fits.

1.1. Object representation

Regular surfaces are represented in implicit form: $\{p \in \mathbb{R}^3 : f(p, a_s) = 0\}$. a_s contains a well-defined set of variables, which uniquely define the surface, such as dimensions, positions, radii, angles, etc. Such a description can be used for minimising $\sum_{p\in P} f(p,a_s)^2$. In order to get a meaningful result, such descriptions are requested, which well approximate the Euclidean distance $d(p,a_s)$. A useful notion is the *faithful* approximation of the Euclidean distance [7]: f approximates d faithfully iff for any fixed parameter vector a_s they are equal up to order one on the surface described by f.

During the minimisation of $\sum_{p\in P} f(p,a_s)^2$ the set of points P is fixed and a_s is modified in each iteration step. To make the iteration *efficient*, f is searched in a special form. When the sum of squares is computed, P should not be 'traversed' in each iteration step, but some quantities (depending on P) should be computed in advance in a preprocessing phase, thus the sum can be calculated in constant time. This is possible if f is given in the form $f_1(p)^T f_2(a_s)$ (f_1 and f_2 are vector valued functions of the same dimension); in this case

$$
\sum_{p\in P} f(p,a_s)^2 \;=\; \sum_{p\in P} f_2(a_s)^T f_1(p) f_1(p)^T f_2(a_s)
$$

$$
\;=\; f_2(a_s)^T \left(\sum_{p\in P} f_1(p) f_1(p)^T\right) f_2(a_s) = f_2(a_s)^T A f_2(a_s).
$$

For a Newton-Raphson iteration, the first two derivatives of f must be computed. If f is in this form, then

$$
\sum_{p\in P} f'(p,a_s) \;=\; 2 f_2(a_s) A f_2'(a_s)
$$

$$
\sum_{p\in P} f''(p,a_s) \;=\; 2\left(f_2'(a_s) A f_2'(a_s) + f_2(a_s) A f''(a_s)\right).
$$

These formulae may still be difficult to compute, so a form where f_2 is as simple as possible is requested. A simple solution is to use another description vector $a_s' = f_2(a_s)$. In that case $f_2(a_s') = a_s'$, and the derivative formulae become really simple:

$$
\sum_{p\in P} f'(p,a_s') \;=\; 2 A a_s'
$$

$$
\sum_{p\in P} f''(p,a_s') \;=\; 2A.
$$

Note that the above transformations must be performed with care: if the dimension of a_s' is much greater than that of a_s, then the size of the system grows, which of course makes computation slower; moreover, it may be difficult to compute the new constraints which must have been introduced.

Later, simple examples will be given to show how efficient representations can be created.

1.2.   Auxiliary objects

Certain constraints can be described just in a very roundabout way, but can be naturally decomposed into simpler constraints by introducing an *auxiliary* object: this is an artificial object not present in the original problem formulation, and it does not need to approximate data points. For example, the constraint 'two cylinders, a cone and a torus go through a common point' may be needed. It can be circumvented by introducing a *point* object and four simple constraints that each surface goes through that point. Coplanarity of some points can be described by introducing an auxiliary plane and prescribing the points to lie on it. The main difficulty with auxiliary objects is that as they do not fit onto data points they cannot be initialised by unconstrained fitting.

1.3.   Simple objects and constraints

Lack of space prevents us to present all 2D and 3D object descriptions and the related constraint definitions (see [2]). Nevertheless, it was felt important to provide simple examples to show how efficient and faithful representations can be formulated. For simplicity's sake, here only 2D lines and circles are described, however, similar formulations can be found after some algebra for the 3D surface elements as well.

The Euclidean distance for lines can be given in an efficient form: if the unit normal is denoted by n and the signed distance from the origin by δ, the signed Euclidean distance of a point p is $\langle p \mid n \rangle + \delta$, which is in the desired form, choosing $f_1(p) = (p_x, p_y, 1)$ and $f_2(a_s) = (n_x, n_y, \delta)$. The normalisation condition $n^2 = 1$ completes the description.

The Euclidean distance for circles (spheres) cannot be written in an efficient form, but a faithful approximation given in [8] can be:

$$|p - o| - r = \frac{(p-o)^2 - r^2}{|p-o| + r} \approx \frac{(p-o)^2 - r^2}{2r}$$

$$= \frac{p^2}{2r} - \frac{\langle p \mid o \rangle}{r} + \frac{o^2 - r^2}{2r};$$

we obtain the desired form by choosing

$$f_1(p) = (p^2, p_x, p_y, 1),$$

$$f_2(a_s) = (\frac{1}{2r}, -\frac{o_x}{r}, -\frac{o_y}{r}, \frac{o^2 - r^2}{2r}).$$

This is still a bit complicated because of the divisions, however, by introducing a new description vector we can get rid of them. $a'_s = f_2(a_s) = (\kappa, o'_x, o'_y, \mu)$. Now $f_2(a'_s) = a'_s$, so computing $\sum f$ and its derivatives is as simple as for a plane. Because the dimension of a'_s is greater by one than that of a_s, a further constraint is needed, which can be derived from the formulae in $f_2(a_s)$:

$$o'^2 - 4\kappa\mu = 1.$$

This description has the advantage that straight lines are just special (and not singular) cases of circles: instead of r being ∞, $\kappa = 0$. In that case o' is the normal of the line, μ is δ, and the normalisation condition is the same as that for lines. From that description the radius and the centre can be computed easily: $r = 1/2\kappa$, $o = -o'/2\kappa$. These formulas are repeatedly used for setting up the constraint equations. For example, the constraint that a line crosses the centre of a circle is

$$\langle o \mid n \rangle + \delta = \left\langle \frac{-o'}{2\kappa} \mid n \right\rangle + \delta = 0$$
$$\langle o' \mid n \rangle - 2\kappa\delta = 0.$$

As mentioned before, there are many more object types and constraints in 3D, these are considerably more complex than the planar entities. The complexity is increased by the fact that cones and tori cannot be represented in a form which is efficient *and* faithful. In [2] we gave both a faithful representation and an efficient representation for these types.

2.   Examples

In this section a few examples are presented to illustrate the importance of constrained fitting in the course of reverse engineering CAD models.

2.1.   Contradictory constraints

This example shows the usage of auxiliary objects and demonstrates how the final solution depends on the order (priority) of the constraints when they contradict each other.

The example consists of a square and three circles in its interior. The constraints are the following:

- the circles have equal radii

- the centres of the circles are collinear, and their line is aligned horizontal

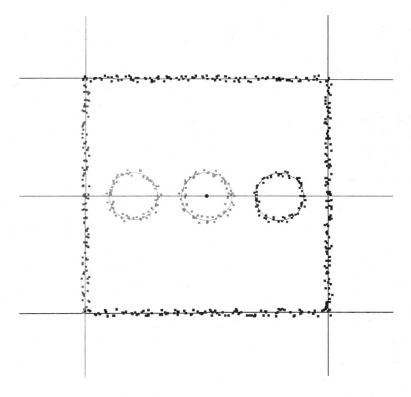

Figure 1. Initial state

- the centres subdivide the width of the square at equal distances of 30 units

- the length of the edge of the square is 100 units

To describe the collinearity of the centres, an auxiliary line object is introduced, and all centres are constrained to lie on it. The initial state does not depend on the constraints, so it is the same for all possible sequences of the constraints; as it can be seen in figure 1.

If the constraints are given in the sequence above, the last constraint will not be satisfied and the length will be 120 units, as shown in figure 2. If the 'centre spacing is 30 units' constraint is given last and ignored, then the result is figure 3. If the 'line of centres is aligned' constraint is given last, the result is shown in figure 4. If the 'centres are collinear' constraint is given last, the result is figure 5.

In practical reverse engineering the number of constraints may be much higher than those of the degrees of freedom, so the well assigned priorities are particularly important to get the right results.

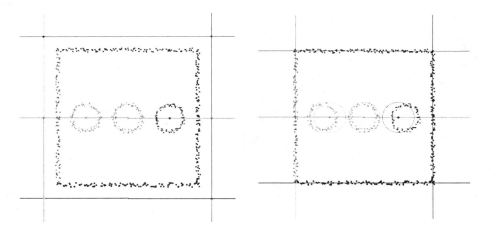

Figure 2.   length of square edge not 100   *Figure 3.*   spacing of centres not 30

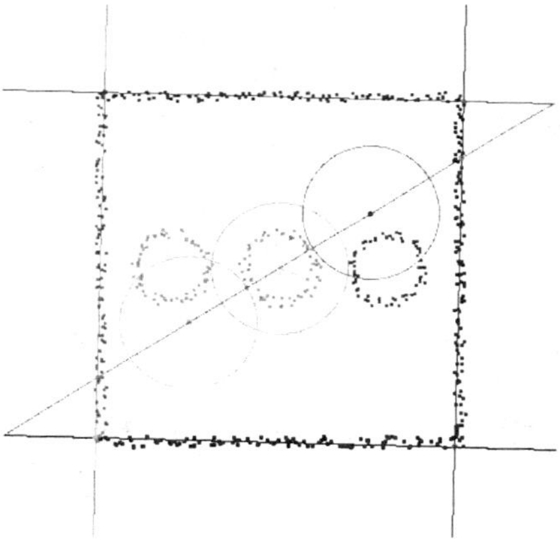

Figure 4.   line of centres not aligned 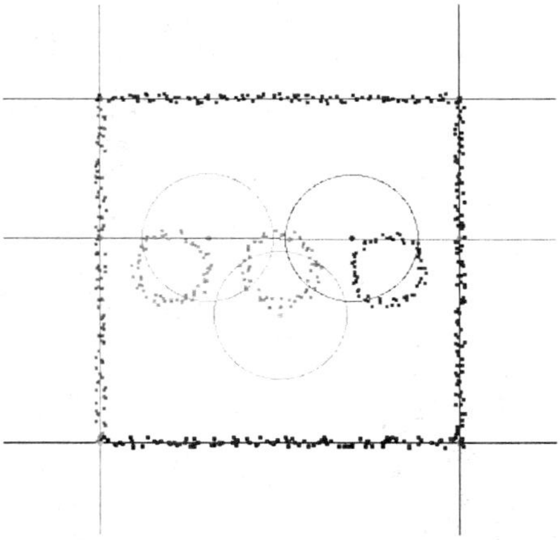   *Figure 5.*   centres not collinear

2.2.   Profile fitting

A very important subproblem in reverse engineering is the reconstruction of linear extrusions and surfaces of revolution. Both involve fitting a profile curve made up of smooth straight line - circular arc sequence to approximate a noisy and 'thick' planar point set. Thickness originates not only from the noise, but also from the inaccurate estimation of the best translational direction or the best rotational axis. Assume the point set is segmented; then circles and straight lines are fitted to the individual segments while adjacent elements are *constrained* to be

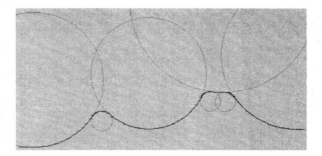

Figure 6.   Rotational
object
                        Figure 7.   Profile of smoothly connected arcs

tangential. We have found that the use of special auxiliary elements is useful for profile fitting as well. We can stabilise the system by forcing each shared meeting point to approximate some data points where the join is expected.

A simple object is presented in figure 6. After determining the best rotational axes, the smooth profile from the bottom of the object has been reconstructed, see figure 7. The thick point set and the full circles are also shown.

2.3.   3D examples

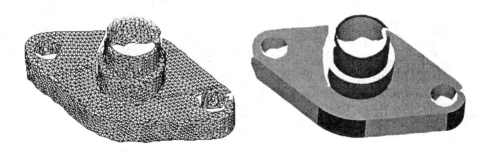

Figure 8.   Decimated point set
                        Figure 9.   Composite smooth region

Finally, two simple 3D test objects have been selected. The first one shows a single view data set, a decimated triangular mesh is given in figure 8. As can be seen, the side is an extruded, composite face set, the separating edges are smooth, these were computed based on the profile of extrusion. The other edges are sharp, and these were determined using surface-surface intersections, see figure 9. This object is a good illustration to show that incomplete models can also be reconstructed in a reasonably robust manner.

In many engineering objects, there are smooth, composite regions, where the boundaries of the primitive surfaces cannot be found directly by locating the abrupt changes of the estimated normal vectors — see also [1]. In such situations again the constrained fitting technique can provide an adequate solution.

The next example is a subset of a larger object. A decimated mesh can be seen in figure 10: there are seven regions which must join smoothly; three planes, three cylinders and a torus. In addition to the above 'face' objects further auxiliary objects need to be used, such as planes of the inner edges, or point-with-directions for ensuring the tangential connection at the innner vertices, or a line to ensure the coaxiality of two cylinders and the torus. Note, that in order to get a topologically consistent B-rep model, we must assure that the two inner vertices with valency four will represent a *single* 3D point. This fact must also be incorporated into the related constraint system together with various constraints including orthogonality and parallelism of planes, parallelism or orthogonality between planes and cylinder/torus axes, equality of radii, and so on. The final result is shown in figure 11.

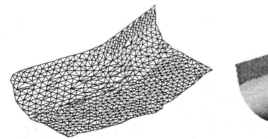 

Figure 10.   Decimated point set   *Figure 11.*   Composite smooth region

Conclusion

After describing a new numerical method to fit surfaces simultaneously to multiple point regions, the significance of this technique in reverse engineering was demonstrated. As it was shown, constrained fitting is necessary to obtain accurate and consistent CAD models, which can later be used in real engineering. Moreover, constrained fitting is necessary for model building: without constraints it would be hardly possible to locate smooth edges and determine coincident geometric entities. The importance of handling contradicting constraints has also been pointed out. Finally, constraints are to be used for 'beautifying' reconstructed models. For example, when 'almost' perpendicular faces or 'almost' coaxial cylinders have been detected, these must be set accurately once the user approves to do so. To recognise and enforce a complex set of constraints in a consistent way is a difficult problem and will be subject of further research.

Acknowledgments

Many thanks are due to Géza Kós (CARI, Budapest) and Ralph Martin (Cardiff University) for inspiring discussions on various problems of constrained fitting.

Pál Benkő is working with the Geometric Modelling Laboratory of CARI, Budapest since 1995. He has just written his PhD thesis supervised by Tamás Várady on reconstructing conventional engineering objects.

Tamás Várady is the head of GML since 1991. He received a DSc degree with his thesis 'Vertex blending surfaces in computer aided geometric design' in 1998. In 1990, he was the cofounder of CADMUS Consulting and Development, of which he is president.

References

[1] P. Benkő, R. R. Martin, T. Várady: 'Algorithms for Reverse Engineering Boundary Representation Models', *Computer Aided Design*, accepted

[2] P. Benkő, L. Andor, G. Kós, R. R. Martin and T. Várady: 'Constrained fitting in reverse engineering', *Computer Aided Geometric Design*, accepted

[3] B. Brüderlin, D. Roller, Eds.: *Geometric constraint solving and applications*, Springer, 1998.

[4] J. Hoschek, W. Dankwort, Eds.: *Reverse Engineering*, B. G. Teubner, Stuttgart, 1996

[5] C. M. Hoffmann: *Geometric & Solid Modelling: an Introduction*, Morgan Kaufmann publishers, inc., San Mateo, 1989

[6] C. M. Hoffmann, R. Joan-Arinyo: 'Symbolic constraints in constructive geometry', *Journal of Symbolic Computation*, Vol 23, 1997, pp. 287–300

[7] A. D. Marshall, G. Lukács, R. R. Martin: 'Robust Segmentation of Primitives from Range Data in the Presence of Geometric Degeneracy', *IEEE PAMI* Vol 23, No 3, 2001, pp. 304–314

[8] V. Pratt, 'Direct least-squares fitting of algebraic surfaces', *Computer Graphics* (SIGGRAPH 87), Vol 21, No 4, 1987, pp. 145–152

[9] T. Várady, R. R. Martin, J. Cox: 'Reverse Engineering of Geometric Models — An Introduction'; *Computer Aided Design*, Vol 29, No 4, 1997, pp. 255–268

[10] N. Werghi, R. B. Fisher, C. Robertson and A. Ashbrook, 'Object reconstruction by incorporating geometric constraints in reverse engineering', *Computer Aided Design*, Vol 31, 1999, pp. 363–399

Volumetric tools for machining planning

G. Renner and I. Stroud*
Computer and Automation Researchc Institute, Hungary
renner@sztaki.hu

*EPFL, DGM-ICAP-LICP, CH-1015 Lausanne, Switzerland

Abstract: The paper presents briefly, as an introduction, the MAT calculation method by subdivision, the so-called «divide-and-conquer» algorithm during which the Delaunay nodes are calculated. The paper goes on to describe how this method can be applied to calculating the MAT of negative objects. The negative MAT provides important information about the exterior of the object for material removal and maximal radii for access. The process of calculating the negative MAT includes the determination of infinite Delaunay nodes as well. The paper presents methods for recombining the Delaunay nodes according to certain rules for object reconstruction, offsetting and object subdivision. Finally the paper describes how these tools can be used for machining and machining planning.

Key words: MAT, offsetting, volumetric decomposition

1. Introduction

Currently the Boundary Representation (B-rep) technique is widely used as a solid model representation technique in CAD and CAM systems. However, one of the drawbacks of this technique is that it is essentially local in nature and there is no implicit notion of volumetric characteristics. For many applications, however, it is necessary to have this information. For example, in order to manufacture an object it would inevitably be necessary to have information on volumetric properties such as thickness, clearances etc. It is possible to test selected points to find material thickness, or other characteristics, but these points have to be selected using some, possibly, arbitrary rules and their computation is complicated and time intensive. In order to circumvent the volumetric reasoning problems this paper proposes the use of techniques based on the Medial Axis Transform surface (MAT). Techniques for calculation of the MAT have been developed over a period of several years now. A recent development is the use of general topological Delaunay nodes for object reconstruction, offsetting and subdivision. This is a development of the multiple start point algorithm described by Renner and Stroud [ReSt96]. Most algorithms mention only Delaunay triangles (in 2D) and tetrahedra (in 3D) but these are only the minimum structures. These may suffice to delimit a MAT vertex, but do not reflect the structure of the object. The generalised node structure is usually more complex.

The paper presents briefly, as an introduction, the MAT calculation method by subdivision, the so-called «divide-and-conquer» algorithm during which the Delaunay nodes are calculated. The paper goes on to describe how this method can be applied to calculating the MAT of negative objects. The negative MAT provides important information about the exterior of the object for material removal and maximal radii for access. The process of calculating the negative MAT includes the determination of infinite Delaunay nodes. The paper presents methods for recombining the Delaunay nodes according to certain rules for object reconstruction, offsetting and object subdivision. Finally the paper describes how these tools can be used for machining and machining planning. The object decomposition method uses rules to recombine the Delaunay nodes into groups corresponding to volumetric elements.

Offsetting technique are used for the computation of tool paths. In contrast with the usual offsetting process, the MAT-based offsetting is much quicker, because the proximity information is pre-calculated, only the recalculation is needed. This makes it possible for a user to use interactive trial-and-error methods to determine the cutter centre surface. The MAT is easy to use for constant value offsetting. This just uses a simple recombination technique to determine the topology of the offset object(s) and then recomputes the geometry. This is analogous to the 2d pocket machining case, where the 2D Voronoi diagram can be used for fast calculation of toolpaths.

2. Divide and conquer algorithm

The divide and conquer algorithm is a method for calculating the Medial Axis Transform surface of a three-dimensional planar polyhedral object, described in [ReSt2000] and [SRX2001]. The method developed by Reddy and Turkiyyah [ReTu95] and extended by Renner and Stroud [ReSt96] is what can be termed a «constructive process». In contrast, the divide-and-conquer is a «divisive» technique, which starts with the exterior of the dual of the MAT and then subdivides it into separate nodes corresponding to MAT vertices.

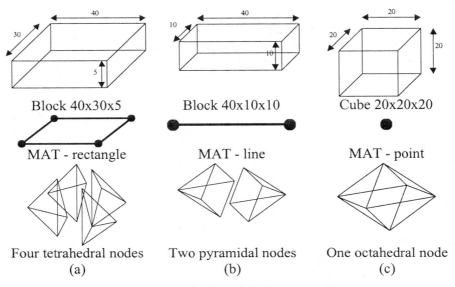

Block 40x30x5   Block 40x10x10   Cube 20x20x20

MAT - rectangle   MAT - line   MAT - point

Four tetrahedral nodes   Two pyramidal nodes   One octahedral node
(a)   (b)   (c)

Figure 1 - MAT and Delaunay nodes

This technique was termed «divide-amd-conquer» because, as the nodes are extracted, the dual structure falls into sub-pieces which are easier to subdivide, in turn, hence the MAT calculation process becomes easier, in theory. Another difference between the divide-and-conquer algorithm and the constructive processes is that the nodes, termed here «Delaunay nodes» determined in the constructive algorithm are assumed to be tetrahedra. Although the tetrahedral node is common it is only the minimum structure. Where there are degenerate parts of the object the node topology can be more complicated. To illustrate this, consider the case of a simple block in Figure 1.

The global structure, that is, topology of the union of the nodes is the same in each case, an octahedron. The different geometries of the three blocks causes the different subdivisions. In Figure 1a the octahedron is broken into four tetrahedra, and the MAT is a rectangle, each corner point corresponds to one of the tetrahedral nodes. In Figure 1b the square cross section of the block means that there are two square pyramidal nodes and the MAT is a line. In Figure 1c the square block means that there is only one single node, an octahedral node, and the MAT is a single point.

The starting point for the divide-and-conquer algorithm is the global structure. This is related to the topological dual of the object, as described in [ReSt2000] and [SRX2001]. For the cube, above, the global structure is exactly the topological dual but the topological dual has to be modified slightly to produce the global structure when there are concave edges and concave vertices in the object. The relation to the dual is because the vertices in the dual represent point touches by maximally inscribed spheres on the face, edge, or vertex. Because of this relationship it is easy to reconstruct the topology of the object, to reconstruct the offset topology of the object or to subdivide the object, as described in section 4.

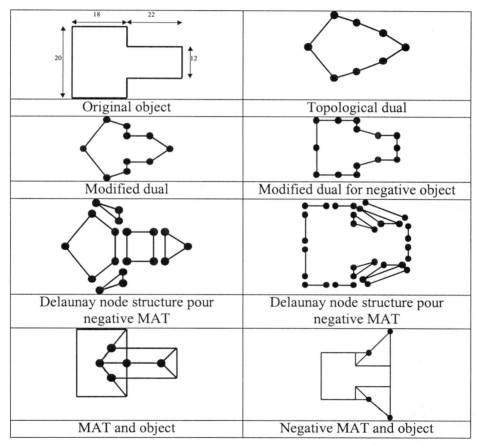

Original object	Topological dual
Modified dual	Modified dual for negative object
Delaunay node structure pour negative MAT	Delaunay node structure pour negative MAT
MAT and object	Negative MAT and object

Figure 2 - MAT and node structures

Figure 2 shows the steps in the creation of the MATs of positive and negative objects with the divide-and-conquer algorithm. For simplicity this is shown with a 2D object, the principle is the same for 3D objects. The first step is to create the dual of the object. This is then modified by adding nodes for the concave edges and vertices.

In 3D this is done by splitting the dual edges corresponding to concave edges of the original object. Concave vertex nodes are added by subdividing the corresponding faces in the dual, but the way in which this is omitted. The modification process is described in more detail in Stroud et al. [SRX2001]. The modified dual is the union of the Delaunay nodes corresponding to MAT vertices.

The calculation of the MAT is done by subdividing the modified dual into nodes, which depends on the nature of the geometry of the original object. The subdivided node structure is the dual of the MAT, each node corresponds to a MAT vertex. Each double edge (or double face in 3D), where the modified dual was subdivided corresponds to an edge of the MAT. The external edges (faces in 3D) of the modified dual correspond to the «wing-edges», where the radii of the maximally inscribed spheres within the object diminish to zero. Briefly the steps of the calculation process are the following:

1. Select any of the faces of the modified dual. The face represents a combination of elements (faces, concave edges and concave edges) forming one limit of a maximally inscribed sphere.
2. Find one or more extra elements, which, together with those determined in step 1, bound the maximally inscribed sphere.
3. Using the vertices corresponding to these extra elements, create and extract the topology of the Delaunay node corresponding to the maximally inscribed sphere.
4. Repeat from step 1 until every group of faces is a Delaunay node.

The calculation process is too complicated to describe in more detail here, the interested reader should refer to [ReSt2000] or [SRX2001] for details. This paper concerns the way in which the MAT information can be used for applications.

3. The negative MAT

Calculating the negative MAT is similar to the calculation of the positive MAT. This is done by negating the object, in effect creating an object-shaped void in a universe of material. The initial step of creating the global structure is done as before, although the actual structure is different, as illustrated in Figure 3. In the global structure for the positive object only the faces are represented as nodes whereas in the global structure for the negative object all the edges and the vertices are represented as nodes, since they are concave edges and concave vertices, respectively, in the negative object.

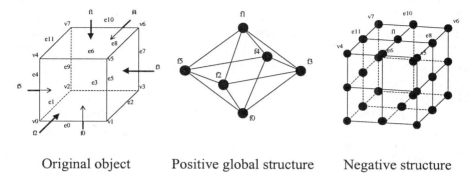

Original object   Positive global structure   Negative structure

Figure 3 - Global structures for positive and negative MATs

There is another difference, though, in the Delaunay nodes for the negative MAT. In the positive MAT there is always an element, which acts as a delimiter for the Delaunay node, hence there is a closed topological structure for the node. In the Delaunay nodes for the negative MAT, however, there are infinite nodes, where the 'inscribed' sphere has infinite radius. With the MAT of a positive object there is always at least one extra element, but this is not necessarily true for the negative MAT. On the right hand side of Figure 2 shows the steps in the creation of the negative mat. The first step is the same as for the positive object, that is the calculation of the topological dual. The second step, the calculation of the modified dual produces a much more complicated structure than for the positive object because, in general, there are more elements for the calculation. The subdivision process creates a set of finite Delaunay nodes and a set of infinite Delaunay nodes, where no extra limiting element is found. An unproved conjecture is that the union of the bases of the infinite nodes is the convex hull of the object. This can be seen by inspection, since if the face is not on the convex hull then a sphere of finite radius will touch the base and also at least one other element, thus creating a finite Delaunay node. The vertices of the MAT of the negative object correspond to the finite Delaunay nodes, the MAT edges join adjacent nodes.

4. Recombination methods

One reason for studying the MAT is to be able to develop improved algorithms for applications. If the algorithms do not improve currently available software then the computationally expensive process of calculating the MAT is difficult to justify. On the other hand, if there are good reasons for calculating the MAT then it is likely that there will be improvements in the calculation methods which will make the MAT more attractive. The purpose of this paper is, therefore, to present MAT-based methods as a comparison for the traditional methods. The techniques described here are not part of commercial software, in fact the only methods for MAT calculation of 3D objects known to the authors are academic methods. The MAT calculation description, above, describes how the Delaunay nodes are created by subdivision. This section describes methods for recombining the Delaunay nodes and how this can be used for applications. It is known that 2D Voronoi methods can be used for the calculation of toolpaths for pocket machining. The methods described here can be used for offsetting and analysis in 3D.

4.1 Offsetting

Object offsetting can be performed relatively easily and quickly using the Delaunay nodes. The Delaunay node subdivision provides information about the interactions between boundary elements. If the object is offset afresh for a set of values then, for each offset value, it is necessary to perform global checks, a sort of Boolean operation, for each value to find the final structure. There is, therefore, a trade-off between the cost of the initial calculation of the Delaunay nodes and the cost of separate offsetting. The MAT-based offsetting method, described here, allows the user to experiment with different values to determine the cutter centre surfaces. In addition to the offsetting process, the radii of the maximally inscribed spheres provide information about the critical offsets, where the topology of the offset object changes. This information indicates the range of tool sizes that can be used for machining different parts of the object.

Figure 4 shows recombination of the Delaunay nodes for offsetting the positive object. The set of the nodes together with the radii of the maximally inscribed sphere is show in the first figure. Column (a) shows the case where the offset is less than four. Column (b) shows the case where the offset is four or more, but less than six. Column (c) shows the case where the offset is six or more, but less than ten. The first line of the figure shows the nodes available for recombination, that is, nodes with associated radius greater than the offset value. The second line of the figure shows the recombined nodes and the last line the offset geometry. In the first column, with the lowest range of offsets all the Delaunay nodes are available for recombination, creating a figure with almost the same topology as the original figure, the only difference being that two circular arcs are added corresponding to the concave figures. The geometry is found by offsetting the geometry of the original object. Edge curves are offset directly, the offset of a vertex is a circular arc. If the offset is zero, though, then the corresponding arc has zero length and is replaced by a vertex. For the second column, with 4 ≤ offset < 6 the two small triangular nodes (nodes A and D) are ignored and the topology of the resulting object is the dual of the union of nodes B, C and E. In effect this means that the circular arcs corresponding to the concave vertices cancel out the two short side edges. In the final figure only node E remains and the topology of the final offset figure is a pentagon.

The offsetting process for the negative MAT is the same, involving a size filter of the Delaunay nodes, node joining and geometric reconstruction, as shown in Figure 5. In effect the external offset is the internal offset of the object shaped void in the infinite object. The infinite Delaunay nodes, represented by lines in the figure have a notional common element at infinity and so are infinite pyramids.

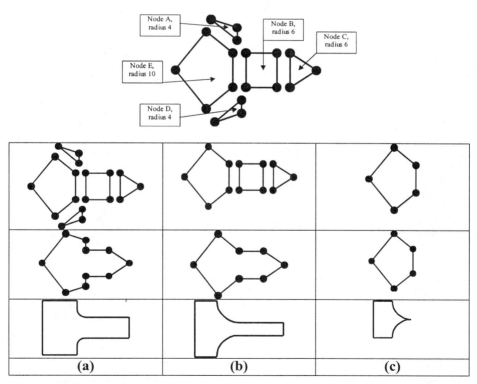

Figure 4 - Delaunay node recombination for offsetting a positive object

The first column shows the case for offsets less than four. In the node recombination step the double edges essentially cancel each other out. The figure produced is similar to the original shape, but circular arcs are introduced for each vertex except the two originally concave vertices. The second column shows the case for 4 ≤ offset < 62.5. As before, the short side edges disappear because the circular arcs corresponding to one of the end vertices cancel out the edge. In the final column, for an offset ≥ 62.5 the top and bottom edges of the side protrusion disappear, leaving the final offset figure.

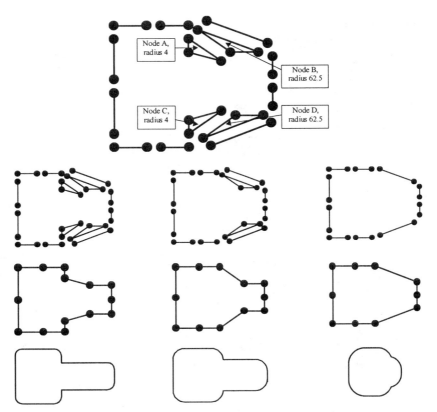

Figure 5 - Recombination for offsetting of a negative object

The above examples are simple 2D examples to show the working of the Delaunay node creation and recombination processes. Figure 6 shows a 3D example created in the same way. In addition the node size information can be used to create offsets where nodes disappear, that is, where the topology of the offset structure changes. The image on the right is the view from above to show these changes. The critical offset values are, for this object: 3.375, 6.75, 10.0, 15.0 and 15.358984.

4.2 Volumetric decomposition

The offsetting procedure uses one set of rules for recombination. Another set of rules gives a volumetric decomposition method, as illustrated in Figure 7.

The Delaunay nodes are joined without regard for their size, but are not joined if one of the vertices of the face corresponds to a concave edge limiting a face of the node. Such a set of simple rules was developed for the object above. More complex rules for general cases are still to be developed.This technique can be used for decomposition of delta volumes, say, to determine volumetric regions to be removed in the machining process.

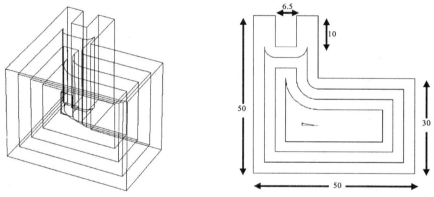

Figure 6 - Topological change offsets in 3D

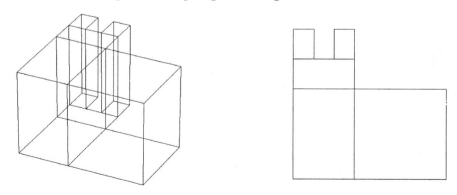

Figure 7 - Recombination for volumetric decomposition

5. Conclusions

This paper describes MAT-based techniques for offsetting and decomposing three dimensional objects. The basis for the method is the divide-and-conquer MAT algorithm for calculating general topology Delaunay nodes which can then be recombined to produce the offset or decomposition. MAT-based applications are still in their infancy and need to be developed further in order to see how the MAT can be used. The offsetting technique presented here is useful, for example, if multiple offsets are to be created. The proximity calculations are calculated once, in the form of a MAT structure, and are then reused.

6. References

[ReTu95] Reddy, J., Turkiyyah, G. (1995), "Computation of 3d skeletons by a generalised Delaunay triangulation technique", CAD, pp. 677-694.

[ReSt96] Renner, G., Stroud, I. (1996) "Medial surface generation and refinement", in "Product Modelling for Computer Integrated Design and Manufacture", ed. Pratt, Sriram, Wozny, pub. Chapman & Hall 1997, ISBN/ISSN: 0-412-80980-

[ReSt2000] Renner, G., Stroud, I. (2000) "Techniques for the calculation of medial surfaces of solids", submitted to CAD Journal

[SAR95]   Sheehy, D. J., Armstrong, C. G., Robinson, D. J. (1995), Computing the Medial Surface of a Solid from a Domain Delaunay Triangulation, in. Proc. Third Symp. on Solid Modeling and Appl., ed. C.M.Hoffmann and J.R.Rossignac, ACM, pp 201-212.

[SPB95] Sherbrooke, E. C., Patrikalakis, N. M., Brisson, E. (1995), "Computation of the Medial Axis Transform of 3-D Polyhedra", in Proc.Third Symposium on Solid Modeling and Applications, ed. C.M.Hoffmann and J.R.Rossignac, pub. ACM, pp. 187-199.

[SRX2001] Stroud, I.A., Renner, G., Xirouchakis, P. "A divide and conquer algorithm for medial surface calculation", in preparation.

A net based solution for lifecycle engineering
Concepts and requirements

A. Bechina, U. Brinkshulte, K.Synnes*, O:E: Artzen**

IPR, Institute for Process Control and Robotics, Universiy of Karlsruhe, Germany,
**Centre for Distance-spanning Technology, Luleå University of Technology, Sweden,*
** *Teknometri, consulting company, Oslo, Norway*

Abstract: A product life-cycle requires product engineering effort starting right from the
product idea and ending for i.e. with support for product recycling. In the
framework of the European project **media**site [10], our study focuses on the
combination of product engineering and Web technologies in order to decrease
the cost. Therefore we defined the term of Integrated Product Engineering
(**IPE**) as the concept integrating people, technology and information into a net
based solution for life cycle engineering. This e-engineering collaborative
platform lead to the concept definition of an extended enterprise. This paper
presents the concepts and requirements for developing a net based solution.
The functionality of a such platform is outlined.

Key words: Integrated Product Engineering, collaborative platform, extended enterprise

1.   OVERVIEW

Today the characterisation of global competition relies on the needs to
decrease the cost, improve the customer satisfaction, time-to market, etc.
Competitive companies today focus on the combination of an Integrated
Product Engineering (**IPE**) effort and World Wide Web use. We define Net
based **IPE** as a management strategy that uses customer requirements, cross-
functional teaming, information process and technology integration to
improve the performance of Product Engineering lifecycles. **IPE** should
provide universal methodologies that answer the fundamental questions
briefly described here:
− *"What to do"* means what activities and results are to be achieved in the
  course of an engineering process.

- "*How to do it*" means how the engineering processes realise the activities and results. This level involves the determination of the appropriated methods in order to perform the activities and results.
- "*With what*" means what technologies and methods have to be used or designed in order to achieve the engineering processes.

This questions reflect the different perspectives that should be taken in account in an **IPE** effort such as functional, behavioural, informational, organisational and operational aspects [1].

The first stage of the project focused on defining a methodological framework on how organisations/enterprises could move toward a totally electronic representation of the product and its processes. The elaborated methods rely on the principle of allowing early data sharing and greater concurrency in the Product Engineering lifecycle. The need to provide a practical approach to share data and support strong structured processes have led to the specification of a technological infrastructure defined as an e-engineering collaborative platform [10].

The goal of this platform is to provide an adequate infrastructure that enables people to collaborate in developing and maintaining product over geographical distances, directly modifying data at the customer's site or at the development centre. Collaboration is a critical issue. Internet and intranets make this function cost-effective and provide the means to support a highly interactive, collaborative workforce.

The challenge in the solution platform definition focuses on :
- the requirements specification of IPE in order to realise collaborative strategies,
- the specification of collaborative functions and their Mapping within the product engineering process in order to be fast and cheap at each stage of the **IPE** effort,
- the specification of a new extended enterprise model seen here as network of manufacturers, partners, customers involved in the **IPE** Processes,
- analysis of the integration between people, technology and information

The next part of our paper introduces the definitions of an Integrated Product Engineering (**IPE**) and the basic principle of the methodologies for lifecycle **IPE**.

The third part proposes an extended enterprise model and focus on the integration process requirement.

The fourth part introduces the e-engineering platform and its main functionality.

The last part outlines some conclusions.

2. INTEGRATED PRODUCT ENGINEERING PROCESS DEFINITION

The **IPE** process covers the process from the idea of a product to its market entry. It is defined as set of activities, concurrent processes which are required to generate the necessary information about the product, its production, logistics, marketing efforts etc.

The basic foundations of the methodology lead on definition of a *generic activity model* and *generic process model* which cover all the phases/lifecycle of **IPE** [2].

2.1 Definition of a generic activity model

A *generic activity model* can be described by a piece of work that forms one logical step within a process [4],[5]. An activity may be

a) *a manual activity*, within an engineering process which is not capable of automation. Such activities may be included within a process definition, for example to support modelling of the process, but do not form part of a resulting workflow.

b) *a workflow (automated) activity,* which is capable of computer automation using a workflow management system to manage the activity during execution of the engineering process of which it forms a part.

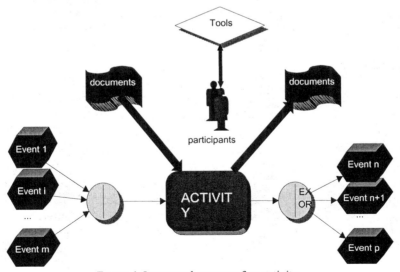

Figure 1. Inputs and outputs of an activity

A workflow requires human and/or machine resource(s) to support process execution. An activity is allocated to workflow participant, where human resource is required. A more detailed and complete definition can be found in [2]

- Each activity has a start date (min, max) and an end date (min, max).
- Each activity has an intermediate result which defines the goals of the tasks.
- For each activity a person or a team is affected
- Document or in wider sense, resources have to be defined in case of use, modification, and producing.

2.2   Definition of Generic Process models

An Integrated Product Engineering process is defined as set of Activities which are required to generate the necessary information about the product and its development [2].

The generic processes applied to a typical Integrated Product Engineering is decomposed into several sub-process called Workpackages and are described as following:

- *Analysis/specification* defines the product requirements in a product concept. This phase is primarily interested in capturing the customers requirements and specifying the technical properties of the product.
- *Product design* phase
- *Quality Function deployment* including the test and simulation. It describes as well a product on the basis of testing procedures and inspection plans that are necessary to ensure quality-related properties.
- *Product development* of manufacturing processes and supporting information and prototyping the product.

For products which are assembled from parts and subassemblies, these functions must be performed for all complex components as well as the final product.

Workpackages allow to structure the whole project and form a hierarchy.

The generic process and activity models allow to plan and create long term or ad hoc extended enterprises that gather all the actors (customers, manufacturers, partners, ..) involved in a common **IPE** process. The sequential or concurrent processes engineering are distributed among the industrial actors and thus lead to definition of an extended enterprise model.

2.3 Process Integration ➜ Extended enterprise model definition

A true net based solution for **IPE** requires the integration of people, business processes and information technology across the life cycle engineering. The technology must support the business/engineering process by enabling the users to get their tasks done and link them to others in the process. This means interfacing application and sharing data across organisational functions as well as with suppliers and customers.

Important issues have been identified such enterprise applications interoperability, security aspects, electronic procurement, distributed engineering/business process management, technologies and standards, etc.

In this paper we focus briefly and mainly on :
- identification and specification of the integration level (information , people, application, ..),
- identification and specification of collaborative functions

The Product Engineering cycle for complex products, four dimensions of integration that must be supported by the technology in an enterprise level **IPE** system could be considered:

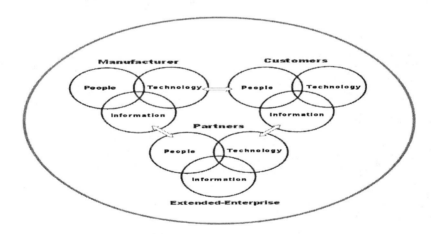

Figure2: Extended enterprise model

1. the integration of the business process (concept to design to manufacturing) for a particular project dealing with a component or subassembly,
2. the integration of the workflow and applications that support the tasks in the project,
3. the multiple projects must be co-ordinated by program or person that deals with the entire assembled product,

4. and finally, the organisations that participate in the implementation of the **IPE** platform, including suppliers, vendors and the manufacturer should be integrated.

The four dimensions of an **IPE** require integration of information technology in three traditional IT domains:

1. *Technical applications* also referred as CAx (computer-aided-everything: CAD, CAE, CASE, ..), tend to be specialised and are used "vertically" within one discipline or function (i.e. design, analysis, testing, process planning, part programming, etc.) within the product engineering cycle. Technical applications may be interfaced horizontally across disciplines, but this is still complicated by the fact that different views of the data occur in different disciplines. Though today some emerging standards such as, EDI , XML, etc. allow to exchange or share the same data.

2. *Management applications* are used to enable structured business processes through cross-functional workflow, budgeting, scheduling and data/document management.(workflow, project management, product data and document management, ...)

3. *Personal productivity tools* promote unstructured or ad hoc linkages along the product engineering lifecycle (word processors, document management, calendar, voice mail, etc. They play an important role in the definition of the collaborative functions.

The integration of applications is complicated by functionality overlap between various categories of applications. And the validity of an extended enterprise model rely on the development of a support integrating people, technology and information.

The first step dealing with integration issues has helped to the definition of the functionality requirement of the technological infrastructure. A net-based **IPE** platform should offer a wide range of functions, including engineering information management, engineering change management, product structure management, etc. However in this paper we introduce only the fundamental aspects concerning mainly the development of collaboration/communication means and the data sharing functions.

The next section introduces the collaborative platform and its needed functionality.

3. AN E-ENGINEERING COLLABORATIVE PLATFORM

3.1 Reference model

Every organisation has different priorities, different implementations strategies, and will, as result require different levels of **IPE** capabilities. However some common function are outlined in this section.

The extended enterprise model rely on a strong need to build new and better relationships with all the economic actors involved [6],[7],[8]. Those requirements could be achieved by connecting all the players in the processes into a common collaborative platform that should encompass the following functions: discussion forum, news, document management, chat, e-meeting, authoring, engineering, training, etc.

In the framework of our work, a reference model of this platform has been outlined and comprise the following components:

– User
– Product Engineering services
– *Operating system services* Each of the services also relies on the standard operating system basic services. These services include: communication (mainly TCP/IP), disk storage, system administration, backup/restore, logging, monitoring
– *Certificate Authority* (CA): plays an important role in a Public Key Infrastructure (PKI). The Certification Authority (CA) is a trusted third-party entity, with well-known and advertised service access points for use by clients. The CA is the authority providing services for creating, administrating, publishing and revoking (cancel) digital certificates.
– Core services described in the next section
– Interfaces

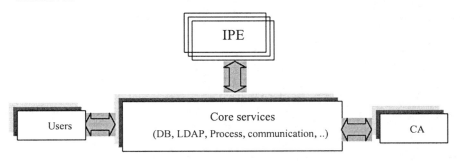

Figure 3: reference model

The next section describes the core services of the collaborative platform.

3.2   Services provided by e-engineering platform

Service			Functions
Core services	Common end user services	User management	Provide the correct user rights
		PostMaster	Can receive different forms of input such as paper documents, email, fax, EDI.
		News	Provides support for individual news groups for each virtual community. Possible to attach relevant documents or links to external information.
		Discussion	Provides support for individual discussion forum. Possible to attach relevant documents or links to external information
		Event	Allow announcements of events such as seminars and courses and the user registration to the events
		E-meeting	Possible to interact with an unlimited number of participants on a personal computer. Supports real-time video, audio, shared application windows, whiteboard, chat and web-based presentations
		Information broker service	Allows the user to define a personal search profile to be used to search for product locally or globally. A product could be a service like flight reservation, a hardgoods like a book or a content

Service		Application/product
Core services / Application support services	Single sign-on	Authentication, (identifying user). Authorisation, (verifying the users rights).
	Session control	Used to transport the user information (user id etc.) between the different services on the e-engineering platform
	Security API	API provides an interface to the authorisation information in the user directory. This API allows applications to control the access rights to a service information. It also allows the web-server to control the access rights to a service
	Directory service for - user data - application authorisation	Provides a structured storage for user related information. Information is based on standard object classes for describing a person. In addition to the user data, data for authorisation is also stored in the directory. LDAP
	Web server	* provides main interface to the end-user * provides the user authentication through SSL. *can forward the user information through the session data. *can control the users rights through the security API. *can launch the services and their corresponding applications
	Relational database	Central storage mainly of runtime data
	Document management	Provides core functions for management of documents
	Messaging service	Support for message distribution between services and users. e-mail, SMS

Most of the functions presented are supplied by applications/tools (document management system (FAM), video conferencing(Marratch pro), virtual community (Net-community) supplied by the consortium of the European project **media**site [10]. However these tools have been be integrated in order to form a common collaborative platform.

The integration platform is still under development and needs to be tailored in order to realise all the functions needed to achieve a true Integrated Product Engineering.

4. CONCLUSION

Thanks to the rapid advances in telecommunications and web technologies, more organisations are involved in the process of an extended enterprise creation. Therefore they meet more easily customer demands for high added-value products and services in a global market. Our e-collaborative platform for Product engineering application represents already a means to increase the communication between partners and thus enhance greatly the productivity. However, The issues generated such security, collaborative strategies definitions, accurate data exchange/sharing, interoperability, etc. cannot be solved by technological means alone, but some socio-economic, legal and policy aspects have to be considered.

5. REFERENCES

[1] A. Bechina,M. Kullmann, B. Keith., (2000). A Description Logic Model for the Management of Automated Guided Vehicles. In Proceedings of the IFAC Conference on Control Systems Design 2000, Bratislava, Slovak Republic,

[2] A. Bechina,A. daum, N. Howind, , (2000). Report on Platform Specification, DeliverableD7.4, Germany, Mediasite project

[3] A. Bechina,A. daum, (2000). Report on Platform Specification, DeliverableD7.6, Germany, Mediasite project

[4] Camarinha-Matos, L. M., Afsarmanesh, H., (eds) (1999), Infrastructures for Virtual Enterprises. Networking Industrial Enterprises. IFIP TC5 WG5.3 / PRODNET Working Conference for Virtual Enterprises (PRO-VE'99), Porto, Portugal, 27-28 October 1999, Boston: Kluwer Academic Publishers

[5] DECKER, Stefan; DANIEL, Manfred, ERDMANN, Michael; STUDER, Rudi: *An Enterprise Reference Scheme for Integrating Model Based Knowledge Engineering and Enterprise Modelling*. Karlsruhe, Universität Karlsruhe (TH), Institut für Angewandte Informatik und Formale Beschreibungsverfahren; (Bericht 365), 1997.

[6] DOGAC, Asuman (ed.): *Workflow management systems and interoperability: proceedings of the NATO Advanced Study Institute on Workflow Management Systems heeld in Instanbul, Turkey* / Heidelberg : Springer-Verlag, 1998. ISBN 3-540-64411-3 .

[7] Edoardo Iacucci(a), Bertil Axelsson(b) (2000). MediaSite: Web-based Services for Virtual Teams and Communities of Practice in SMEs. ITBM 2000: Information Technology for Business Management, China

[8] GROISS, Herbert; EDER, Johann: *Bringing Workflow Systems to the Web*. Institut für Informatik, Universität Klagenfurt, Austria.

[9] Wognum, N., Thoben, K.-D., Pawar, K. S. (1999), Proceedings of ICE'99, International Conference on Concurrent Enterprising, The Hague, The Netherlands, 15-17 March 1999, Nottingham: University of Nottingham, ISBN 0 9519759 86.

[10] Mediasite EU project: http://www.mediasite.nu

Integration Issues on Combining Feature Technology and Life Cycle Assessment

H.E. Otto, K.G. Mueller and F. Kimura
Dept. of Precision Engineering, The University of Tokyo
Bunkyo-ku, Hongo 7-3-1, Tokyo 113-8656, JAPAN
tel. +81-3-5841-6495 - fax. +81-3-3812-8849
e-mail:[otto,kmueller,kimura]@cim.pe.u-tokyo.ac.jp

Abstract   Within recent research trends and an increased interest in targeting not only product modeling but product life cycle modeling as a whole, for feature technology used to support and enhance mainly computer-aided design and computer-aided manufacturing, new dimensions of requirements arise. Further demands on higher automation, more flexibility, enhanced man-machine interaction and increased built-in knowledge of application systems, require sophisticated (system) interfaces and perhaps even local re-design of feature frameworks currently developed. To combine and relate data from both a product model and a life cycle inventory data base, currently several (in most cases manual, non optimized) methods were employed. However, they usually all ignore the potential of pre-defined structures and soft technologies now-a-days integrated within a product model. Within the scope of the work described in this paper, the focus will be on studying and discussing the approach to extend a feature model in order to explicitly utilize feature technology within product life cycle assessment. In particular methods and structures for two assessment approaches namely discrete and continuous life cycle assessment will be introduced and discussed.

Keywords   Feature technology, model and framework integration, product design, life cycle inventory parameters, life cycle assessment

1   INTRODUCTION

Work on feature technology as it evolved in the past two decades grew from its origin to identify part regions of interest for manufacturing of products to a powerful technology with many frameworks and system architectures developed as well as implemented. However, with recent research trends and an increased interest in targeting not only product modeling but product life cycle modeling (LCM) as a whole, for feature technology used to support and enhance mainly computer-aided design and computer-aided manufacturing, new dimensions of requirements arise. Feature technology is a powerful tool that at certain representation levels, can be successfully used to capture the relationship of shape and functionality in many ways. An important aspect if one recalls the most labor-intensive work within life cycle assessment (LCA), which is to determine the life cycle inventory (LCI). On the other hand, approaching automation of extracting LCI data elements from already available digital models such as product models and CAD models, requires efficient interfacing between life cycle assessment software and these models. However, within this context as outlined, to properly utilize functionality of feature technology on a broad base, framework and entity concepts developed up to now, need to be extended. Within the scope of the work described in this paper, the focus will be on studying and discussing the approach to extend a feature model in order to explicitly utilize feature technology within product life cycle assessment. In particular methods and structures for two assessment approaches namely discrete and continuous life cycle assessment will be introduced and discussed.

2 FEATURE TECHNOLOGY AND LCA

2.1 Background and problems

Early work on features was done during the late 1970's and early 1980's within problems of identifying regions of part geometry important to manufacturing. After half a decade of untiring efforts, when feature technology left its initial embryonic state, integration of feature models and geometrical models was targeted with the basic goal to integrate different processes such as manufacturing and design. Although, from the very beginning, the basic idea of features was to associate functional information with shape information, early feature frameworks and implementations of feature systems displayed a low degree of integration between a feature model and a geometrical model. These individual approaches were focused on either a more function/meaning related approach [4,14] or a more geometry related approach [16,6].

Besides integration and consistency problems partially mastered in frameworks developed in the late 1980's and early 1990's [22,21,23,17], successful modeling of relationships and control of feature properties across features as well as across different feature classes represent a further important milestone in the historical development of feature technology (cf. [24]). Extensions of feature modeling domains not only towards geometry in particular [5,1], but also towards further applications within assembly modeling [18,9] and product life cycle modeling [13,19,12] are further indicators of efforts to translate into practice and integrate the potential of FT within today's rapidly evolving fields of product life cycle assessment and modeling.

The first approaches, to what today is called life cycle assessment, were developed in the 1960's. The studies of that time dealt mainly with energy issues [2], which was related to the general interest in the scarcity of raw materials and fossil fuels. Formalized LCA tools developed in the USA were mainly based on raw materials and energy analysis, such as the REPA method by [10]. The development of LCA procedures began in Europe in the early 1980's, especially in Switzerland, where the Ministry of the Environment initiated research lead to the development of the Critical Volumes Method, and the Ecopoint Method [7]. Several European research groups began publishing on LCA methods in the 1990's, often supported by national agencies (e.g. CML classification method in the Netherlands by Heijungs et al. [8]), Environmental Priority Strategies (EPS) method in Sweden [26]). Related methods, such as the KEA method [28] and Material Intensity Per Service-unit (MIPS) [25] were also developed in Europe. In the recent past the International Standards Organization has issued a framework for conducting LCA's [11].

The difficulty of LCA is not based on the complexity of the methodology, but on the complexity of problem. Each single step is very simple and can be understood easily but the often thousands of steps required to assemble a complete LCI make a accurate LCA extremely difficult and expensive. To give an impression of the scale associated with compiling an LCI, Bretz [3] estimated that an analyst may require some 10^4 to 10^5 or more numerical data elements for a complex product which will consist of thousands of unit processes.

Although, as can be seen from the brief survey given above, progress in various different directions within the fields of feature technology and life cycle assessment has been achieved, essential concepts and structures that support an efficient integration of both from the viewpoint of information/knowledge modeling and processing within the entire product life cycle as discussed above, are still missing.

2.2 Scope and objectives

As pointed our earlier, in life cycle assessment the most labor-intensive work is to determine the life cycle inventory. This is also because it is the most data-intensive part and extremely difficult to automate. A natural, straightforward solution to this problem would be to automate the extraction of LCI data elements from CAD drawings to make LCA a viable analysis tool during the design process. However, this can only be achieved by efficiently interfacing product model software and life cycle assessment software.

An example of a previous approach is an experimental assessment software which consists of evaluation modules integrated with a commercial CAD system, extended for handling form features on a limited base. This prototype system integrates information on possible manufacturing methods, production costs and eco-toxicological data. It is primarily aimed at optimizing the manufacturing method of the embodiment design, so that the emissions and costs due to manufacturing are minimized. The system extracts geometrical and feature information from both the geometrical model and the feature model and so evaluates manufacturing costs and energy requirements during the manufacturing stage [15]. The disadvantage of this implementation is that it is based on an inflexible architecture, resulting in a 'closed system', thus there is no possibility of interfacing this system to exchange the product's pre-use inventory data with existing LCA software, which would allow a more complete life cycle assessment.

An alternative, better structured and more flexible approach introduced and discussed in the next section is to utilize the inventory information already available in product and process models for determining the LCI in an automated fashion. This requires the linking and integration of different models and data bases on a system and module based firm footing. To overcome shortcomings of related work discussed above, system component interfacing and information integration are realized by employing a true feature technology supported open system architecture.

However, within this approach two different directions can be chosen, depending on the type of LCA, i.e. the time range and frequency, LCI parameters are updated and eventually computation of LCA is carried out during computer-aided design sessions. Within the scope of the work described, the focus will be on evaluating continuous LCA, a progressive type of computation and discrete LCA, an a-posteriori type of computation.

3   APPROACH

3.1 Outline and general framework

To integrate different system modules and models as outlined earlier, one requires system components with an open system architecture that provides an application programming interface (API) for each relevant model to be accessed by a LCA module. For example as shown in Figure 1, this requires at least one API for the process model and also at least two for the product model, providing a means to connect the geometrical model and the feature model to individual LCA modules.

Any local or explicit access to models other than from within its own modeler must not be allowed, to guarantee a clean and modular architecture where access to individual models is only permitted through a dedicated API. Especially in the case of geometrical models, due to their central role within a product model, a well-designed, attribute supporting API is indispensable.

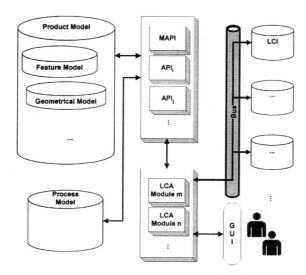

Figure 1 *Overview of the system architecture*

Unfortunately in practice there are still many commercially available CAD systems and geometrical modelers in use, which do not support any of these characteristics. Due to the fairly wide range of data types, the context of API parameters needs to be set a-priori by individual LCA modules. For example, in the case of the API for the feature model, measurement units of parameters such as feature dimensions or feature volume need to be fixed, to guarantee the proper interpretation and integration of data from different sources. Attributed parameters also help to keep the interfacing API transparent, since the number of functions and parameters required can be kept small, while only the range of individual interface components is modeled with appropriate attribute values. A design that keeps the API flexible, while fixing the data context on a dynamic base.

Another aspect of flexibility is the modular architecture for the API of a feature model. Due to the nature of features being both application dependent and user extensible, the feature API needs to have a modular internal interface that can be extended according to newly created feature types and feature attributes. These modular feature API (see MAPI in Figure 1) also operating on the basis of attributed parameters, are then flexible enough, to provide on a feature-based ground efficient access to product and process information.

3.2 Discrete life cycle assessment

From a computing system and data structure point of view, as pointed out earlier, there are two different approaches, to support LCA, which will be introduced in the following and discussed in detail in the next section. The main difference of these two approaches lies in the method and degree, how LCI parameters and LCA models are integrated with other product relevant models utilizing feature technology. LCI parameters are seen as those parameters that relate to the LCI directly or indirectly, many of which are already included in some explicit or implicit form in a product model, but only need to be linked in a correct and efficient manner to determine the LCI. In Figure 2 a selection of those parameters is given. Note, for illustration purposes three levels of detail are distinguished, i.e., the product level, the part level / assembly level and the feature level.

One approach to integrate these parameters with product information would be to provide a set of integrated LCA modules that contain partial LCI / LCA models which can be linked to product information and a LCI database, to externally compute as well as store LCA results. Since this approach is, from the viewpoint of computer-aided product design, somewhat discrete being applied only from time to time when a design solution seems to be worth being analyzed in respect to its product life cycle qualities, it is named discrete life cycle assessment (dLCA). Within dLCA the entire data retrieval and access, data analysis and computation of relevant LCI parameters is specified and carried out within integrated LCA modules.

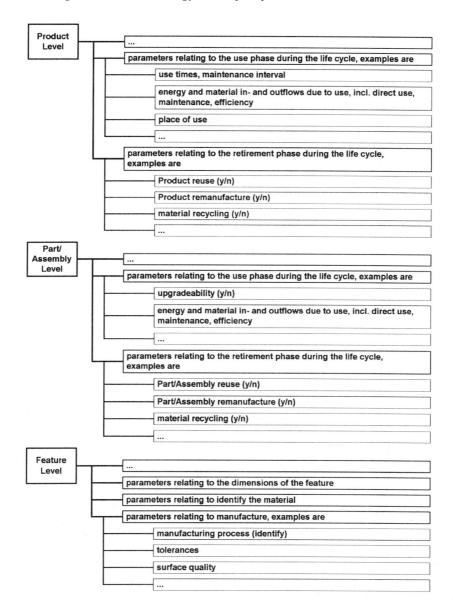

Figure 2 *A selection of LCI relevant parameters*

3.3 Continuous life cycle assessment

An alternative approach would be to explicitly integrate LCI relevant parameters, partial LCA models and results of LCA within models of a product. Within such a system architecture, during different design stages of a product, each change in any model immediately would be triggering a recalculation and consequently an updating of all LCI parameters and LCA results as well. Within this approach, besides LCI relevant parameters as shown in Figure 2, further parameters, containing LCA results of individual entities such as features, parts, assemblies or even a complete product, need to be included. A representative selection of those parameters may be a set as shown in Figure 3.

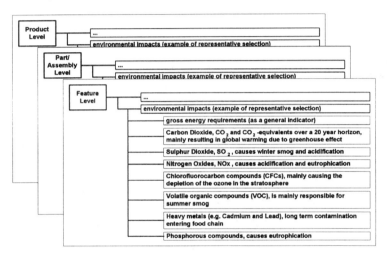

Figure 3 *Representative selection of parameters containing LCA results on environmental impacts*

Within this approach, all product data, LCI parameters and LCA results to be computed are explicitly interrelated on a real-time base. Since any design change is triggering an update of all relevant entities connected across all models being included in the framework, LCA is, from a computer-aided product design point of view, carried out on a progressive, continuous base, thus resulting in the name continuous life cycle assessment (cLCA). Within cLCA, the majority of data access and retrieval, data analysis and computation of relevant LCI data is specified and carried out internally in data and computation structures which are - as encapsulated entities - an integrated part in all product related models. LCA modules integrated in the framework as outlined earlier act here mainly as an external support to efficiently interface to further related databases such as the LCI database, etc.

4 ASSESSMENT AND EVALUATION

4.1 Theoretical aspects

For the dLCA approach, since computation of the LCA is mostly executed externally, a fast and efficient access to all LCI relevant data is necessary. As can be seen in Figure 2, many LCI parameters at the product level and part / assembly level can be represented with basic numerical values, logical values and a few string values. However, to compute LCA parameters also geometrical and geometry related product data are required. Here integrated feature models represent a great benefit from a data structure point of view. All parameters related to dimensions, manufacturing, local tolerances, surface finish, etc. are already present in a feature model that in turn is integrated with other product model components such as the geometrical model the part / assembly model, etc. Also procedures to access these data are already available within feature-based operators, which only need to be made available to the LCA modules through a MAPI as shown in Figure 1. For the cLCA approach, from a data structure point of view, the feature model represents even an essential part of the product model, for without features, it would be, due to efficiency reasons (as will be shown later), very difficult to pursue this approach. Since besides relevant LCI parameters all sets of computed LCA parameters (cf. again Figure 3) are included in an extended product model, one needs a set of basic elements of a product, where internal LCA parameter computation can be associated with. Here features represent a natural, almost atomic base, where geometry related details of all parts of a product were stored and maintained in a manner consistent to all other, non-geometry related product data.

From a computation structure point of view, requirements and benefits of the two approaches are quite different. In the case of dLCA, the specification of partial LCA models and LCA parameter computation resides entirely within integrated LCA modules, which also contain structures and descriptions on how to access an LCI database. Since partial LCA models are externally specified within LCA modules and computation of LCA parameters is asynchronuous to changes in a product model, no special requirements regarding computation or parameter updating are necessary. In the case of cLCA, both results of computed LCA parameters and specification of their computation reside (on a real time base) within an extended product model. In other words, the functionality of LCA modules within dLCA as described earlier needs to be encapsulated within extended entities at different levels of a product model. Moreover, sophisticated mechanisms are required that guarantee a synchronized bi-directional bottom-up as well as top-down data flow among all integrated LCI and LCA parameters along all model levels. They are necessary to provide for a consistent parameter updating each time a design change has been committed in the product model.

Taking a brief look at the computational expense of both approaches, it comes at no surprise, that the cLCA approach is fairly expensive in terms of data throughput and computation required. Since cLCA needs to re-compute and update all related parameters, any time a change in the product model is affecting at least one of them, computational expenses increase tremendously with the model size. It is

estimated that with completion of each design phase described in [20] as conceptual
design, embodiment design and detail design, the data volume is exponentially
increasing (see Figure 4) in a similar manner as costs do when changing design
parameters. The cost associated with changes to the design increases by a factor of
10 for each design stage and changes in the manufacturing stage can cost more than
1,000 times the amount that would have been required if the decision would have
been reached during the concept stage [27]. It has to be realized that a LCA carried
out at the later stages can be more accurate due to the higher product definition, but
the increased data throughput and computation required also increases the effort for
the LCA. This means that when an assessment of the LC leads to changes in the
design, then it is more cost-effective to both assess and change early on during the
design process.

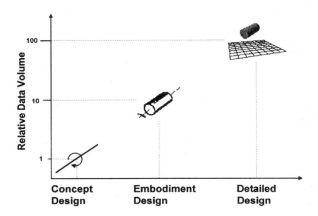

Figure 4 *Estimated order of magnitude of product data required in each design phase*

Although dLCA is much cheaper in absolute terms of both data throughput and
computation, we estimate the order of magnitude of expense to retrieve and access
all parameters before executing another LCA calculation within dLCA is in average
similar to one sequence of updating and data forwarding within cLCA. Here FT
helps to significantly reduce through its well structured model and efficient access
to geometry related data the otherwise cumbersome access to LCI relevant product
parameters. An aspect that is also relevant for the dLCA, though compared to
cLCA less dominant, since actual LCA is done much less frequently than within
cLCA. In general, if frequency of computing LCA parameters increases within
each approach, FT's role as outlined above becomes a dominant factor to reduce
substantially computational expense for both approaches.

    From a computer-aided product design point of view, the scope possible of
supporting individual design phases as mentioned earlier depends for both
approaches on the data structure and models used within a system. Referring to
each design phase, dLCA and cLCA are limited by the smallest, most abstract but
complete entity available, which in the framework as described, is the feature. In

other words, within both approaches, limits are lined out by the feature model used. In general one can assume, since features can only be abstracted as far as there is actually at least concrete geometry of a spatial region of interest available, currently LCA in both forms discrete and continuous can be supported only from embodiment design on, not earlier. Although, LCA is most effective the earlier it is applied during product design, at the time of writing, feature models supporting conceptual design are not known to the authors.

4.2 Practical aspects

From the practical viewpoint of implementation dLCA is a straight forward approach that can be pursued without major obstacles, assuming an open system architecture as outlined in Figure 1 is given. A modular system component structure and APIs for all relevant models provide an ideal, well-structured environment to implement dLCA without heavy requirements on the programming environment itself, though attributed structures, a sound typing scheme and if possible, object-oriented approaches should be supported. In the case of cLCA, implementation requirements are heavier. Besides an attribute-like structure extension to allow for storing additional LCI and LCA parameters in an extended model, the entire computation of the latter needs to be somehow encapsulated. Moreover at run time, a bi-directional data flow between all LCI and LCA parameters needs to be established and synchronized. In terms of supporting implementation, FT provides only moderate help for dLCA. Besides reducing specification / programming expenses by providing through an API already available feature-based operators for navigating and accessing geometry related data, there is not much additional support FT can provide for dLCA. Different is the case for cLCA. If feature models are not included in a product model, similar constructs or support structures need to be implemented first. It would be quite expensive from a computation point of view to retrieve and compute all geometry related LCI parameters by utilizing a geometrical model and a part / assembly model only.

Finally to complete the assessment and evaluation of practical aspects, the scope of detail of carrying out LCA shall be discussed. In the case of dLCA, it is relatively easy to explicitly control the scope of computation related to e.g. each life cycle phase of a product or a specific part of the product. Simply by setting the focus of retrieval and access of relevant LCI parameters to certain product model levels and respective entities, thus selectively limiting the data input for the LCA, computation can be directed in individual LCA modules towards individual parts and life cycle phases as demanded by a designer. A task that is explicitly supported by FT in respect to all structural and computational aspects as discussed earlier. However, in the case of cLCA this is quite different. Due to the encapsulated structures that prevail throughout all portions of a product model deep down to the feature model, explicit control of the scope of computation regarding the range of LCI parameters used for the LCA is not possible. For example, exclusive local LCA of a part of a product, regarding its pre-use and post-use is not possible within cLCA without additional changes of components in the system architecture as outlined earlier. Therefore, current LCA methods as carried out in practice now a days, computing LCA parameters once for an entire product for all three life cycle phases and afterwards computing only new values of critical parts regarding their pre-use and post-use phase - neglecting assumed small changes incurred in the use-phase - while product design proceeds, are difficult to support with cLCA. To support such methods, additional data structures and computation structures are required, to allow for parameter selection and data stream re-direction of once during implementation specified and encapsulated structures related to LCI and LCA. Here FT may provide some structural help only, if it can adopt to these requirements within its model and operators.

5   CONCLUSIONS

LCA is becoming more important as an assessment tool within the design process. To support interfacing of computer-aided design and LCA, in a first attempt, the problem of extending the architecture and product data related models of current systems utilizing FT has been targeted. In particular methods and structures for two assessment approaches, i.e. dLCA and cLCA, haven been introduced and discussed. Considering the current state-of-the-art of CAD systems and computer hardware available today, requirements of dLCA clearly present a reasonable match regarding data throughput, data structures and computational power required. Also from both a theoretical and a practical point of view, until more LCI data with higher precision and reliability become available, a less precise dLCA with respect to small changes during design - reflected also in modifications of a product model - seems better to be in balance with current LCI databases than the costly cLCA. On the other hand, if new generations of software and hardware platforms significantly increase in efficiency, quality and power, it might be not far fetched, to assume that cLCA shall be the proper choice to base future research and development on.

6   REFERENCES

[1]   Au, C.K. and Yuen, M.M.F. (2000) A semantic feature language for sculptured object modeling, *Computer-Aided Design*, **32**, 1, 63-74.

[2]   Boustead, I. (1995) Life cycle assessment: An overview, *Energy World: The Magazin of the Institute of Energy*, 230, 7-11.

[3]   Bretz, R. (1998) SETAC LCA workgroup: Data availability and data quality, *International Journal of LCA*, **3**, 3, 121-123.

[4]   Descotte, Y. and Latombe, J.C. (1984) GARI: An Expert System for Process Planning, in: *Solid Modeling by Computers* (Picket and Boyse, eds.), Plenum Press, New York, USA.

[5]   Fontana, M., Giannini, F. and Meirana, M. (1999) Free form feature taxonomy, *Computer Graphics Forum*, **18**, 3, 107-118.

[6]   Gossard, D.C., Zuffante, R.P. and Sakurai, H. (1988) Representing Dimensions, Tolerances, and Features in MCAE Systems, *IEEE Computer Graphics & Applications*, **8**, 2, 51-59.

[7]   Habersatter, K. (1991) Report No. 132: Ecobalance of Packaging Materials State of 1990, Swiss Federal Office of the Environment, Forests and Landscape (BUWAL), Berne, Switzerland.

[8]   Heijungs, R., Guinee, J., Huppes, G., Lankreijer, R. M., de Haes, H. A. U., Sleeswijk, A. W., Ansems, A. M. M., Eggels, A. M. M., Duin, R. V. and de Goede, H. P. (1992) Environmental life cycle assessment of products, guidelines and backgrounds, Technical report, Centre of Environmental Sciences (CML), Leiden University, The Netherlands.

[9]   van Holland, W. and Bronsvoort W.F. (2000) Assembly features in modeling and planning, *Robotics and Computer Integrated Manufacturing*, **16**, 4, 277-294.

[10]   Hunt, R., Sellers, J. D. and Franklin, W. (1992) Resource and environmental profile analysis: A life cycle environmental assessment for products and procedures, *Environmental Impact Assessment Review*, 12, 245-269.

[11]   ISO 14040 ff. (1997) DIN EN ISO 14040 ff: Umweltmanagement - Ökobilanz, Beuth, Berlin, Germany.

[12]   Krause F.-L., Kind, Ch. and Martini, K. (1997) Application of Feature Technology in a Disassembly-Oriented Information Technology Infrastructure, in: *Proc. 4-th CIRP International Seminar on Life Cycle Engineering*, June 26-27, Berlin, Germany, 345-355.

[13]   Krause F.-L., Stiehl, Ch. and Martini, K. (1998) New Applications for Feature Modeling, in: *Proc. Israel-Korea Bi-National Conference on new Themes in Computerized Geometrical Modeling*, February 18-19, Tel Aviv, Israel, 1-9.

[14]   Luby, S.C., Dixon, J.R. and Simmons, M.K. (1986) Creating and Using a Features Database, *Computers in Mechanical Engineering*, **5**, 3, 25-33.

[15]   Müller, D.H. and Oestermann, K. (1997) Simultaneous consideration of ecological and economical parameters in early stages of product design, in: *Proc. International Conference On Engineering Design*, August 19-21, Tampere, Finland, Vol.1, 337-342.

[16]   Ostrowski, M.C. (1987) Feature-based Design using Constructive Solid Geometry, Internal Report, General Electric Corporate Research and Development, Schenectady, New York, USA.

[17]   Otto, H.E., Mandorli, F., Cugini, U. and Kimura, F. (1994) Domain-Oriented Semantics For Feature Modeling Based On TAE Structures Using Conditional Attributed Rewriting Systems, in: *Proc. ASME International Design Engineering and Computers in Engineering Conference,* September 12-14, Minneapolis, MN, 13-27.

[18]   Otto, H.E., Kimura, F. and Mandorli, F. (1998) Support of Disassembly / Reassembly Evaluation within Total Life Cycle Modeling Through Feature Neighborhoods, in: *Proc. of ASME International Design Engineering and Computers in Engineering Conference,* September 13-16, Atlanta, GA.

[19]   Otto, H.E. and Kimura, F. (1998) Towards a framework for unmitigated integration of feature technology within life cycle modeling, in: *Proc. of JSPE 1998 Annual Conference*, March 18-20, Kawasaki, Kanagawa, Japan.

[20]   Pahl, G. and Beitz, W. (1996) Engineering Design: A Systematic Approach, Springer-Verlag, New York.

[21]   Pratt M.J. (1989) A Hybrid Feature-Based Modelling System, in: *Preprints International Symposium on Advanced Geometric Modelling For Engineering Applications*, November 8-10, Berlin, Germany, 177-189.

[22]   Rossignac, J.R., Borrel, P. and Nackman, L.R. (1988) Interactive design with sequences of parametrized transformations, Research Report RC 13740, IBM Research Division, Thomas J. Watson Research Center, Yorktown Heights, New York, USA.

[23] Shah, J.J. and Rogers, M.T. (1989) Feature Based Modeling Shell: Design and Implementation, in: *Proc. ASME Computers in Engineering Conference*, July/August 30-3, Anaheim, CA, 255-261.

[24] Shah J.J. (1991) Assessment of features technology, *Computer-Aided Design*, **23**, 5, 331 - 343.

[25] Schmidt-Bleek, F. (1997) Wieviel Umwelt braucht der Mensch? Faktor 10 - das Maß für ökologisches Wirtschaften, dtv, Taschenbuchausgabe, Germany.

[26] Steen, B. and Ryding, S. O. (1994) Valuation of environmental impacts within the EPS-system, in: *SETAC: Integrating Impact Assessment Into LCA*, SETAC-Europe, 155-160.

[27] VDI-Richtlinie 2210 (1975) Datenverarbeitung in der Konstruktion; Analyse des Konstruktionsprozesses im Hinblick auf den EDV-Einsatz, VDI-Verlag, Düsseldorf, Germany.

[28] VDI-Richtlinie 4600 (1995) Kumulierter Energieaufwand - Begriffe, Definitionen, Berechnungsmethoden (Entwurf), VDI-Verlag, Düsseldorf, Germany.

Integrating Life Cícle Aspects Within Product Family Design: An Example for SMEs

H.E. Otto, K.G. Mueller and F. Kimura
Dept. of Precision Engineering, The University of Tokyo
Bunkyo-ku, Hongo 7-3-1, Tokyo 113-8656, JAPAN
tel. +81-3-5841-6495 - fax. +81-3-3812-8849
e-mail:[otto,kmueller,kimura]@cim.pe.u-tokyo.ac.jp

M. Germani and F. Mandorli
Department of Mechanical Engineering, The University of Ancona
Via Brecce Bianche, Ancona I-60131, ITALY
tel.: +39-71-2204969 - fax.: +39-71-2204801
e-mail: germani@popcsi.unian.it, ferro@ied.unipr.it

Abstract: Recent concern for the environment and high competition in globalized markets pose eminent requirements to the producing industry, including small and medium sized enterprises. Envisioned goals of a eco-design combined with a drastical reduction of product development time and increase of customized features of products require new strategies and tools. One solution subject to international investigation is the development of modular products. This approach we enhance by integrating life cycle assessment. Work done in this field was concentrated on developing methods to define either a product's platform architecture or relationships between product modules and documented customer needs within the design analysis of product families. Unfortunately, all of these approaches failed to include a framework to enable also the integration of life cycle assessment. Within the scope of the work described in this paper, we target fundamental problems within the design of product families regarding the integration of life cycle assessment and economic application of information technology, suitable in terms of cost and performance for small and medium sized enterprises. The application field, where first case studies were undertaken and real product models implemented was within the design of parallel-shaft speed reducers.

Keywords: product family design, eco design, life cycle inventory parameters, life cycle assessment

1   INTRODUCTION

Life cycle assessment (LCA) is becoming increasingly more important as an assessment tool within the design process. LCA may be used in an attempt to minimize certain environmental impacts of a design during the whole life cycle, and as a result weak points and areas requiring special attention of a design are highlighted. The downside of LCA is that a large amount of data is required. The data is difficult to assemble alone due to the volume and even requires periodical updates, which makes LCA relatively expensive. The cost of carrying out an LCA can often not be justified in small and medium sized enterprises (SMEs), since savings on the product side cannot be guaranteed to compensate the expenses on the development side.

Realizing that many SMEs specialize on a product, which is produced within a product family, it seems reasonable to prepare an industry specific, parameterized LCA for the product family. The approach taken here is to utilize existing affordable software products and to develop software links in an appropriate manner, resulting in a greatly reduced effort for both the layout of the varied design within the product family and the LCA evaluation.

Since an LCA firstly requires the life cycle inventory (LCI) of the product, the approach described here is to parameterize the LCI using sizing parameters (design parameters) for the product family. A database with LCI data on relevant generic raw- and engineering materials and energy products and a database with site specific manufacturing LCI data are both linked to the simplified product inventory, thus providing a product specific LCI. Actual values for the sizing parameters, which are used to determine the LCI, are extracted from a digital spreadsheet used during the design process.

2 PRODUCT FAMILY DESIGN AND LCA

2.1 Current situation in practice and research

Today's world wide practice of mass production mostly targeting low product costs determined by high productivity at minimum costs is about to change due to several reasons, as outlined earlier. New approaches to product design and development such as modular product development and eco-design are among the most progressively pursued.

The need for modular product development has increased substantially over the recent years and therefore considerable research attention has been dedicated to solve related problems. Work done in this field was first concentrated on developing methods to define a product's platform architecture [1,2,3]. An approach where a modular product is based on a structure (rules and relations), that is called product platform, on which the designer can assembly different modules developing new product variants to satisfy individual customer requirements. Obviously, a right definition of the product's platform architecture permits an easy configuration of variants (high interchange ability of modules). Other important research issues are the definition of modules and modularity, and the relative identification in an existing product. In several approaches one can find function-based methods (c.f. [8]) that have been derived from function diagrams originally developed by Pahl and Beitz [6]. Modules, in this case, are defined as physical assemblies of components that have a correspondence with functional groups defined in the functional structure. Analysis of the function diagrams then allows for the evaluation of modularity and the subdivision of modules. To determine relationships between product modules and documented customer needs as described in [4,9] represents yet another approach within the design analysis of product families. An approach that allows to automate the product configuration on the basis of customer requirements.

'Integrating' LCA into the design process is still often seen as simply carrying out LCA in parallel to the design process. The material requirements are extracted from the design manually and entered into an LCA program together with data on the companies manufacturing processes and information on the use and post-use phases. This becomes more difficult with increasing complexity. For LCAs at the embodiment and detailing phase of the design process, methods for greater integration (avoiding data redundancy, by reducing the amount of LCI relevant information that is repeatedly determined and input) have been proposed, though non have been implemented in commercial software. They have in common that only the pre-manufacturing and manufacturing phase is assessed, while other phases are neglected.

One trend is the automated extraction of material requirements from CAD-drawings, which can be used as an input to LCA programs. The information does not include manufacturing data, which have to be entered manually. A step further is the utilization of a feature technology based CAD approach. Here the geometrical representation of the product is obtained by defining simple geometric objects through a predefined structure, which includes design features (e.g. geometry, dimensions) as well as manufacturing features (e.g. material, machining process). The objects are set in relation to each other to form a part, and the parts are arranged to define a product. Thus the product model includes structured inventory and inventory related information, that can be automatically extracted to determine an overall manufacturing and pre-manufacturing inventory. A specific example can be found in [5]. The disadvantage of this implementation is that it is a 'closed system', thus there is no possibility of interfacing the program to exchange the product's pre-use inventory data with existing LCA software, which would allow a full life cycle assessment. Another example is the method proposed by Birkhofer's group (c.f. [7]), which aims at integrating process and product development by integrating, what they call 'environmental knowledge', into the design process. The main feature of the computer-based system is the interface to product and process development (CAD) and to internal and external environmental information. The environmental information is previous LCA data of materials, machine elements, sub-assemblies, etc. The method is aimed at the embodiment stage of the design process and does not include a systematic approach to include the use-phase of the product.

In summary it may be concluded that current LCA software represent standalone products and research approaches concentrate on specific phases of the life cycle, often eliminating use and post-use phases and only assessing manufacturing and pre-manufacturing phases. These approaches seem to be too inflexible, complex and expensive for especially SME's, where often there is no dedicated LCA team looking at the life cycle optimization of products.

Despite international efforts in research, latest results in areas discussed above are adopted and put into practice by industry quite slowly, and if so, mostly in big enterprises. However, SME industries still remain somewhat isolated ignoring necessity and benefit of these new approaches, though it seems only a matter of time until they too are forced by intensifying global markets and new the environment and natural resources protecting laws and regulations to adapt to current practices of product development methods.

2.2 Targeted problems and overview

Within the work of introducing LCA to SMEs as described in this paper, the scope of this first attempt was to provide for an affordable LCA that although being somewhat simplified, is helping to reduce efforts and costs to design environmentally friendly product families. Since SMEs usually specialize on a product that is produced within a product family, an industry specific, parameterized LCA for a particular product family can be realized.

Since an LCA firstly requires the life cycle inventory (LCI) of the product, the approach described here is to parameterize the LCI using sizing parameters (design parameters) for the product family. A database with LCI data on relevant generic raw- and engineering materials and energy products and a database with site specific manufacturing LCI data can then be both linked to the simplified product inventory, thus providing a product specific LCI. Hereby to accommodate to budget and resources available to the SME, the development of a framework employing easy to handle, inexpensive commercially available hardware platforms and software products was given high priority. To test in how far such an approach can be translated into practice, development and implementation of a fully operating prototype system is targeted.

3   ECO-DESIGN AND PRODUCT FAMILY DESIGN

3.1 Outline of the approach

Basic requirements - among other things - stemming from the profile of our industrial partner (SME industry) resulted in an approach, providing a framework that allows for transparent, flexible, inexpensive and easy to handle implementation. To design product families and carry out computation of basic life cycle inventory parameters, easy to handle graphical user interfaces are provided to access all central system components (see detailed discussion below), which employ only inexpensive commercially available software that can be used on basic windows NT workstations.

In the next following sections details of methods used in the framework will be discussed together with an outline of the system architecture and tools used for implementation. For evaluation of both the approach and the computer-aided generation of design solutions regarding applicability, performance and future innovative improvements, a first implemented prototype system has been provided for both sides, our research team and our collaborating SME industrial partner.

3.2 Framework and structures

Work on the framework was basically aimed at supporting the definition of functional structures by providing well-documented studies of typologies for an application field, subject to investigation. Problems of evaluation of modularity of products can be solved with a functional decomposition of different product typologies as discussed in [4]. The identification of how many common functional groups are present in different products represents the parameter that allows for the definition of commonality in a specific production. Once the modularity is determined, the same functional groups previously identified are then candidates, which become the modules suitable for new products. The correspondence between modules and functions allows for an easy assembly of new configurations to satisfy different customer requirements.

Results of such an analysis are libraries of modules for product designers. However such modules may have different configurations, while performing the same function. Two types of differences, due to technical specifications, shall be mentioned explicitly. Namely, different components dimensions and the use of different components. In the first case, the dimensions can be managed easily with relations and formulas derived from structural constraints, catalogues of standard components, etc. In the second case, it is likely to be more difficult, due to the required designer skills, automating the new configuration.

At present, our prototype system supports only the first case. However, in the near future we are going to develop an interface to an integrated KBE system to include and manage the designer's experience.

3.3 System architecture and prototype implementation

Currently the implementation of our prototype system supports the configuration of single modules and their assembly. Design parameters are identified with a selection, related to determine the product geometry, which is processed in a digital spreadsheet. This system component in turn is linked to a three-dimensional parametric solid modeling CAD system. To compute a set of life cycle inventory parameters, used to for a basic life cycle assessment, several LCA modules which combine relevant product data and LCI related parameters are integrated in the prototype system. Within this basic framework, a semi-automatic parametric design of product families can be achieved. In a further step, currently underway, the integration of a KBE system and a design knowledge base are prepared (see Figure 1). This shall give the framework and its implementation enough functionality, to organize, relate and process available design parameters, obtained from various sources such as product functionality, customer needs, LCI database, etc.

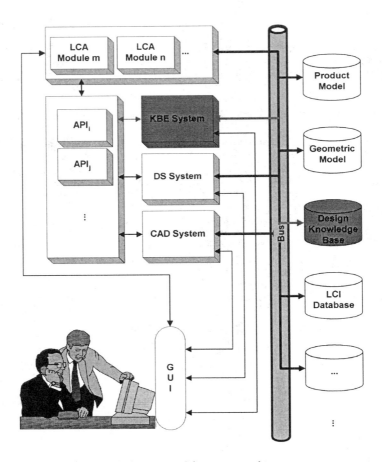

Figure 1 *Overview of the system architecture*

An important aspect of the system is the link between the digital spreadsheet and the integrated CAD system. Parameters and relations currently implemented within the digital spreadsheet contain all basic design information such as simple design rules, dimensioning relations, dimensions of standard components of a product, etc. During an interactive design session using the implemented prototype system this design information, abstracted as design parameters and relations, turns then together with the user input into the driving force to create as well as control the final overall product dimensions and configurations of module assemblies. Since design parameters are linked to the CAD system as well, further analysis of spatial interference and gaps between components, or studying a better way to execute an assembly, etc. are supported employing the exact geometrical model of a designed part or product. This structured access to the geometrical model is also important for determining some life cycle inventory parameters, when part or product data related to exact geometry such as material volume are required. The open system architecture allows also for an easy interfacing with an FEM system or a CAPP system. At the end of a design cycle results of evaluations can be fed back into the system through modifications of design parameters and relations stored in the digital spreadsheet to generate new configurations. In future versions of the system this step of data feedback shall include also complex design knowledge and newly gained experience abstracted as rules of an integrated knowledge base expert system. Eventually the technical documentation (drawings) and bills of material to manage the production processes can be obtained automatically.

Tools and hardware platforms used for implementation and experimental work consisted of windows NT operated workstations and of an object oriented application development environment (VisualStudio) and a digital spreadsheet (Excel) both from Microsoft, a CAD system (SolidEdge) from Unigraphics solutions (UGS), a technical computing environment (MatLab) from MathWorks Inc. and an LCI database.

4   EXAMPLE

4.1 Test data and industrial partner

We have applied the method described above to the design of a product family of bridge crane speed reducers. In the case study described below subject to examination is a speed reducer for a bridge crane with main requirements of a customer as following: Two different lifting speeds (0.75 rpm and 7.43 rpm) for lifting 63t with two different powers 3.3kW and 33 kW. Additionally the overall dimensions of the entire product should be as small as possible. Our industrial partner within the European Union was an Italian SME (Omme S.P.A.) which is an enterprise working in the field of design and manufacture of custom-made medium and large sized gears and gear trains, respectively. Details of part and product geometry are shown below (cf. Figures 2 and 3) as rendered screen dumps created with our implemented prototype system.

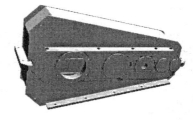

Figure 2 Examples of the housing with and without the gear train

Figure 3 Examples of individual parts and modules

4.2 Scope and procedure of generated solutions

The sample outcome of the LCA study of the gearbox is presented in this section. The scope of the LCA is to demonstrate, how with firstly the existing product data, secondly the manufacturing inventory data, thirdly a LCI database on basic energy conversion processes and engineering materials and fourthly the assumptions on the use phase the importance of the different product phases can be compared to each other.

The SME gains from the LCA that weak points of the product are highlighted, but also it is an easy step to perform an assessment of the life cycle costs and these can be minimised once the problem areas are determined. This is an advantageous standpoint, which helps to sell products, especially as many clients are aware that life cycle cost (and the related total cost of ownership) carry significant weight in a sales situation.

The assumptions on the use phase are listed in Table 1. The average power requirement was determined from the fact that the crane is utilised for loads between 20% and 100% of its full load and assuming a symmetrical load distribution the average load would be 60% of the full load condition and with constant lifting speeds and lifting distances, the power requirement is proportional to the load.

For the purpose of this paper the parts inventory is simplified and summarised in Table 2. The various types of steel used in the gearbox are not distinguished, as the level of detail is not required. The manufacturing inventory could not be determined in time for this paper and therefore had to be neglected.

The sizing parameters required for the LCI are in essence the parameters listed in Tables 1 and 2, the most relevant being the usage characteristics, the efficiency of the gearbox and the parts inventory of the gearbox. These are the parameters that the SME can influence and so optimise the life cycle of its product.

Table 1: *Assumptions on working conditions of gearbox*

	High-Speed	Low-Speed
Working hours per year [h]	2000	2000
Operating per working hour	50%	50%
Operation ratio (high and low speed)	90%	10%
Operating time per year [h]	900	100
Design Life [a]	10	10
Avg. power requirement [W]	19,800	1,980
Electric motor efficiency	94%	94%
Gearbox efficiency	91%	87%
Net Energy Input [kWh]	208,324	2,421

Table 2: *Simplified summarised parts inventory of gearbox*

Part	Material	Mass [kg]
Bearings	Steel	30.35
Gears	Steel	475.70
Shafts	Steel	158.50
Housing	Steel	200.00
		864.55

4.3 Evaluation of results computed

It can be seen that the low-speed operating conditions only require about 1% of the net energy required from the grid while the other high-speed operating conditions accounts for 99%. This would indicate that the gearbox efficiency of the low speed condition does not need specific attention, while for the high speed condition a 1% increase in the gearbox efficiency would result in about a 1% decrease in total energy requirement.

In Table 3 a selection of LCI data is listed for demonstration purposes. Nevertheless, this simplified LCI data highlights that for most emissions the use phase dominates the life cycle to such an extent, that it can be generally concluded, that the design of heavily used gearbox requires a focus on increasing the efficiency for its predominating operating condition. In this specific case, from the point of view of the SME producing the gearbox, the gearbox efficiency of 91% for the high-speed condition (cf. Table 1) should be attempted to being increased in order to have the greatest effect on the whole life cycle of the gearbox. From a practical point of view there would be a point in monetary terms where the additional cost to increase the efficiency would be greater than the energy cost saved.

Table 3: LCI data, for demonstration purposes only (using The Boustead Model, Boustead Consulting, UK)

LCI Data Resources	Unit	PreUse	Use	PostUse	Total	Use/Total
Gross Energy Requirement	MJ	66,000	2,456,000	-	2,522,000	97%
Water	t	24	65	-	89	73%
Air emissions						
CO2 Equivalent (20 year horizon)	t	4.6	145.6	-	150.2	97%
SOx	t	0.04	1.92	-	1.96	98%
NOx	t	0.00	0.88	-	0.88	100%
Metal	kg	0.02	1.12	-	1.14	98%
Hydrogen Chloride (HCl)	kg	0.2	4.4	-	4.6	96%
Water Emissions						
Chem. Oxygen Demand (COD)	kg	0.60	1.26	-	1.86	68%
Biochem. Oxygen Demand (BOD)	kg	0.02	1.13	-	1.15	98%
Lead (Pb)	g	0.6	0.0	-	0.6	0%
Solid Emission						
Mineral Waste	t	0.8	1.9	-	2.7	69%
Mixed Industrial Waste	kg	30	113	-	143	79%
Slag/Ash	kg	280	515	-	795	65%
Inert Chemical Waste	kg	7.5	0.0	-	7.5	0%

5 CONCLUSIONS

The paper's aim was to demonstrate that low-cost specialized solutions for product family design and life cycle assessment can be developed for SMEs, tailored to the specific needs. The cost are kept to a minimum by using standard inexpensive packages and linking them with tailor made API's.

It was shown that the sizing within the product family was carried out by defining relevant design parameters and the linked software packages can semi-automatically produce solid models, engineering drawings and bills of materials. The bills of materials and the use-phase parameters were used as input parameters for a LCA, which in the example showed the dominance of the use-phase on the life cycle. The most important factor the SME can influence was shown to be the gearbox efficiency in its predominating operating mode (high speed condition).

6 REFERENCES

[1]   Allen, R. K. and Carlson-Skalak, S. (1998) Defining Product Architecture During Conceptual Design, in: *Proc. of 1998 ASME Design Engineering Technical Conference*, September 13-16 , Atlanta, GA.

[2]   Kohlhase, N. and Birkhofer, H. (1996) Development of Modular Structures: The Prerequisite for Successful Modular Products, *Journal of Engineering Design*, 7, 3, 279-291.

[3]   Martin, M. and Ishii, K. (2000) Design for Variety: A Methodology For Developing Product Platform Architectures, in: *Proc. of 2000 ASME Design Engineering Technical Conference*, September 10-13 , Baltimore, MD.

[4]   Mc Adams, D., Stone, R. and Wood, K. (1999) Functional Interdependence and Product Similarity Based on Customers Needs, *Research in Engineering Design*, 11, 1-19.

[5]   Müller, D. H. and Oestermann, K. (1997) Ökologische und ökonomische Bewertung technischer Produkte mit Hilfe eines featurebasierten Konstruktionssystems, in: *Features verbessern die Produktentwicklung: Integration von Prozessketten*, VDI-Gesellschaft Entwicklung, Konstruktion, Vertrieb, VDI-Verlag, Düsseldorf, Germany, 365-380.

[6]   Pahl, G. and Beitz, W. (1996) Engineering Design: A Systematic Approach, Springer-Verlag, New York.

[7]   Schott, H., Grüner, C., Büttner, K., Dannheim, F. and Birkhofer, H. (1997) Design for environment - computer based product and process development, in: *Proc. of the IFIP WG5.3 4th International Seminar on Life-Cycle Engineering, LIFE CYCLE NETWORKS*, Berlin, Germany, Chapman & Hall London.

[8]   Stone, B. R., Wood, L. K. and Crawford, H. R. (1998) A heuristic Method to identify Modules from a Functional Description of a Product, in: *Proc. of 1998 ASME Design Engineering Technical Conference*, September 13-16 , Atlanta, GA.

[9]   Tichem, M., Andreasen, M.M. and Riitahuhta, A. (1999) Design of Product Families, in: *Proc. of Int. Conf. on Engineering Design ICED 99*, August 24-26, Munich, Germany, 1039-1042.

Evaluation of Used-Parts Quality
for Integration of the Inverse Manufacturing Process[*]

Michiko Matsuda[*1], Hiroto Suzuki[*2] and Fumihiko Kimura[*3]

*1Department of Information and Computer Sciences, Kanagawa Institute of Technology
1030 Shimo-ogino, Atsugi-shi, Kanagawa 243-0292, Japan
telephone:+81-462-91-3213, fax:+81-462-42-8490, e-mail: matsuda@ic.kanagawa-it.ac.jp
*2Arthur D. Little (Japan), Inc.
*3Graduate School of Engineering, The University of Tokyo

Abstract: To complete the product life cycle, one has to include an inverse flow from products to materials or parts. In the inverse manufacturing process, products are decomposed into several parts after the use phase. Some of these parts are re-used at the production stage and during the maintenance/up-grade process. In this paper, the evaluation system of decomposed parts' quality is proposed in order to connect the output from the inverse process with the input to the forward manufacturing process. To represent the deteriorated status of used parts, geometrical quality description is introduced into the product model. Behavior-mechanism relation tables are used for connecting functional behavior of the product with corresponding deteriorating parts in the evaluation process. Final output from the evaluation process is geometrical deteriorated status of the geometric element of the parts. These data can be used at planning the stage for reusing parts in the forward flow of product life cycle.

Key words: computer-aided integrated manufacturing, inverse manufacturing, product life cycle, product modeling, used parts quality evaluation

1.   INTRODUCTION

An active discussion on product life cycle has recently ensued, driven by an increasing interest in the environmental impact of manufacturing. Product design is the initial stage of the product life cycle. The product design

---

[*] Funding support: Grant-in-Aid of Scientific Research (C) by The Ministry of Education, Science, Sports and Culture (Japan)

process has to be supported with considerations of environmental aspects. Furthermore, an approach is now made to not only consider the product design and manufacturing preparation concurrently, but also to design the total product life cycle as a whole from product planning, through product design and manufacturing, to product usage, maintenance and re-use/recycling/disposal. This approach is called *"inverse manufacturing"*, stressing the controllability of the re-use/recycling process, where closed cycles of manufacturing activity are pre-planned and controlled [1,2].

The trends mentioned above demand the necessity to make the production process more energy efficient as well as to save resources. They require that the manufacturing preparation system should estimate not only the production time and cost but also energy usage, and incorporate the reuse/recycle of used-parts and materials wherever possible, and so on.

In this paper, mechanical products are considered that are produced with large varieties of types and very small production volumes. In this type of production, it is important that used parts from a product can be accepted as materials for another product in the manufacturing phase [3]. Therefore, it is important to support the reuse phase by constructing the additional description of representing the deteriorated status in the product model and by providing the evaluation method for the geometrical quality of each part using this model. As a result, this support connects the inverse flow with the forward flow in the product life cycle and completes the product life cycle lines.

2.   INVERSE / FORWARD PROCESS INTEGRATION

2.1   Life cycles of product

The life cycle of a certain product includes various phases such as product design, manufacturing, usage/maintenance, disassembly and reuse/disposal. In these phases, the reuse phase can be divided into two types. One is the reuse of each part. In this type of reuse, decomposed parts are refurbished and/or exposed to another machining process in order to adjust their form and accuracy to a new usage. The other type of reuse is material recycling by melting down parts. In this paper, only the former type of reuse for each part is considered.

As seen from the manufacturing phase, various kinds of products originating from various and separate life cycles are processed on one common production line. Figure 1 shows this. To represent a complete product life cycle, one has to include an inverse flow from products to parts. In the inverse process, products are decomposed into several parts after the use phase. Some of these parts are reused at the manufacturing stage and during the maintenance/up-grade process.

In the manufacturing phase considering parts reuse, several types of parts from different products have to be processed in the same manufacturing process. The manufacturing process can be divided into machining process and assembly process. Many of the mechanical products, which are considered in this paper, have already been intended to be used repeatedly by refurbishment if necessary [4].

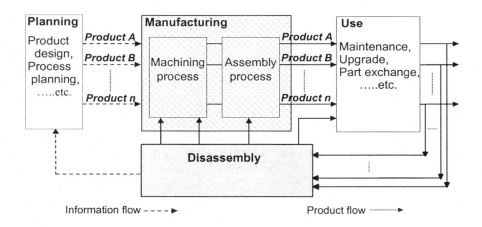

Figure 1. Product life cycle and Inverse manufacturing

2.2   Introduction of evaluation process

After the usage/maintenance stage, the product is disassembled into several parts in the inverse manufacturing process. By reusing these parts, the inverse process is integrated into the traditional forward manufacturing process. If only blank materials are used in the manufacturing process, only one machining process plan can be designed. However, if used parts from different products are also treated as candidates of machining in a material stock, several processes should be considered. Because the deteriorated status of each used part can be different, one part may need to be refurbished while others may be able to be used as assembly parts without refurbishment. Also some parts may be used as repair parts in the maintenance process. Here, refurbishment means the additional machining required for remaking the parts into the parts required for another product. Figure 2 shows this integration.

When reusing used parts, their quality needs to be evaluated based on their usage history data, since used parts may have changed their size and accuracy due to deterioration such as abrasion and deformation during their former use. In order to support this evaluation process from the product information processing, the structure of the product model and the new evaluation procedure should be provided.

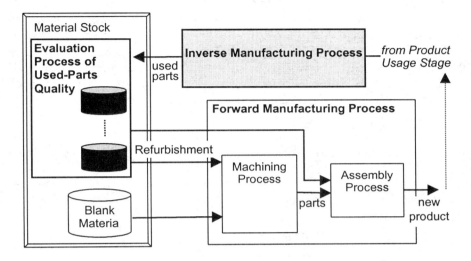

Figure 2. Integration of inverse manufacturing process and traditional process

3.   PRODUCT MODEL FOR EVALUATION

3.1   Role of product model in evaluation process

For reusing parts, the product model has to be able to be used in the quality evaluation of used parts before their reuse. For this evaluation, the product model has to fulfil the following three requirements [5].

The first requirement is the ability to apply the prediction process of the geometrical deteriorated status using a computational method such as a behavior simulation. As considered, a geometry-oriented product model is mainly applied to predict how a certain geometrical property, such as accuracy of form may have an effect on the product behavior. In the product model, these geometrical accuracies are represented in the geometric model, and should be treated as parameters of the input model in the calculation process.

The second requirement is the ability to add the result of the evaluation into the original product model. In the present product model, it is hardly supposed that the information about geometrical quality should be added for the inverse process. In order to add the result of the evaluation process, another attribute should be added to the normal product model.

The third requirement is the ability to be used for production planning including refurbishment. Since such machining is done for each part, these geometrical accuracies have to be assigned to each geometric element from the refurbishment viewpoint. The deteriorated condition of each used part is evaluated as a change in accuracy of form based on the usage history. An altered product model based on the evaluation of deterioration is added to a material library as a candidate for machining. If it matches requirements for the parts of another product, refurbishing process data for the selected candidate parts are generated based on the selected product model. Then this product model is again circulated as a new product model in the next life cycle.

3.2   Geometrical quality description

The geometrical quality description is to relate the mechanical behaviors and the geometrical accuracies. Figure 3 shows the structure of geometrical quality description [5].

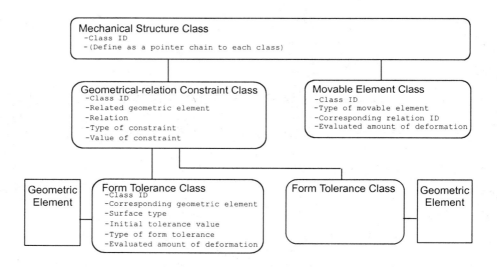

Figure 3. Geometrical quality description

Form tolerances are defined for each geometric element. The type of form tolerance is determined according to the surface type of the geometric element. A geometrical-relation constraint is defined for the geometrical relation between geometric elements. If the geometrical-relation constraint is defined between parts and any relative motion occurs in the relation, a movable element has also to be defined for the geometrical-relation constraint. As commonly used movable element descriptions have already been constructed, the appropriate movable element is applied to the objective movable relation. The whole connection between each description in the product is described as a mechanical structure description. In this description, the connective relation is represented in the form of a pointer chain between each description.

3.3   Product description for evaluation

In the mechanical product, many of the product functions are realized through the mechanical behaviors. These behaviors are made up of the mechanical capability of each part as well as the part's role in the overall structure. Therefore the geometrical accuracy given by the geometrical variations of each part and its relation can have an effect on the behavior.

Figure 4 shows the additional description required in the product model for evaluating used parts' quality. Functional constraints for satisfying the product function are defined as the geometrical accuracies of assembly relations, non-interference with the motion envelope of moving parts, and so on.

Corresponding to each functional constraint, behavior-mechanism relation tables are generated from the product model. In this table, the possible sets of the movable elements in which deteriorations occur are given corresponding to a functional constraint to be achieved.

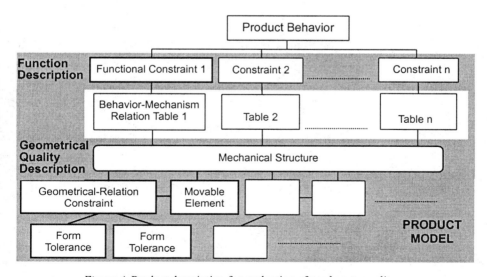

Figure 4. Product description for evaluation of used-parts quality

4.   EVALUATION OF USED-PARTS QUALITY

4.1   Evaluation process

The evaluation process for geometrical quality of used parts has two phases. The first phase is the preparation of the behavior-mechanism relation table, and the other phase is execution of the evaluation. The Evaluation process using the product model is shown in Figure 5.

Usually, the behavior-mechanism relation tables can be prepared at the product design stage. The behavior-mechanism relation tables are generated corresponding to each functional constraint. The value of the corresponding predicted geometrical accuracy in the behavior-mechanism relation tables are obtained from the product mechanical-behavior simulation.

At the execution phase of the evaluation, the product behavior at the product usage stage is compared with functional constraints which the product should be satisfying. According to the result of this comparison, the deteriorated status of each part which is a component of the product, is evaluated using product model and behavior-mechanism relation tables. The final output of this evaluation system is the predicted deformation value for each geometric element of the parts.

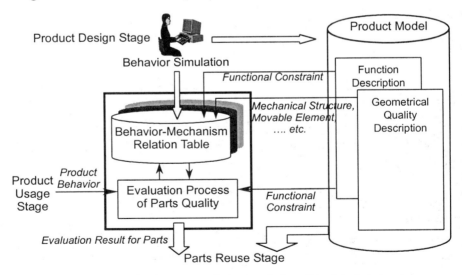

Figure 5. Evaluation process for geometrical quality of used-parts

4.2   Preparation of the behavior-mechanism relation

In the preparation process which will be included in the product design stage, each set of deteriorated status given by the behavior-mechanism relation tables are tried with the behavior simulation as shown in Figure 6.

First the behavior-mechanism relation table is generated with an entry corresponding to each functional constraint. Then, by looking up the mechanical structure description in the product model, a corresponding movable element is found.

In the behavior simulation of the product, the value of the corresponding geometrical-relation constraint in the movable element description is interpreted as a dimension of the clearance between parts. By superimposing the result of behavior simulation, the effect of that set of deteriorated status on the behavior can be calculated as the threshold value of the corresponding predicted geometrical accuracy. These results are added to the behavior-mechanism relation table based on the result calculated by product behavior simulator.

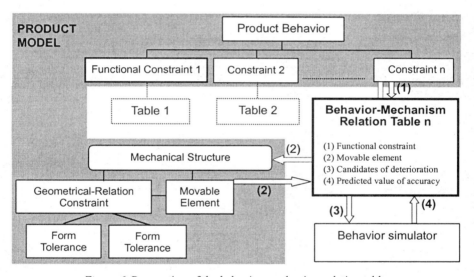

Figure 6. Preparation of the behavior-mechanism relation table

4.3 Evaluation process

Evaluation process of the geometrically deteriorated status of the parts is executed using product model and behavior-mechanism relation tables. Figure 7. shows the procedure of the evaluation.

Usage history is also added to the product model during the product use stage. This history mainly describes the functional status of the product corresponding to the functional requirement on the behavior-mechanism relation table. For the evaluation, as the first step the usage history is compared with the threshold value of each functional requirement on the behavior-mechanism relation table.

If a functional requirement to be achieved is found to be satisfied from the usage history, the behavior-mechanism relation tables corresponding to the functional requirements can be selected for evaluating the deteriorated status of the related movable elements. When several behavior-mechanism relation tables are selected, tables are overlapped to determine whether the same movable element are related or not.

Using this methodology, the deteriorated status for each movable element is estimated. Then in each movable element, the deformation is assigned to the form tolerance of each geometric element involved in the movable element. The relationship between geometrical-relation constraint and form tolerance is used for this assignment.

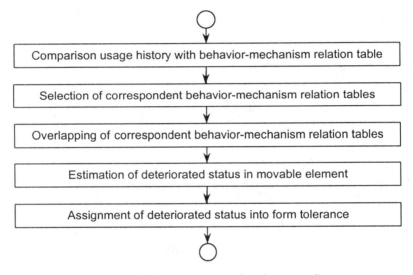

Figure 7. Evaluation process of used-parts quality

5. EXAMPLE

The proposed evaluation method is applied to a slider-crank mechanism. Figure 8 shows the mechanism (left) and solid model (right) of the example model.

Based on this solid model, both form tolerances and geometrical-relation constraints are added. Figure 9 shows the part of the geometrical quality description for this example model. In this example, a hinge element class is defined between the cylindrical shaft of part1 and the cylindrical hole of part2. By repeatedly generating geometrical quality descriptions for all parts, the mechanical structure class can be defined as shown in figure 11.

Then, a behavior-mechanism relation table is generated based on the mechanical structure and a functional requirement as shown in Figure 10. In this case, the positional error of the evaluation point on part3 is supposed to be the requirement. As there are three movable elements involved in the mechanical structure class, seven possible sets of deteriorated states are given for the behavior simulation. Then each possible set is checked in the calculation process. In this process, behavior simulation is executed in order to estimate the maximum deformation value for each case. After calculating all cases, the maximum deformation value for each movable element can be detected. Here, ADAMS which was developed by Mechanical Dynamics, Inc., is used as a behavior simulator.

In the evaluation process, if this function requirement is not satisfied, the estimated deformations are assigned to the form tolerance of each geometric element. In this example, the clearance of the above-mentioned hinge element is assigned to the cylindricities of the shaft and hole. As a result, the change of cylindricity of both parts is evaluated as the true cause for the change of behavior.

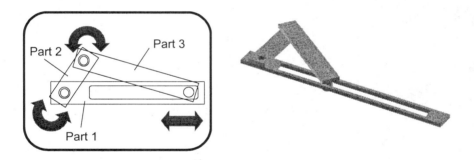

Figure 8. Example parts (slider-crank)

6. CONCLUSIONS

In this paper, an evaluation system for used-parts geometrical quality is proposed as one method to integrate the inverse manufacturing process with the traditional forward manufacturing flow. To structure this system, the product model with geometrical quality description and the behavior-mechanism relation table are introduced.

The product model with geometrical quality description is provided by expanding the definition of the present geometric tolerance. Therefore, the parts deformation due to deterioration could be interpreted into form

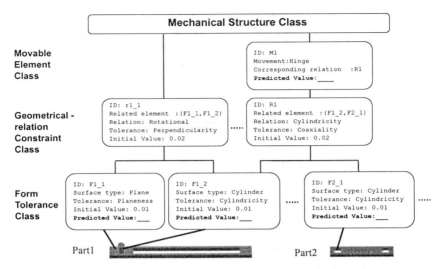

Figure 9. A part of geometrical quality description for example parts

Relation Table				
Mechanism Slider–Crank				
Functional Requirement: FR01				
Error Parameter: Radial Error of Movement				
Start Element: g12				
Evaluated Element g22				
Thereshold Value: 0.5				
hinge1	hinge2	slider	Error	Judge
1.00	1.00	1.00	1.2135	0
1.00	1.00	0.50	1.2248	0
1.00	1.00	0.10	1.6987	0
0.50	1.00	1.00	0.9410	0
0.50	1.00	0.50	0.8520	0
0.50	1.00	0.10	1.2742	0
0.10	1.00	1.00	0.6136	0
0.10	1.00	0.50	0.5879	0
0.10	0.80	1.00	0.7009	0
0.10	0.80	0.50	0.7565	0
0.10	0.80	0.10	0.7370	0
0.10	0.75	1.00	0.4966	1
0.10	0.75	0.50	0.6530	0
1.00	0.50	1.00	0.9685	0
1.00	0.50	0.50	0.9161	0
1.00	0.50	0.10	1.2648	0
0.50	0.50	1.00	0.6550	0
0.50	0.50	0.50	0.5746	0
0.50	0.50	0.10	0.8188	0
0.10	0.50	1.00	0.5976	0

Figure 10. A part of generated behavior-mechanism relation table for example parts

accuracy for geometrical element of physical parts. The behavior-mechanism relation table provides a function to connect product behavior with parts deformation.

An example part is applied to the proposed evaluation process from the generation of the behavior-mechanism relation table to the prediction of deterioration status. This trial shows the effectiveness of the proposed geometrical quality description and evaluation method using the behavior-mechanism relation table.

The output of this evaluation process will be useful for planning of parts reuse. Especially, form accuracy data for geometrical element of parts will be useful for the refurbishment planning phase.

7.   ACKNOWLEDGEMENT

The authors are grateful to Dr. Udo Graefe of the National Research Council of Canada for his helpful assistance with the writing of this paper in English.

REFERENCES

[1] Kimura, F. (1997). Inverse Manufacturing: From Products to Services, The 1st Int. Conference Managing Enterprises – Stakeholders, Engineering, Logistics and Achievement (ME-SELA'97).

[2] Hata, T., Kimura, F. and H. Suzuki. (1997). Product Life Cycle Design based on Deterioration Simulation, CIRP 4th Int. Seminar on Life Cycle Engineering, pp 197 -206.

[3] Matsuda, M., and Kimura, F. (1999). A Model Integration Framework for Life Cycle Engineering from the Machining Process Viewpoint, Proceedings of the CIRP 6th International Seminar on Life Cycle Engineering, pp. 332-341.

[4] Suzuki, H., Matsuda, M. and Kimura, F. (2000). Prediction of Used Parts Deterioration Based on A Tolerance Model, Proceedings of the CIRP 33rd International Seminar on Manufacturing Systems, pp. 149-154.

[5] Matsuda, M., Suzuki, H. and Kimura, F. (2001). Tolerance Modeling Using The Feature Based Product Model for Prediction of Used Parts Deterioration, Proceedings of International IIFIP Conference Feature Modeling in Advanced Design for the Life Cycle Systems FEATS2001, in print.

A Telecooperation Management System for Secure Data Conferencing in Distributed Product Development Processes

Prof. Dr.-Ing. F.-L. Krause,
Dipl.-Ing. H. Jansen,
Dipl.-Inform. R. Schultz,
Cand.-Inform. H. Gärtner
Fraunhofer Institute for Production Systems and Design Technology – IPK Berlin
e-mail: [first name]". "[last name] "@ipk.fhg.de "

Abstract: Telecooperation is a technology which fulfills the communication requirements arising from global distributed product development. Implementations of the T.120 standard take the heterogeneous environment into account but do not sophistically solve security and ergonomic problems. The Telecooperation Management System (TCMS), developed by the Fraunhofer Institute for Production Systems and Design Technology, enhances T.120 implementations for their successful use in distributed product development.

Key words: distributed product development, process chain, computer aid, telecooperation, teamwork, application sharing, data security, Telecooperation Management System (TCMS)

1.   SITUATION OF THE DISTRIBUTED PRODUCT DEVELOPMENT

As a result of global competition, international consolidation and cooperation of companies is leading to the distributed development of products. Comprehensive use of computer aided technologies in all phases of the product development process and worldwide availability of networks allow the sharing of digital information between different company sites /1/.

These assumptions allow the application of information and communication technologies (ICT), particularly computer supported cooperative work (CSCW) and telecooperation systems. Examples of these systems are electronic mail, chat, white boarding, application sharing, and video conferencing.

Because of the optimised exchange of information between different team members in the development process, the organisational benefits of telecooperation are /2/:
- reducing the number of changes,
- reducing development costs and time,
- increasing the quality.

In particular, the use of application sharing systems will bring benefits to the product development process, because these systems allow the common, distributed use of any computer program. If a CAD program is shared in a data conference, all members can see the same view of the model displayed, they can mark areas and modify the model. Therefore, a flexible and comfortable cooperation between team members is independent of their location in the world.

But, high requirements from application sharing systems on the network infrastructure lead to several problems. Reasons for the problems are:
- heterogeneous hard- and software in different networks,
- interoperability of the telecooperation systems used,
- different security requirements and infrastructure as well as
- lack of user friendliness of the systems.

The first two points are the result of company specific guidelines during the construction of their network infrastructure. Only the consideration of standards allows the communication between systems of different hardware manufacturers. A promising way to handle these problems is the use of TCP/IP based systems programmed in Java. This is due to the fact that TCP/IP and Java are designed to work in a heterogeneous environment. For application sharing systems, the most common standards are known as H.323, H.320 and T.120 /3/. The H.32x standards include the T.120 standard and are designed for video and audio conference via ISDN (H.320) or IP based networks (H.323) /4/. The T.120 standard defines the communication for application sharing, white boarding, remote pointing, chat and file transfer /5/.

Platform-independent use of T.120 systems is possible because Microsoft (NetMeeting), SUN Microsystems (SunForum), Silicon Graphics (SGIMeeting) and IBM (DC-Share) offer implementation software and most of them are available for free. With several million downloads, these systems are wide spread, but are not used for inter-company communication.

The third point in the list of problems concerns the security policy of the companies. During product development, exchanged data is very important for the companies. For example, the models represent development know-how or the style of the planned product. Therefore every company defines a security policy where, among other things, the security requirements for telecooperation are specified.

Telecooperation is only possible between companies, if the required resources are permitted in all companies involved. Figure 1 shows the generally available resources for telecooperation and the three sets (A, B and C) of company specific resources. Telecooperation between companies A and B is only possible with the resources in the intersection A ∩ B. Therefore, telecooperation systems must allow company specific configuration on one hand, but enable flexible communication with other systems on the other hand.

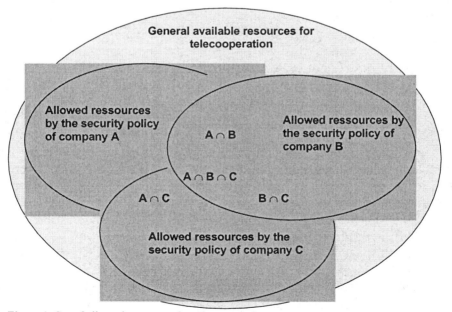

Figure 1: Set of allowed resources for telecooperation

The point to point connection between the conference computers due to IP based teleconference systems leads to problems concerning the security infrastructure. These issues include:

- the definition of permissions for every possible connection results in many open ports for many IP-addresses in the firewall,
- the use of unofficial or dynamic IP-addresses is not possible,
- the administration of the firewall requires a lot of time and
- many and not well ordered entries are the reason for loosing the overview.

From an ergonomic viewpoint, the use of telecooperation systems is complex for an engineer. If the user wants to establish a conference, he has to know the IP address of the computer of his conference partner. He must obtain the IP address and is, therefore required to call his contact for this information. Figure 2 shows the IP based invitation of a conference partner with the T.120 system MS NetMeeting.

Figure 2: Invitation to conference with NetMeeting

Nowadays, the third and the fourth point are not sophistically solved by the T.120 implementations. The reasons for this are, on the one hand the required end-to-end-IP-connection between conference computers which results in the creation of great holes in the firewalls. On the other hand, the

IP-based invitation of conference partner can not be supported by address books, because the information is always changing if the chosen person works in a pool or the dynamic host configuration protocol (DHCP) is used.

This paper describes the concept of the telecooperation management system (TCMS) which provides a solution for several teleconference protocols. The exemplary supported standard is T.120 for two reasons: in the product development process, pure application sharing is the better solution compared to application sharing with video conferencing. Video reduces the space on the screen, where the CAD window displays the model. Moreover, video and audio occupy the bandwidth on networks and have a negative effect on the performance of the shared program.

2.   ARCHITECTURE OF THE TELECOOPERATION MANAGEMENT SYSTEM

To fulfill the requirements of distributed product development, the solution must be platform independent, scalable, and must support the company specific security policy. Therefore the concept of the TCMS requires a TCMS on both sides and is implemented in JAVA. Figure 3 shows how the TCMS splits the data stream in an internal and an external part.

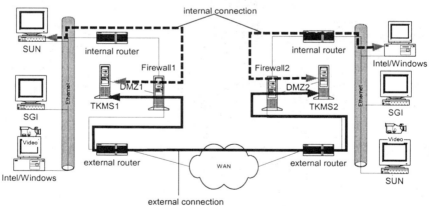

Figure 3: Telecooperation with the TCMS

The communication between every internal user workstation requires only connections to the demilitarised zone in the firewall. The external part of a teleconference requires only one connection to the external TCMS trough the firewall for every possible conference.

In order to keep the company security and network configuration secret, both TCMS negotiate the available resources before a conference and decide by conference parameter if it is allowed. The different modules of the scalable TMCS are shown in Figure 4.

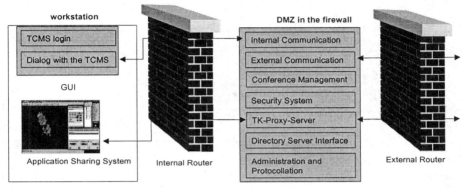

Figure 4: Architecture of the TCMS

The left side of the picture represents a user workstation with the TCMS user interface and the right side the demilitarised zone in the firewall with the modules of the TCMS.

The graphical user interface (*GUI*) is implemented using web technologies. The dialogs for login and for establishing a conference are either HTML pages or JAVA applets. The advantages are, on the one hand, the low installation expense because, apart from a web browser on the user computer, no additionally software is necessary. The user interface and the TCMS have to be installed only one time for the entire company. On the other hand, is the ability of using a common browser wide spread, whereby the user qualification is reduced to the functionality of the system.

The *Internal Communication* realises the communication between the TCMS and the user interface. Login and conference requests from the internal network are received and conference invitation from external partners are delivered.

The *External Communication* realises the communication with other external TCMS. Because of security aspects, only conferences between TCMS are possible. During the communication, both TCMS exchange the conference parameter and negotiate the coding algorithms and ports for the communication.

The main module of the TCMS is the *Conference Management*. This module controls the required resources for incoming and outgoing conferences and decides by conference parameter if a conference will take place or not. These parameters are:
-   internal user and internal conference computer,
-   external user
-   required conference system and
-   date and time of the conference.

The *Security System* realises the verification of conference parameters and the coding of all transferred data between internal and external TCMS and user workstation and TCMS.

For every required teleconference protocol, a *Proxy Server* can be implemented and integrated in the TCMS. The advantages of this solution are the independence of teleconference protocols used, the hiding of the internal network infrastructures and the reduction of holes in the firewall to one port for the IP address of the proxy server. The implemented proxy server supports the T.120 protocol on the platforms MS Windows, Sun Solaris, SGI IRIX and IBM AIX.

The *Directory Server Interface* allows the use of information stored in a companies own directory services.

The *Administration and Record* allows the definition of conference permissions and the comprehensive recording of all conferences, attacks, and errors.

3.   ESTABLISHING A TELECONFERENCE WITH THE TCMS

The TCMS makes telecooperation between two companies or sites in the distributed product development possible because it allows the realisation of company own security policy by enhancing the T.120 standard by several security features. The proxy server reduces the open ports in the firewall, allows network address translation and encrypts the transferred data using the secure socket layer (SSL) technology.

The *Conference Management* allows the invitation of a conference partner by using his e-mail address, shown in figure 5. The conference request is transferred to the internal TCMS where permission is verified. If permission is given, the internal TCMS starts a connection with the external TCMS and both negotiate the resources for the conference. Finally, the external TCMS starts a connection to the invited user, because links between user and the IP address of the computer are defined during the user login.

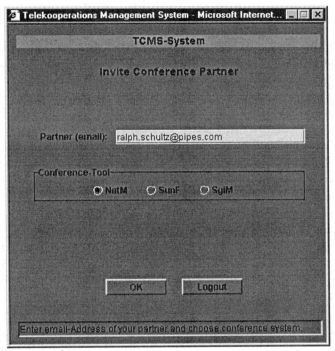

Figure 5: Inviting a conference partner

If the invited conference partner accepts the invitation and permission for the conference is given, a window is shown to the initiator of the process so that he should start the chosen teleconference system and connect to the internal proxy server of the TCMS, shown in figure 6. If a domain name service is available in the network, the IP address of the proxy server can be replaced by the name of the computer where the TCMS is running, for example TCMS.supplier.com.

Figure 6: Invitation accepted

The internal proxy server is waiting for the connection from the user computer and forwards all incoming data to the external proxy server which forwards the data to the workstation of the conference partner. A teleconference is now established. If any error occurs before or during a conference, the TCMS will provide the administrator and the user with context specific information to solve the problem.

4.   SUMMARY

Distributed product development in global markets requires efficient and secure communication between partners. This demands a small amount of time and money for the installation, adaptation and maintenance of the necessary infrastructure. Not only small and medium sized enterprises need to be able to communicate with several partners via low budget networks like the internet, but the transport of secret information on public networks leads to additional costs for the integration of high security technologies to save the data.

Teleconferencing with TCMS solves all the mentioned problems, because available features are:
- transparent establishment of a conference (only e-mail is required),
- reliable user and server authentication before a conference,
- hidden infrastructure with a minimum opened firewall
- secure data transmission (secured by SSL or IPsec),
- conference parameter based release of resources,
- realisation of company specific security policy,
- easy to install (zero maintenance solution),
- easy to configure and administrate and
- easy to use.

The Telecooperation Management System provides a significant reduction of requirements for flexible and secure teleconferences. Platform independence and comprehensive configuration possibilities allow the integration of the TCMS in any existing infrastructure to fulfil the company specific security policy. Up to date authentication and coding technologies allow conferencing over public networks. The modular architecture and the proxy concept allow the easy adaptation of the TCMS forced by changed telecooperation protocols.

5. REFERENCES

/1/   *Spur, G.; Krause, F.-L.:* Das virtuelle Produkt – Management der CAD-Technik. Carl Hanser Verlag, München, Wien 1997

/2/   *Krause, F.-L.; Schultz, R.:* Integration praxisorientierter Telekooperationssysteme in verteilte Produktentwicklungsprozesse eines Automobilhersteller/-zulieferer-Netzwerks. In: VDI-Berichte 1537: Der Ingenieur im Internet - Erfolgreiche Anwendungen in der Industrie. Tagung Karlsruhe, 27./28. März 2000. VDI Verlag, Düsseldorf 2000, S. 145-168

/3/   N.N: International Telecommunication Union. URL: *http://www.itu.int/home/index.html*, 2000

/4/   *N.N* A Primer on the H.323 Series Standard. URL: http://www.databeam.com/h323/h323primer.html, 2000

/5/   *N.N* A Primer on the T.120 Series Standard. URL: http://www.lotus.com/products/sametime.nsf/standards/8DDB25B6C08E70 E5852566640072FCD2, 2000

6. BIOGRAPHY

Prof. Dr.-Ing. Frank-Lothar Krause, born in 1942, studied Production Technology at the Technical University of Berlin. In 1976, he became Senior Engineer for the CAD Group at the Institute for Machine Tools and Production Technology (IWF) of the TU Berlin and earned his doctorate under Prof. Spur. Since 1977, he has been Director of the Virtual Product Development Devision at the Fraunhofer Institute for Production Systems and Design Technology (IPK Berlin). He earned the qualification as a university lecturer in 1979 and has been University Professor for Industrial Information Technology at the IWF of the TU Berlin since 1990.

Dipl.-Ing. Helmut Jansen, born in 1946, studied Electrotechnical Science at the Technical University of Berlin. In 1977 he joined the department of Industrial Information Technology at the Institute of Machine

Tools and Manufacturing Technology of the Technical University of Berlin. In 1980 he joined the Virtual Product Development Division of the Fraunhofer Institute for Production Systems and Design Technology (IPK Berlin). Since 1985 he is head of the department Information Technology and has worked in several national and international research projects in the field of Man-Machine-Communication, Pattern Recognition and Virtual Product Creation.

Dipl.-Inform. Ralph Schultz, born in 1965, studied Computer Science at the Technical University of Berlin. In 1995 he joined the department of Industrial Information Technology at the Institute of Machine Tools and Manufacturing Technology of the Technical University of Berlin. Since 1999 he is joining the Information Management Department at the Virtual Product Development Division of the Fraunhofer Institute for Production Systems and Design Technology (IPK Berlin). He has worked in several national and international research projects in the field of telecommunications and security in distributed product development.

Cand.-Inform. Hendrik Gärtner, born in 1971, studied Computer Science at the Freie Universität of Berlin. Since 1999 he is joining the Information Management Department at the Virtual Product Development Division of the Fraunhofer Institute for Production Systems and Design Technology (IPK Berlin). He has worked in national and international research projects in the field of telecommunications and security.

New Product Development Process:
Proposal for an Innovative Design Modelling Framework Including Actors Evaluation of Innovation Costs and Value

Sechi N. [1], Lawson M. [1], Soenen R. [2]

nathalie.sechi@renault.com; mlawson@univ-valenciennes.fr; rene.soenen@bat710.univ-lyon.fr
[1] *Laboratory of Industrial and Human Automatics and Mechanics (LAMIH) –CNRS, UMR 8530, University of Valenciennes – France, Renault S.A.*
[2] *Lab. of Information Sciences, Graphics, Image et Modelisation (LIGIM), University of Lyon*

Abstract: Firms are facing very short and important innovation cycles, particularly in IT and Telecommunication sectors. Then a question appears: why do some innovations succeed whereas other fail. From offer's point of view, a way could be to evaluate impacts of a decision to innovate for each of the actors involved in this product trajectory. Therefore the goal of such an approach is reducing high Innovation development risks by integrating the diverse stakes of life cycle actors and by helping design teams to integrate the evolution of some key environmental processes. We introduce in this paper the characteristics of the Innovation Process and Engineering Design Phase for high level innovations. In this framework, we propose an Innovation Valuation Model integrating strategic and tactic impacts in term of value and cost.

Key words: Innovation process, Innovative Design, process approach, product trajectory, innovation valuation

Introduction

Firms are facing very short and important innovation cycles, particularly in IT and Telecommunication sectors. Thus, a fundamental question appears : why do some innovations succeed whereas others fail. To try to answer to this question we may adopt the environmental or global point of view (for instance technologies selection by environment) or the innovation process actors' points of view and particularly the offering firm's vision. We choose to study the latter in the light of the life cycle actors' diverse stakes.

Moreover, our research particularly will be supported with the life-cycle of Automotive Industry and a specific case, the integration of e-services in a vehicle.

In this way, we have to define the innovations we study and their Innovation Process' characteristics (Part 1) . In a second part, we shall focus on the key phase, the expectation phase: the Innovative Engineering Design Process. In fact, we argue that the specificity of 'high level Innovation' Processes conducts to cast doubt over design traditional performance analysis in benefit of an approach using Innovation impacts' estimation. So we propose to introduce an instrumentation of this evaluation model for innovation actors (Part 3). In a fourth part, we present our validation framework, the development of innovative technological-intensive services (Part 4), and finally we conclude (Part 5).

1.   Analysis Framework

We focus on high level innovations we need to define (part 1.1.). Moreover, we argue that the corresponding Innovation Process also integrates high level innovations' specificity (part 1.2.).

1.1. Innovations studied : high level innovations

The term of Innovation includes products which have very different logic and risk for our economic system's evolution. We here focus on high level innovations, as they imply high changes in all the processes of the product's life cycle and in the firm's evolution. On the base of (Tidd 97), we can locate innovations studied in the following representation (figure 1.1).

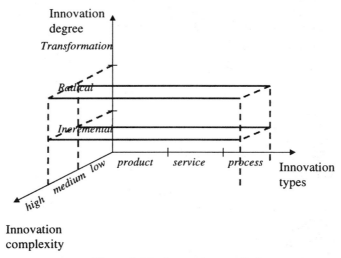

Figure 1.1.1 : Innovations studied

We then concentrate on 'Radical' Innovation, referring to Abernathy's definition (Abernathy 78). On the other hand, those high level innovations tend to be a combination between product, service and process and have a relatively high technological level. Typical high level innovations can belong to Information Technology and Communication sectors.

In this framework, we shall define innovation actors as the agents who have stakes in the innovation development. In this 'actors network' we can find actors of the product's life cycle – as final users, the offering firm, organisation partners - but also actors affected by the firm's results, as shareholders and public authorities. Referring to Normann and Ramirez (Normann 93), we call this group of agents a 'Value Constellation'. The authors remind us that the actors of this value creation-system co-produce the product.

Moreover, we have to note that we can name and define innovations only in reference to an initial state – a product we call As-Is. The As-Is is a commercialised product plus its associated life-cycle processes, introduced by the firm or competitive firms. We then have to observe the product-process As-Is and its 'performance trajectory' (Bower 95) including performance evolution of the product. At the opposite the Innovation is called To-Be.

1.2. The Concurrent Innovation

Innovations are often developed conjointly in parallel to others to be finally joined in a final determined time.

As G. Segarra (Segarra 01) explains us, the concurrent service / product development is defined in analogy with the concurrent engineering approach : it will consist in developing simultaneously the product and the services which will be enabled by the product. Following the author, his iterative approach (the product development impacts the service development which will impact the product development (see the figure 1.2 below)) must allow to develop the product in such a way that it will be enough open in order to facilitate the constant development of new services with a reasonable effort.

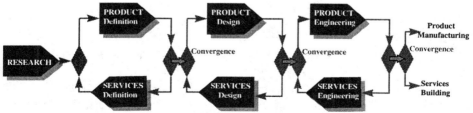

Figure 1.2 : Concurrent product - services development (Segarra 01)

The transition from one step to another (i.e. from definition to design) will be conditioned by the convergence of the product - services actors toward a common view point.

1.3. The Innovation Process

As technology intensive innovations in IT and telecommunication sectors show us, the success of an innovation process for high level innovations particularly depends on technologic and competitive environment changes. In addition, the process implies the intervention of diverse complementary agents and resources and generates important exchange of information and knowledge.

This complexity pushed us to choose a process approach in order to represent components of the innovation process and links between them. In this way (cf. the following 1.3 figure), the innovation process depends on offered resources – by the extended firm as by the environment – and drives us to final and intermediate results by the mean of stages. Information and knowledge exchanged and created are defining our solution. Moreover, the process can be viewed as a process highly reactive to its environment. At every process' stage, environment choices – corresponding to information and knowledge flows – curb the product trajectory. At last, we cannot neglect the role of the firm's strategy: its rules have the role of a canal, they conduct the process orientation.

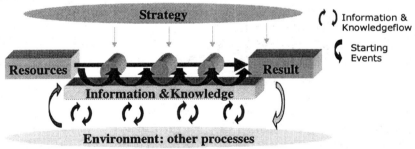

Figure 1.3.: an environment reactive process

In consequence, we may define three levels in environment processes:
the internal processes of the firm, some processes of the firms' partners, some processes of the global environment.

Processes which determine the Innovation Process thus have three levels. At a first level, we distinguish all concurrent processes from the extended enterprise. They include on one hand the firm's internal processes – as technology landscape, production processes, other new development processes, knowledge creation process – and on the other hand some processes of partners which interact on our product development process. Moreover, in the light of evolutionist studies (Nelson 82), some processes from the global environment such as external technologies selection processes or knowledge and competence selection processes (Cohendet 98) should also have an impact upon decisions about product development.

In conclusion, in this first part we have presented the type of innovations we will focus on, as well as the characteristics of Innovation Process. This permits us to bring out our study's scope : high level innovations and the way to develop and manage them, from the offering firm's point of view. More precisely, in next paragraph we will emphasise a key process : innovative engineering design process, and hence introduce our contribution about conducting this process through the estimation of innovation impacts.

2. Objectives and methodology

2.1. A representation of Innovative Engineering Design

At the heart of Innovation Process, Innovative Engineering Design process is a complex activity implying intervention of actors owning different levels of knowledge and information upon the solution. This process stretches from an innovative idea selection as far as the synthesis of the innovative solution, in such a way that it fulfils life cycle conditions.

Of course, many definitions and descriptions of design processes have been proposed both by academic and industrial authors. As an illustration, a good list of design methods has been realised in (Cross 94). For our part, we are choosing to bring the different visions of stakeholders , and the progressive convergence of these visions, to light. Each actor estimate so his perception of the innovative possible solutions' impacts upon their respective interests and is involved in a negotiation process through impacts estimation process

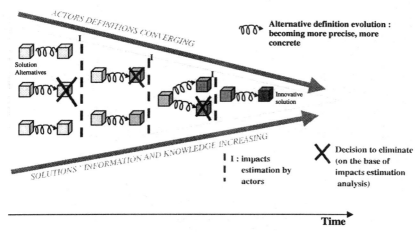

Figure 2.1 : Innovative Engineering Design Process as a Convergence process

This impacts estimation process, based upon the different alternatives definitions, will therefore help decision-makers to validate or eliminate some alternative. Then, from a product development point of view, design process can be seen as a funnel : from the beginning to the end of the process, some product features are chosen whereas others are rejected, until only one product concept is selected (cf. figure 2.1.).

2.2. Innovation valuation model methodology

The convergence process inside Engineering Design is conducted by actors' collective evaluations. We propose to pilot those decisions by actors' estimation of value and costs of the innovation. Consequently, one of the original points of our contribution is that the evaluation criteria we choose are the economic value and costs of Innovation for the actors. We define the economic Value as an individual or collective judgement on the product (Mouchot 95) ; (AFAV 89), based here not only on financial impacts but on real impacts too, such as enterprise knowledge development and competencies evolution.

Then we need to know which components are involved in such a decision model based on impacts evaluation. Referring to the 'Rational' Decision Model introduced by B. Walliser (Walliser 86), and the well known IDEF 0 formalism, we have consequently chosen to define the steps of an innovative solution's choice integrating value and cost impacts as follows (figure 2.2.):

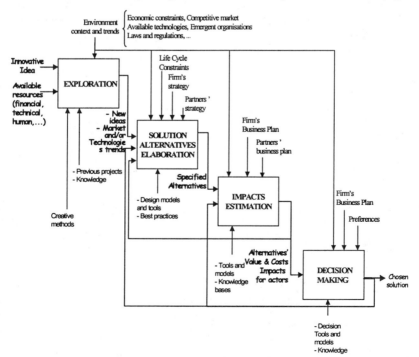

Figure 2.2.: *Steps of an innovative solution's choice using value and cost impacts*

On the base of system approaches we can consider initial innovative idea and available resources as inputs and the final result – our choice – as the output. The first phase we define is the exploration of the design space, regarding, as following steps, the environment context in terms of market and technologies' trends, available resources or capabilities from the extended enterprises, etc. Then, a second step will be the elaboration of different technical solutions by Engineers also constrained both by technical and economic elements, from the future product's life cycle issues. Each of these solutions, to which are associated some technical specifications, are the inputs for the third phase, impacts estimation (our main topic). The aim of this latter is to provide the global value and cost for actors, referring to our chosen representation models. Finally, the last phase equals to the decision model. Different possible decision models could be used, such as a simple discretionary decision to a collaborative informal choice or a metrics formalised model. We must notice that this overall process model is not sequential : at each step it includes possible returns to a previous step.

The decision process we propose is based on the innovation economic value estimation by the actors and the manner to extract information from the environment so as to feed this decision process. For each stage of the development process, Innovative solution values for the actors network are built in a converging manner.

After locating impacts of the possible solutions inside of Innovative Engineering Design Process, it would be interesting to enter inside the impact estimation's box. We shall thus present in the following part the different types, dimensions, and determinants of value and cost impacts.

3. Innovation Valuation Model : critical impacts from a decision to innovate

3.1. Impacts levels and dimensions

Each decision, along the Innovation Process, generates impacts on every Innovation Actor belonging to the 'Value Constellation' (Normann 93). In fact, as Normann and Ramirez underline it, innovating today implies an active intervention from all innovation stakeholders on account of final decision impacts on their interests. Therefore, we are facing a value creation system or network which builds a solution through information and knowledge flows. Each decision has an effect on each actor's interests. Let's note that the impact horizon is relatively large, and accordingly we choose here to limit our study to the most impacted actors.

Innovation success has interested several authors since the 1970's (e.g.(Madique 86)). These authors notably tried to identify the key factors for success. However today we can observe that their main weakness was that their studies focused exclusively on financial results. Then appeared since the 1980's large studies showing the important role of information relations between system's actors : in firms relations, employer-employees relations, manager-shareholder,... Since the beginning of the 1990's, the dominant tendency has been to take mainly into account knowledge issues (works on knowledge creation (Nonaka 96), Knowledge Management, and more traditionally, knowledge based firms relations ((Richardson 72), (Prohalad 90) ,(Teece 94),...). Otherwise, new economic models show us how brand value can determine firm value and is becoming a real study center. Our point of view is that all those dimensions of decision impacts are components of an Innovation success.

In conclusion, we define in our paper five dimensions of impacts for a solution choice, relatively to As-Is impacts : financial impacts, informational impacts, knowledge and competence impacts, reputation impacts, time impacts.

Moreover, one can observe that an innovation's development has different levels of impacts. We cannot reasonably integrate impacts on the evolution of a enterprise in terms of core competencies (Prohalad 90), market orientation and effects of product's usability. Accordingly, we just distinguish two levels of impacts for actors : strategic and tactic impacts.

As Hamdouch (Hamdouch 97) reminds us, Evolutionist Economists showed that an organisation follows a path, owns a story. There is an interdependency between its past, present and future choices. On the other hand, this trajectory is conducted by an evolution strategy defined by firm decision-makers at several levels. We touch here an important evolution factor of the firm that the product trajectory has to respect, but also can curb.

Elsewhere, one obviously observes tactic or short-term impacts which integrate direct effects of the innovation's life-cycle phases on actors. In fact, in this part we are not focusing on the product trajectory but on the direct relation of respective actors with the product.

3.2. Our solution representation

The decision to innovate implies choices on product characteristics and nature of relations between innovation actors. We shall study the both respectively.

In order to give a receiving structure for an innovative solution's different representation models proposed by actors, we propose a modelling reference framework (Soenen 01) , which is CIMOSA inspired (AMICE 89). This framework's aim is to enable a better circulation of data and information between actors. Their different models are then classified according to three axes as explained in the following figure 3.2.

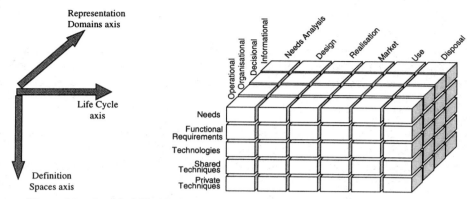

Figure 3.2 : Our Modelling Reference Framework for innovative solution representation

In the life cycle axis, co-designers are regrouped inherently to the life cycle phase they refer to. Thus Needs Analysis column deals with needs expression and solutions tracks proposals in terms of news functions or new technologies to be integrated. For example, realisation column is about models and tools used to define and specify forms, surface textures, etc. of product elements and their assembly features. In order to mark out some steps in the product's definition process, we adopt the five definition spaces from (Jacquet 98). Then, covering down this definition spaces axis, product definitions become more embodied, more concrete. About representation domain axis, we distinguish operational, organisational, decisional, and informational domains, according to the product's aspect they represent.

The industrial organisation is a determinant point at the time of solution choices. The type of competence asked define a correspondent type of actor, the nature of contract between complementary firms – buying, cooperating, long term contract,...– as well as the type of contract between offer and customers have significant effects on all dimensions of impacts. Those points have been developed within several points of view. The types of organisational relationships are what we note External Industrial Organisation (EIO) – forms of partnership between firms –, Internal Industrial Organisation (IIO) – relations between firm's internal departments –, and the Customer Relation Management (CRM).

3.3. Strategic impacts

We can analyse strategic impacts of offer from two points of view : objectives view (1) and global position view (3).

On base of N. Malhéné's study (Malhéné 00) about firms' evolution system, we situate the Innovative Product's trajectory in a Strategic Period, as we are going from a base position – the As-Is – towards an expected result – the To-Be target. Moreover, the firm's target objectives are defined in a

Strategic Horizon and are periodically casted in doubt over. So expected results of our product trajectory are supposed to be an answer to a part of target objectives. Then a first level of strategic impacts is this answer degree.

Secondly, we can compare product trajectory's choices and environmental (public authorities visions, legislative directives, competitive firms,...) global choices, notably in term of technologies evolution.

3.4. Tactic Impacts

Tactic impacts include all the direct effects of a decision to innovate for life-cycle actors – users, leader department, internal customers, partners – in the several cited dimensions. As we said previously, the decision to innovate is characterised by a solution representation. So concerning for example a final user, in the life-cycle usability phase, we can ask ourselves what impact does the product's level of technologies have, from a learning or reputation point of view. In addition, has he gained time or money through this transaction and how much the gain could it be? Does the offer receive more frequent information returns by commercialisation of that new product; does it increase its reputation capital ?

Through tactic impacts estimation we are therefore trying to expect as much as possible the direct and short term value and cost of an innovation for life-cycle actors.

VALIDATION FRAMEWORK

In this part, we focus on our research IT Innovative solutions that will be integrated in a vehicle. We shall firstly show that those innovations are high level ones, according to our precedent definition. We shall secondly see what particular questions we can consequently ask ourselves in this framework. We shall also be able to notice the interest of our approach's integration.

4.1 Innovations in IT and communication

Innovations in IT and communication are characterised by: a high level of technologies, a constantly changing environment, a high level of dependence between technologies, a both services and physical products.

Therefore, we are facing new types of services that we can call technology-intensive services (see Segarra (Segarra 01)). In fact they are services which need technological product innovations to support them. They are particularly present in new technologies industries, but more and more in manufacturing industries, and auto industry. In such services, the innovation includes both a new technological product and accompanying services to be delivered to the customers. So high changes in services require frequently changes in product configuration and in the overall infrastructure (i.e. IT and Telecommunication). On the other hand, new technologies boost changes in services. One can observe a mutual interdependence between services and technology, which is important in the creation of new services delivery possibilities. In this way, innovative technology intensive services are a particular example of multiple potential strategic and tactic impacts in term of value and cost.

4.2. The automotive industry: a complex products and actors system

Introducing new services in an existing commercialised product – in occurrence a vehicle – implies both concurrent product-innovative services management and new competencies management. If we refer to G.Segarra's study (Segarra 01), we can apply the representation of the concurrent product-services development to innovative technology-intensive services (see figure 1.2). In our framework, the innovative developed product would be the trajectory of a particular vehicle. It includes other concurrent R&D projects which will be effectively introduced in the initial vehicle- the As-Is. Therefore, we need to integrate our results step by step in the global design. So, technology-intensive services' Innovation Process and Vehicle's Innovation Process are converging at each phase to permit a continuous common view point.

On the other hand, as G. Segarra shows us too, new actors will be involved in the convergence process. These new actors will be all the partners of the services development.

Conclusion and perspectives

We argue that traditional economic performance's evaluation are not sufficient for all types of innovations. Our particular experience of IT innovative solutions in the Auto Industry also shows us that innovations characterised by an high level and high complexity generate important impacts for all innovation actors. It thus would be interesting to estimate them during the Engineering Design process.

In this framework, the Innovation Process must be particularly vigilant to movements of its environment and strategic objectives of the offer. Otherwise, the Innovative Engineering Design process is characterised by the intervention of all innovation actors – the 'Constellation Value'- which converge to a final unique solution. We have proposed to pilot this convergence process by introducing impacts' estimation. Strategic and tactic impacts are defined in terms of value –positive returns- and cost-negative returns-, may have several dimensions –financial, informational, knowledge, reputation, time -, and are determined on the base of a solution representation.

Posterior steps of our research – that will be the object of a doctorate thesis – are to enter more in details in the instrumentation of the innovation valuation model and, conjointly, to validate our proposal in an industrial case.

References

Abernathy W.J. et Utterback J., 'Patterns of industrial innovation ', in Tushman M.L. et Moore W.L., *Readings in the Management of Innovation*, 97-108, HarperCollins, New York.

AMICE 89, Amice Consortium, 'Open System Architecture for CIM', 688 ESPRIT project research report

Association Française pour l'Analyse de la Valeur (AFAV), *Exprimer le besoin, applications de la démarche fonctionnelle*, coll. Afnor Gestion, AFNOR, 1989.

Bower Joseph L., Christensen Clayton M., ' Disruptive Technologies : Catching the Wave *Harvard Business Review*, January- February 1995, pp. 43-53.

Cohendet P., « Informations, connaissances et théorie de la firme évolutionniste », *L'économie de l'information, les enseignements des théories économiques*, Sous la direction de P. Petit, ed. La Decouverte, Paris, 1998, 406 pp.

Cross Nigel, 'Engineering Design Methods', Strategies for Product Design, ed. John Wiley & sons, 1994

Hambouch Abdel-Illah, « Les frontières fonctionnelles de l'entreprise : esquisse d'une approche évolutionniste » , *Working Paper du CRIFES-METIS*, Paris 1 University, n°88, décember 1997, 20 p.

Jacquet L., 'Contribution à l'élaboration d'une démarche de spécification fonctionnelle', PhD, LAMIH, University of Valenciennes, 1998

Maidique M.A. et Zirger B. J., « The New Product Learning Cycle », Research Policy, December 1985.

Meinadier, *Ingénierie et Intégration des systèmes*, ed Hermes, Paris, 1998, 543 pp.

Mouchot Claude, *Les théories de la valeur*, ed. Economica, coll. Economie Poche, 1994, 112 p.

Nelson R.R., Winter S.G., *An Evolutionary Theory of Economic Change*, Cambridge, Harvard University Press, 1982.

Nonaka Ikujiro, Takeuchi Hirotaka, *La connaissance créatrice : la dynamique de l'entreprise apprenante* , ed. De Boeck, 1997, 303 p.

Normann R., Ramirez R., ' From Value Chain to Value Constellation : Designing Interactive Strategy ', *Harvard Business Review*, july 1993.
Phahalad C. K. & Hamel Gary, ' The Core Competence of the Corporation ', Harvard Business review, 1990.

Richardson G.B., ' The organisation of industry ', *The Economic Journal*, 1972

Segarra G., Sechi N, 'From Product to Service engineering: Innovative technology-intensive services in Automotive Industry', FEATS 2001 Conference, Valenciennes, june 2001.

Soenen, Lawson, Sechi, 'From Needs Expression to Design', Proceedings of MICAD 2001 Conference, 2001

Teece D.J., Chesbrough H.W., ' When is virtual virtuous ? organizing for innovation', *Harvard Business Review*, January-February, 1996.

Tidd J, Bessant J, Pavitt K., ' Managing Innovation : integrating technological, market and organizational change', ed. Willey, 1997, 377 p.

Walliser B, *Anticipations, équilibre et rationalité économique*, ed Calmann-Levy, 1985.

Towards the Workflow-enabled Civil Construction Enterprise Integration

Mário Paulo Teixeira Pinto(a), João José Pinto Ferreira(b)

(a) Superior School of Industrial Studies and Management. Polytechnic Institute of Porto
Av. Mouzinho de Albuquerque, 32 4490 - 409 Povoa de Varzim - Portugal
email: mariopinto@eseig.ipp.pt
(b) Engineering Faculty of Porto University / Electrical Engineering & Computers Dpt.
INESC Porto - Manufacturing Systems Engineering Unit
Rua José Falcão , 110, 4050 - 315, Portugal
email: jjpf@fe.up.pt, URL: http://www.fe.up.pt/~jjpf

Abstract.   This paper presents the result of a research project. In the course of this project we demonstrated the application of business process management concepts in the extended construction enterprise, by using a Workflow Management System. The civil construction enterprise presents, usually, a distributed organisational structure (strategic enterprise, construction sites, warehouses, head-offices, etc.), and operates in a transient and variable environment, according to the collection of works in progress. A common scenario is the construction process that involves the co-ordination of many smaller firms (subcontractors), normally coordinated by one contractor. In this context, the paper starts with the presentation of the civil construction enterprise, by briefly describing the organisational, functional and process models, and concentrates its discussion in the supply chain business process. The rationale for this process is further presented based on so-called *construction enterprise basic building block*, capable of being replicated for each construction entity. Follows the presentation of a prototype illustrating the above concepts and demonstrating the value added of such approach. Workflow technology will definitely become the cornerstone technology to support B2B *E-Business* scenarios in areas such as Construction. In this paper we will further discuss this topic by bringing new construction enterprise requirements in the event of such as the usage of new mobile terminals such as UMTS, PDAs, and so on.

1. THE CIVIL CONSTRUCTION ENTERPRISE

The construction enterprise presents, usually, a distributed organisational structure (strategic enterprise, construction sites, warehouses, head-offices, etc.), and operates in a transient and variable environment, comprising the concurrent management of far apart construction sites. Construction industry uses a very unstable and transient workforce. The result of a construction project is usually a custom-built product from numerous raw materials and components, drawn from several sources. Unlike advanced manufacturing industries, concepts such as controlled working conditions, parallel working, standardisation, and innovation, are not embedded in the construction industry.

Another familiar scenario to civil construction is widespread usage of subcontracting, transferring to these enterprises the responsibility to execute some specific construction activities. To this end, the construction process typically involves the co-ordination of many smaller firms (subcontractors) controlled and monitored by the contractor. The effective co-ordination and communication of these smaller firms by the principal contractor is essential to the success of any project. This clearly demands the usage of adequate decentralised business process co-ordination infrastructures, allowing firms to contribute to common process integration [7].

Basic functions and interactions in the civil construction enterprise are illustrated in figure 1 (derived from [8]). The information, material, management and financial flows are also illustrated. This description further includes all the extended enterprise participants (main contractor, subcontractors and suppliers) and locations (construction sites, warehouses and had offices).

A construction enterprise is, typically, structured in the following organisational areas:

- Management enterprise, including the strategic planning and the administrative area;
- Technologic planning, with activities related by engineering like I&D, projects, works budgeting;
- Works planning, responsible to determine the needs of materials, workmanship and equipments, supported in MRP ("material requirements planning) and CRP (Capability requirements planning); Works control, in three aspects:
- costs, chronogram and quality standards; Construction sites, with the production development, material and quality control, and measurement of works realised;
- Equipments parks (and head-offices), responsible to management and maintenance of machines;

- Warehouses, responsible to material management, and so to the materials flows to the construction sites.

These organisational areas are responsible for the execution of an extensive number of business functions [5]. These functions (and organisational areas) are in fact replicated in several nodes of the extended enterprise. As examples we would highlight examples such as construction sites, equipments parks and warehouses. The number and location of these functional units is in fact variable in time, and depends on the actual work in progress.

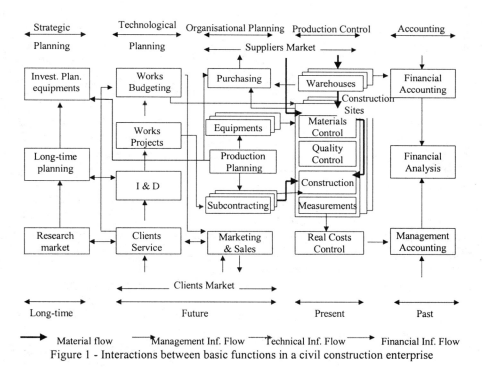

Figure 1 - Interactions between basic functions in a civil construction enterprise

2. THE CIVIL CONSTRUCTION SUPPLY CHAIN

2.1 The Supply Chain (SC) building block

A common scenario is the construction process that involves the co-ordination of many smaller firms (subcontractors), normally coordinated by one contractor. From this perspective, and by analysing the construction company supply-chain, the extended construction company building block was derived (figure 2). This basic construct is therefore replicated for each contractor and sub-contractor recursively.

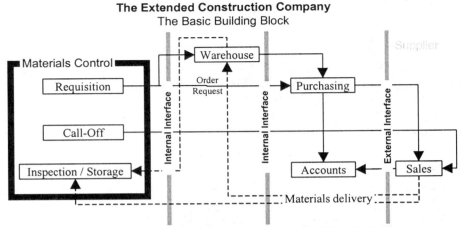

Figure 2 - The Extended construction company building block [3]

For each construction entity, we identified four distinct environments, with interfaces enabling the communication among them: construction site, warehouse, administrative area and suppliers. The materials control, in the construction site, allows the local materials management. At the warehouse, materials management is performed for several construction sites according to their respective production plans. The warehouse is also an intermediary between the construction sites (getting materials requisitions) and the purchasing area (release purchasing orders). The administrative area is responsible for accounting and for processing purchasing orders. Suppliers deliver materials to the construction entity, either at the warehouse or at the actual construction site. We can analyse with more details some one of these functionalities described in figure 2:

- *Inspection/Storage* is a functionality that provides the materials quality control.
- *Call-off* enables a contact with supplier to ask for the next shipment for a previously negotiated (for this type of orders, we have for example agreed phased deliveries at the construction site, allowing a lower local material stocks).
- *Requisition* is an order request that is sent from construction site to the warehouse or purchasing area
- *Warehouse* The global materials management is performed at the warehouse. Products are received from suppliers and delivered at construction sites.
- *Purchasing* is the department responsible for receiving material requests from construction site or/and warehouse, and for producing

purchasing orders to suppliers. It also makes deals and agreements with suppliers (namely procurement), to get the best conditions.

- *Accounts* processes the financial and management accounting information.
- The *Sales* department is part of any supplier. Receives purchasing orders and provide the materials delivered ether at the warehouse or at construction site according the established contract.

2.2 Interoperability across SC members

The relationships between contractors and sub-contractors were further identified as being contract based. This interaction, as illustrated in figure 3, features a complex and dynamic relationship [3]:

- The need for the contract emerges from the actual production plan
- Contract manager handles the detailed contract elaboration and starts the process
- From the actual moment when the sub-contractor starts its own activity, the contractor starts the regular monitoring of the work in progress. This monitoring activity is a corner stone to feeding back information both for accounting and for construction management.

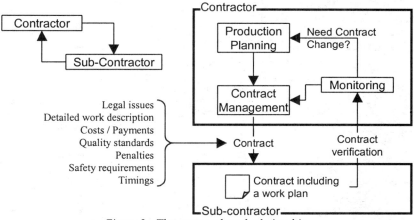

Figure 3 - The contract-based relationship

The relationship between contractor and sub-contractor works in both directions, and demands coordination and interoperability between processes. Inside a hierarchic level, the sub-contractor could be analysed like a contractor, establishing and executing a work plan. He will have to perform its own resource management and, if necessary, manage its sub-contractors.

3. WORKFLOW SUPPORT

3.1 Rationale

The approach to enterprise integration rests in the usage of a workflow backbone to support interoperability and integration with legacy systems, and the construction of a scalable and networked solution. On the other hand, the aim is to provide tools for the management of change by building on business process model-based co-ordination solutions to support daily construction management operations.

The inherent geographical distribution and number of acting entities (namely the contractor and sub-contractors as well as the other participants in the supply chain), clearly demands the usage of stage-of-the-art workflow management tools and the need of interoperability [1]. Moreover, only a distributed infrastructure supporting the so-called extended construction organization will enable innovative communication and collaboration across the networked organization. The rationale for the authors approach is further illustrated in figure 4, picturing the WFMS (Workflow Management System)-based business process model enactment [3].

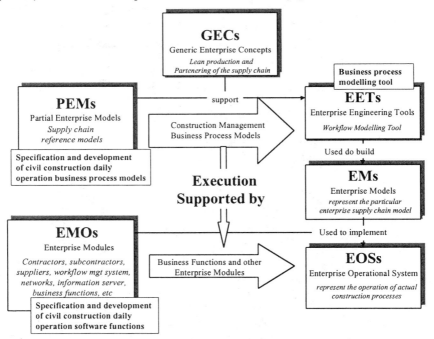

Figure 4 - Enterprise Integration Approach [4]

In this scenario, a distributed workflow management environment is responsible for enabling co-operation and inter-operation of business processes across enterprise boundaries in the construction extended enterprise.

3.2 Case Study

The distributed model execution of business processes are based on the so-called *Move* and *Produce* primitives derived from the Activity Production Control Architecture [2]. These primitives enable business process execution by describing process activities and the corresponding functional entity responsible for its execution, be it a software module, machine or human resource. The *Move* primitive supports information or material flows, and the *Produce* primitive supports the actual task performance. Execution primitives support therefore the interchange of orders and status reporting between the workflow engine and the functional entity performing the actual operations.

The supply chain model process execution is based the execution of successive *Move* and *Produce* activities, according a particular production plan. In this scenario, the entity responsible to execute one activity, receives jointly instructions and information necessary to start the activity. Upon its conclusion, the involved entity will make available the resulting product, namely to the following entity.

Figure 5 illustrates the requisition of materials (in the construction site) and the processing of a purchasing order (in the warehouse). The *Produce* primitive enables the form filling for material requisition comprising information such as the product type, quantity and delivery deadline. The *Move* primitive then supports the material requisition transfer from construction site to the warehouse. The following activity, performed in the warehouse, receives information (filled requisition form) and if necessary prepares a purchasing order to fulfil the received material request.

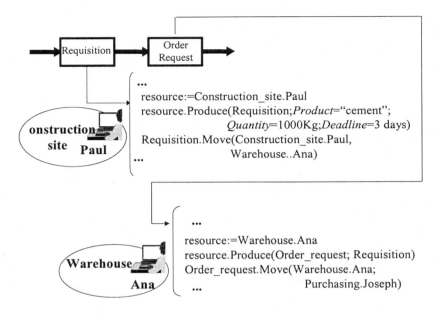

Figure 5 - Supply chain conceptual model

4.   CONCLUSIONS

Unlike in advanced manufacturing industries, the concepts of controlled working conditions, parallel working, standardisation and innovation, are not truly embedded in the construction industry. It is, therefore a most interesting green field to bring in new electronic business concepts. The challenge in this area derives from the multiple existing physical and organisational boundaries of such company. Supply chain management in the construction industry seems to be a critical area to driving innovation and to sustaining incremental and continuous improvement. Effective implementation of supply chain integration leads to performance improvement in the shape of reduced costs and better quality.

Workflow technology could become a key to support business process integration in areas such as construction, providing management tools and interoperability between processes. It offers modelling tools and supports co-ordination in a distributed environment. To this environment, we would bring in new requirements in a construction enterprise to improve a communication and collaboration across the networked organisation. Particular attention should be given to the convergence of wireless networks and INTERNET. In such a transient and variable environment new roles will for sure be given to 3rd Generation mobile terminals in the construction environment.

References

1. [Aalst]
"Interorganizational Workflows". Department of Mathematics and Computing Science, Eindhoven University of Technology

2. [Bauer]
Bauer et. Al; "Shop Control Systems: From Design to Implementation"; Chapman & Hall; ISBN 0-412-36040-3;

3. [Ferreira]
João José Pinto Ferreira, David Proverbs, Manuel Pintor, Mário Pinto: "The Extended Construction Enterprise", INESC Porto Internal Report 1999.

4. [GERAM]
"Generalised Enterprise Reference Architecture and Methodology", Version 1.5, 199709-27 IFIP - IFAC Task Force

5. [Pinto]
Mário Pinto: "Workflow Management System in Processes Co-ordination in the Extended Construction Enterprise " MSc Thesis 2000

6. [WfMC]
Lawrence, Peter: "Workflow Handbook 1997"; John Wiley & Sons Lda.

7. [Vernadat]
Francois B. Vernadat: "Enterprise Modelling and Integration, Principles and Applications", Chapman & Hall

8. [V. Rembold]
V. Rembold, B.O. Nnaji, A. Storr: Computer Integrated Manufacturing and Engineering, Addison-Wesley

Bulgarian Experience in Use of Internet for Production and Business Applications

Todor Neshkov -assoc. prof., Dean, Faculty of Mechanical Engineering

Julian Pankov - postgraduate student in TU - Sofia

Abstract: This paper is directed to the problems liked with presentation of production and trade enterprise in Internet. Specialised WEB site with a lot of searching facilities was developed with several models for collecting and presentation data for Production Company.

Key words: IT, WEB design and WEB presentation, Internet for Business application

1.   INTRODUCTION

1.1   Internet in production and business in Bulgaria

At the present moment when preparations for integration to EC in Bulgaria are made, problems for use of Internet in production sector, including mechanical engineering, are becoming more and more substantial.

Brief analysis of present situation in Bulgaria, shows that use of Internet in production scope is still very weak. As a illustration of that is the following: only 12% of the companies participating in 55-th International Technical fair in Plovdiv'99 in Mechanical Engineering Pavilion, had in their advertisements emails and only 2 % had a web page.

The percentages mentioned above have risen tremendously for the last two years only and now they are: 65% are the companies with e-mail and 38% companies with WEB pages (fig. 1).

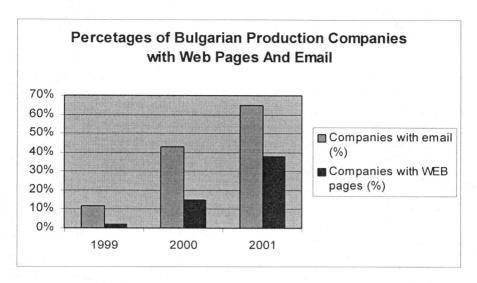

Figure 1. Percentages of Bulgarian Production Companies with Web Pages and Email

At the same time, peoples involved in small and medium business are facing many problems with finding markets for their production, ensuring basic materials for production, participating in actions etc. Internet technology is ideal for all of these, but unfortunately, it is still not used effectively. Why? The answer is: lack of information, lack of habits and abilities for use of enormous opportunities of new Internet technologies in production, trade and services . Simultaneously main part of WEB pages and sites are entertainment oriented.

The existing infrastructure in Bulgaria, especially in small towns, brings up additional problems as: slow speed connections to the Internet, outdated technical equipment etc.

Although there are a big number of search engines in Bulgarian WEB space, most of them are with general orientation. Main disadvantages of that approach are: large amount of search – result pages, lack of technical information in the returned data, limited possibilities for refining the user's query .

Analysing that, project for developing of specialized Internet databases, containing information for products and services of mainly production companies in Bulgaria began 3 years ago. Lecturers from Technical University - Sofia, generated the main idea with financial support provided by private company and Ministry of Economics. This led to development of specialized site - BGCATALOG.COM

2.   OUR TECHNOLOGY

The structure of BGCATALOG enables database search in several ways depending on customer's preferences through:

1. Search by product / activity
   This search is not only keyword oriented, but also has an advanced feature for choosing the parameters of the searched product / activity. With this option the user can refine his request that is guarantee for fast search and quality results.
2. Search by company's name
   New option for choosing the type of the company - trading, production, service, importers/exporters, Ltd., Plc., etc. is under development. Wildcards also can be used.
3. Search by company's address
   The user can use in this search not only the name of the city/village, but also the name of the street, phone number, e-mail etc.
4. Search by branch of the company
   Three-level tree structure according to Unified Classification of Republic of Bulgaria was developed, for assisting user's queries.
5. Combination of the mentioned above
   For example, ' Production Company for hinge bolts for VW in Sofia '. Wildcards can be used in every kind of search mentioned above. Logical combinations like 'AND' 'OR' 'NOT' also can be used.

BGCATALOG is organized on two databases - one is the main database with catalogues of the companies and the other is a database with templates for presentation of the company's data. This structure is very flexible and allows different individual design appropriate with demands of each firm and transparent embedding of information from main database in company's WEB-pages. For achieving hi-speed, we use dedicated database - server for access to both databases.

We have developed three different applications for maintain and support of BGCATALOG ' s databases -fig.2 - WEBASSIST, WEBDESIGN and WEBSEARCH

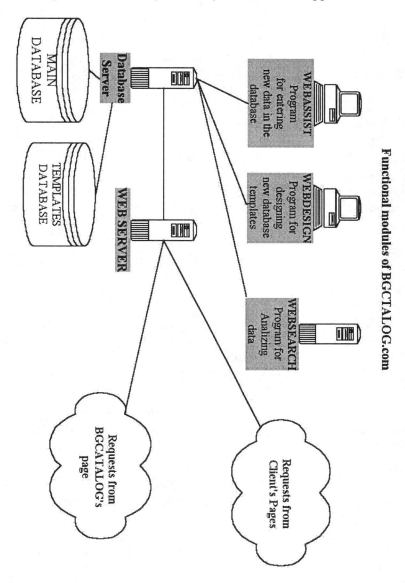

Figure 2. Functional modules of BGCATALOG.COM

WEBASSIST is a program for editing and inserting new information in the main database. It is oriented on six basic models we have developed for collecting information. The program allows speeding - up the process of including and refreshing company's catalogues in the database with it's user friendly interface and table organization with respect to every product and service in each company's list.

WEBDESIGN is the other application we have developed and it is main purpose to assist designers in making new templates. These templates can be used for full control over presentation of each product/service from the main database. The program is HTML - oriented with many graphical elements and easy of use.

Third program in this range is called WEBSEARCH. It is used for processing inputted data to be suitable for searching. It performs actions like extracting terms in particular branch from company's data, basic morphological analysis of Bulgarian language, spell checking etc. This additional operation on company's data speeds up following WEB -based searches and is used in server-side scripts for increased accuracy of finding.

Final elements of BGCATALOG are shared server-side scripts, which are used for actual WEB - presentation of the companies. These scripts are used both from personal company's WEB pages and from BGCATALOG' s WEB PAGE. They provide different behavior depending on the context thus providing best appearance of the data over INTERNET.

3. APPROACH FOR COLLECTING INFORMATION FOR DATABASE FOR BGCATALOG

We have developed six basic models for product presentation for assisting Bulgarian producers in shaping their materials for Internet catalogue.

3.1 MODEL 1 – Model of WEB page with links to databases

WWW pages could contain additional information for the company as:

- Name of the company
- Complete address of the company and its representatives - city, ZIP code, street, flat, phone number etc.
- Production opportunities and history of the company
- Participation in international projects, trading with international companies etc.
- Organization structure of the company, company profile etc.
- Links with other WEB pages
- Future plans of the company

The following models describe presentations of objects in a database and are very useful for manufacturing companies.

3.2   MODEL 2 - Textual presentation

This model is suitable for products and services that do not require additional graph materials as picture, images, charts etc.
Information could be structured as follows:
- Branch of the products or services according to the Unified Classification of Republic of Bulgaria
- Brief description of every product or service

3.3   MODEL 3 - Table presentation with graphics, which is visualized by a user request

Information could be structured as follows:
- Branch of the products or services according to the Unified Classification of Republic of Bulgaria
- Brief description of every product or service
- Tables with technical parameters for every product or service
- Graphical materials for every product

This type of presentation in the databases is appropriate for products with characteristics, which can be classified in tables. Graphical information for a particular product is viewed on a user request. This model is very suitable for machine building, textile, chemistry and other industrial's products.

3.4   MODEL 4 - Table presentation with graphics material

Information could be structured as follows:
- Branch of the products or services according to the Unified Classification of Republic of Bulgaria
- Brief description of every product or service
- Tables with technical parameters of every product or service
- Graphical materials for every product

In this model, graphic material for every product is visualized in the table with parameters. The model is suitable for the same products as mentioned in the previous model.

3.5   Model 4 - Textual - graphical presentation

Information could be structured as follows:
- Branch of the of the products or services according to the Unified Classification of Republic of Bulgaria
- Brief description of every product or service
- Graphical materials for every product

This model is appropriate for products and services, where tabular presentation is not applicable. Each product is accompanied with graphics and text.

3.6   Model 5 - Summarized graphics for the products, presented in a table form

Information could be structured as follows:
- Branch of the products or services according to the Unified Classification of Republic of Bulgaria
- Brief description of every product or service
- Tables with technical parameters of every product or service
- General graphic materials

The model is for a big number of products of the same class.

Each model ensures a standard mode for describing the web page's structure and enables variations and individual appearance of every single company.

4.   OUR TARGET GROUP

Due to our three-year experience, we have gathered huge amount of information for visitors on BGCATALOG ' s web page. This gives the idea for the target group of our users. Collected information confirms the original orientation to Bulgarian business users. As it may be seen from fig. 3, most of the hits to the WEB page are during working hours of the day in Bulgaria.

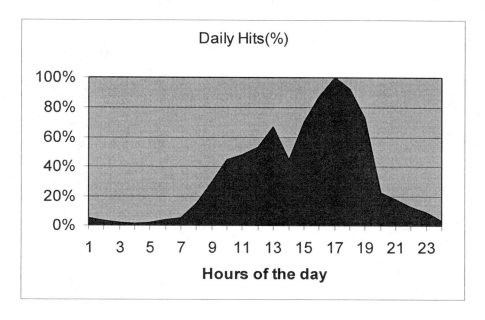

Figure 3. Daily Hits of BGCATALOG

Another confirmation of the business target group is shown on chart 4 – it is weekly statistic of hits of the WEB page.

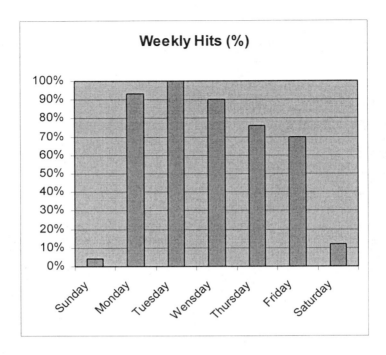

Figure 4. Weekly Hits of BGCATALOG

Another valuable information we have collected for nearly two years are the rate of keywords, which our users have been looking for. This gave to the project new direction for experimental new methods for speeding up the searching process.

The first method of searching is a model without any preliminary processing of the data and it is based on a full-text database search. The speed of the online queries depends entirely from the speed and realization of the searching of the particular database server. Our experiments have shown that this approach ensures stable work, but unfortunately is very slow for big amount of data, especially when this data is structured in many different tables, which is the most common case in technical information.

The next approach is based on a simple preliminary word extracting as a base for subsequent online queries. Here we have small amount of initial work to do before the data is ready for online searching. Several indexes can be added to tables with keywords depending on keywords statistics, alphabetical order etc. The method shows good online speed and minor of preliminary processing.

The last one is more complex - extracting of words followed by their classification as terms by branches. This gives a completely new direction of the options for searching as finding related words from particular keyword in specific technical field, pointing the main terms in certain area etc. Disadvantage of this method is the bigger amount of preliminary processing.

Experiments have shown that there is link between speeding up online queries and increasing the time for pre - processing of the data. For an optimal speed of specialized database is very important to find the balance between introductory work over the data and complexity of the online query.

5.   CONCLUSIONS

Internet technologies have wide application in production sphere. In the near future, they will be used not only for information, advertisement, production catalogues and searchable databases, but also for gathering information data for automation production by use of CNC and PLC etc. Databases with production standards are going to be developed. Huge systems for summarizing of requests and distribution of production over departments, suppliers, stock production etc. through Internet and Intranet are already realized in some big companies.

We consider that good experience of production practice over Internet technologies in machine building should be propagated by:

- Populating of WEB addresses of specialized pages
- Realization of European projects
- Short term qualification courses for personnel and exchange of education materials

All of the above will extend competing of SME, increase quality of production and realization of plans of the company in contemporary fast changing world.

Local min.-energy, F E based formulation for well-path design

Kailash Jha
IIS, Scientific Computing Ltd.
New Delhi, India

Abstract: A new algorithm has been developed in this work to generate an optimal well path for drilling applications. Minimum energy based finite element technique has been used for interpolating well path. The straight portion of the well path at beginning and end of segment has been maintained. The dog-length severity has been maintained in present interpolation technique.

Present algorithm will give total length of the well as well as maximum value of the dog-length severity. These two parameters are useful to estimate the cost of drilling.

Key words: B-splines, interpolation. minimal energy, chord length parameterization, dog-length severity, well planning, horizontal well, directional drilling.

1. INTRODUCTION

In present work minimum energy based curve fitting has been done for horizontal well path design (see Joshi [1]). A brief overview of horizontal well design is given in reference [1]. In early days of oil drilling, only vertical wells were drilled. These types of wells can not be optimal wells because of poor flowability and their simple geometry. Poor flowability results in low productivity and simple geometry restricts the approach to complicated reservoir. If there is any obstacle in direction of drilling then drilling process has to be restarted. This leads to a huge wastage of money and time. These difficulties have been overcome in directional drilling. Directional drilling may be of horizontal shape. In general directional drilling can take any shape. Obstacles in the direction of drilling may be due to presence of salts or faults. If there is salt or fault in direction of drilling it is very hard for drill bit to cross it. This is true for any types of drilling. Another importance of directional drilling is to enhance the reservoir contacts and better productivity. A long path in direction of reservoir gives a larger contact area and hence enhances the injectivity. There are basically two types of non-vertical wells

(1) Deviated and (2) Horizontal

Deviated drilling applies to a well, which is designed to reach a particular point in a reservoir, which is substantially different from the surface location

(2) Horizontal drilling applies to a well, which is designed to enter a reservoir roughly parallel to formation boundaries and remain there for some distance.

A non-linear optimization for well design has been discussed in reference [2]. In this work, optimal parameters have been calculated and compared with actual trajectory of the well.

In the present work an attempt is made to get the optimal well path by iterative design. An optimal bore well path has minimal length for pre-specified minimum radius of curvature as well as minimum total energy. Total length of the well for a given minimum radius of curvature has been calculated in this work. Total length of well includes the straight portions and the curved portions of the well. In any segment of well path, if the minimum value of radius of curvature is less than given minimum value, the energy parameters are rectified to get the required condition. Dog-length severity is measurement of magnitude of radius of curvature.

Drilling parameters can be changed to get the required minimum radius of curvature. By iterative technique optimal drilling parameters can be obtained. Total length and minimum radius of curvature are very important to calculate total cost of drilling. There are four coefficients, which have been used to control smoothness of the well path. For each segment of the well bore, pre and post hold lengths are specified. In present energy based technique an attempt is made to achieve the straight-line requirement of the borehole at starting and end of the segment by current interpolation technique.

In the proposed research work, points on well path, pre-hold and post-hold lengths are the prime inputs. Tangents are optional. If tangents are not given, 3-point method [3] is used to get tangent values. Points and tangents give geometry and direction of well path for a segment. Pre-hold and post-hold lengths give the straight portion of the well path. Cubic B-Spline curve fitting based on local minimal energy has been developed to get required well path. This technique assures matching of input parameters. For each segment of the well path, three elements have been considered to get finite element based minimum energy formulation. First and third elements of a segment are straight lines where as middle portion (second element of the segment) has curved shape. For each element different coefficients can be considered according to the requirements. In straight portion, energy coefficients are put to zero. In curved portion of the well suitable values of coefficients have been used. Effect of these coefficients is studied in this work to get an optimal well.

While designing a well path, it is required that the well profile should be smoother, radius of curvature should be more than given value and total length should be as minimum as possible. Higher order derivatives also can be considered in constrained equation.

This paper has been organisation as follows:
Section-2 contains a brief literature survey of energy based curve fitting as well as well path design. Current work has been explained in section-3. In section-4, results of implementations have been discussed along with comparison with previous works. Conclusions have been made in section-5.

2.   RELATED WORK

There are number of researchers who worked on the principle of minimum energy to fit a curve through set of points [4-12]. Early researchers have tried to approximate total energy of curve/surface by equation-1

$$\int_t K^2(t)\, dt \qquad \text{-----------------------------(1)}$$

Where K(t) is curvature function and t is parametric value.
Kjellander[4] and Lee[6] used approximation of eqation-1 for splines. Kallay [5] used discrete approximation to solve equation-1 for given length. Cheng and Barsky[7] used energy-based spline fit to pass through specific regions. This is useful when some of fitting points are uncertain. Vassilev[9] used energy technique for fairing curve and surfaces. Result obtained by approximating curvature function (K (t)) does not approximate the curve correctly (see Wang *et al* [8]). It may be suitable for certain specific requirements. Wang et al [8] has given different approximations and compared the results. They have used different parameterisation techniques like (a) uniform, (b) chord length and (c) centripetal to get different results.

They found that energy has a bigger impact in generation of curves and surfaces than parametric technique in curve or surface interpolation. They also found that approximation of energy would not approximate the curve/surface properly. A brief description of energy based formulation is given in Park *et al.* [10]. Well design concept has been explained in Joshi [1]. In this reference a brief description of horizontal well technology has been done. Comparison with vertical well has been performed effectively in this work. Concept of optimisation is explained in Helmy [2]. In this work non-linear optimisation has been done to get optimal drilling parameters.

I adopted similar terms as used in references [10-12] in my energy function. In present work, one more energy term has been used than reference [10], which is responsible for smoother curvature (see equation-2). This term

Fig. 1(conventional S-type well)

involves the third derivatives. In present approach local minimization technique has been adopted to ensure the present requirement of fix pre and post hold lengths. Pre and post hold lengths are the straight portions of well segment which are shown in the Figure-1. ab and cd are the straight portions of the well segment shown in Figure-1 which are called pre and post hold lengths. Curved portion of well segment is bc. Points and tangents are defined at segment points a and d respectively. Tangent directions for this segment are ab and cd. Proposed work solves quadratic functional minimization where a resulting curve can be obtained by solving a linear set of equations see [10].

3.   CURRENT WORK

In this paper an energy based interpolation technique has been developed to get a smooth well path in such a way that it has minimum energy.

Total energy considered in this formulation is given by equaition-2, which is given below:

$$E = \int_t \alpha(C_t(t) \cdot C_t(t)) + \int_t \beta(C_{tt}(t) \cdot C_{tt}(t)) + \int_t \gamma(C_{ttt}(t) \cdot C_{ttt}(t))$$
$$+ \gamma 1 \sum_{i=0} || P_i - C(t_i) ||^2 \text{ -----------------(2)}$$

Where α, β and $\gamma 1$ are non-negative value called stretching, bending and fitting coefficient respectively.

γ is a new non-negative coefficient of energy responsible for having smooth curvature.

C (t) is required cubic B-Spline curve.

P_i denotes given points.

C_t, C_{tt} and C_{ttt} are first, second and third derivations of curve C(t).

t is parametric value. These energy terms have been explained in [10-12]. Effect of these coefficient have been studied in present work. I adopted a quadratic functional minimization technique, which results in a curve by solving linear set of equations see Park *et al* [10]. Assuming equation-3 given below to be solution of equation-2.

$$C(t) = \sum N X^T \text{ --(3)}$$

Where N represents basis functions and X represents Control points (see Piegl and Tiller [3]) in equation-3. Equation –2 can be rewritten as given below:

$$X^T (\alpha N_t N_t^T + \beta N_{tt} N_{tt}^T + \gamma N_{ttt} N_{ttt}^T) + \gamma 1(AX - P)^T(AX - P) \text{ - ---(4)}$$

$$\Rightarrow X^T(\alpha K1 + \beta K2 + \gamma K3)X + \gamma 1(AX - P)^T(AX - P)$$

Where N_t, N_{tt}, N_{ttt} represents first, second and third derivatives of Basis Function see Piegl and Tiller [3]. K1, K2 and K3 are 4x4 matrices see [8,12]. A and P are representing coefficient matrix and input points respectively. Constraints are given by equation-5 below:

$$C_k X = D_k \text{ --(5)}$$

Where k varies from 0 to one less than number of constraints
The final matrix equations can be written as given below (see[10]):

$$\begin{pmatrix} K+\gamma 1 A^T A & C^T \\ C & \phi \end{pmatrix} \begin{pmatrix} \\ \end{pmatrix} = \begin{pmatrix} \gamma 1 A^T P \\ D \end{pmatrix}$$

Where V is a vector storing Langrange Multiplier.

K is a segment stiffness matrix with size (nC x nC). nC indicates number of control points.

C is (nconstrats x nC) matrix.

nconstrats indicates number of constraints given.

Above matrix equations can be solved by Gauss-elimination to get control points.

The interpolation technique is applied locally to maintain linearity requirement at the start and end of the segments and uses the contribution of energy from the third derivative (see Fang and Gossard [12]) which is given in equation-2. In current work linearity requirement for segment is overcome by putting the energy coefficients (α, β and γ) for straight element as zero. The term local interpolation indicates that all points for a segment are derived from input parameters (points, tangents, pre and post hold lengths) for a segment.

In the present implementation, for each segment, eight points have been considered along with four tangents at start and end of the straight portions of each segment. Three elements have been considered for each segment. First and last elements are straight lines. Energy coefficients are kept zero to maintain linearity for straight element of the segment. Points are exactly interpolated in this work and minimum value of the radius of curvature is maintained in each segment as well as for entire well path. If any segment violates the minimum radius of curvature it will be indicated by present formulation. Once this mismatch occurs, the energy coefficients for curved portion are modified. $\gamma 1$ value reversibly affects the smoothness of the curve. Magnitude of the tangents is assumed as total arc length for the segment, which may effect the nature of curve. Parameterization is also effecting the nature of curve see[8,9]. This formulation is useful in drilling applications where the well path is highly directional. There may be situations when a drill bit will not move forward due to presence of faults or salts. In such situations, diverting the well path is required in real time while still meeting some overall requirements. The simple output of the program is shown from Figure-2 to Figure-9. Final output of the program will give a well survey points. Real wells also have been tested in this work.

4.   RESULTS AND DISCUSSION

Results for different wells have been shown from Figure-2 to Figure-9 using OpenGL on a PC using Windows operating system. Figure-2 to Figure-6 shows results for test example-1. Figures 7 and 8 show results for test example-2 and Figure-9 shows result for test example-3.

Boreholes are approximated as small solid cylinders. Well paths are shown as lines. Curvature is plotted against the parametric value. Y-axis shows magnitude of curvatures magnified by 2000 and X-axis shows parametric values with magnifying factor 1000. Current observation shows that the affect of third part of energy term is also important along with other terms in energy based curve fitting. Fang and Gossard [12] have expressed its importance. It has been found in current work that as the value of the γ increases the maximum fitting error and smoothness increase and total length of the well path decreases (see Table-1 and first two rows of Table-2). Table-1 shows twenty sets of coefficients (α, β, $\gamma1$ and γ) in favor of above statement about γ for a segment shown in Figure-7. In well industry, it is

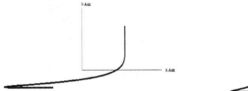   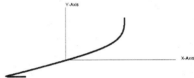

Fig. 2(well ($\alpha=\beta=\gamma=0.0$ and $\gamma1=1.0e^6$))   Fig 3(same well as Fig-2, except $\gamma=.0455$)

important that a well should be of minimal length as well as maximum possible radius of curvature. Figure-2 and Figure-3 show the borehole for a segment, for two different sets of coefficients. Figure-4 shows the curvature plots for these bore holes. In Figure-2, values of coefficients (α, β, $\gamma1$ and γ) are 0.0. 0.0, $1.0e^6$ and 0.0 respectively. In Figure-3, these

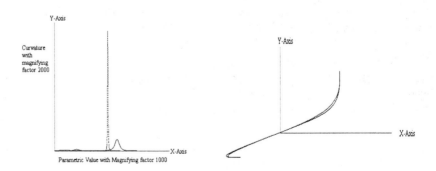

Fig. 4 (Curvature plots for figures 1,2) Fig. 5 (well paths to show the effect of α/β)

coefficients are 0.0, 0.0, $1.0e^6$ and .0455 respectively. Black solid line in Figure-4 shows curvature plot for the borehole shown in Figure-2. The blue dashed line in Figure-4 shows the curvature plot of the well bore shown in Figure-3. If ratio of α to β is greater than one then stretch on the well path will be effective otherwise bending will be effective (see Figure-5). If ratio of α to β is more than one it increases fitting error as well as smoothness of the curve and decreases total length of the well path. If it is less than one, it decreases the fitting error and smoothness of the curve as well as increases total length. Figure-5 shows two well paths for different set of coefficients for the segment shown in Figures 1 and 2. Third and fourth rows of Tables-2 show the effect of α and β. Effect is measured in terms of total length, maximum curvature and maximum fitting error. The dotted blue line shows the stretch effect. In this case, α, β. $\gamma 1$ and γ have value 1.0, .2, $1.0e^6$ and 0.0 respectively. In the same figure with black line shows the effect of dominance of β value, which results more bending. Figure-6 shows the curvature plots of the well paths shown in Figure-5.

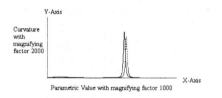

Fig. 6 (curvature plots for different α/β)   Fig. 7(bore well for test example-2)

Smooth behavior of dotted blue line shows the effect of dominance of α on β. In this case curvature plot are smoother. In the same figure black line shows that, higher beta dominance results in comparatively non-smoother path. It has been found in present observation that there should be a maximum limit for α to β ratio for a smooth interpolation. I have achieved a good result up to a maximal value 5.

Figure-7 shows another well segment whose curvature is plotted in Figure-8 to show the effect of $\gamma 1$(fitting coefficient). Blue dotted line of Figure-8 shows the curvature plot for a segment having α, β, $\gamma 1$ and γ values 1.0, .2, $1.0e^4$ and .0455 respectively. The black solid line shown in the same figure, shows the curvature plot for the same segment having α, β, $\gamma 1$ and γ values 1.0, .2, $1.0e^6$ and .0455 respectively. Curvature plots show that as the $\gamma 1$ value increases, fairness of the well path decreases. As the $\gamma 1$ value increases maximum fitting error decreases and total length increases (see 5th and 6th rows of Table-2). Figure–9 shows a real well with coefficients 0.0,

Fig. 8(curvature plots for example-2)   Fig. 9(real well bore example-3)

0.02, $1.0e^6$ and 0.0 respectively. Table-2 (7th and 8th rows) shows comparison of this well with well obtained by least square fit. Real effect of energy terms can be realized well in isolation of the particular coefficient. As the value of beta increases total length as well as fitting error increases and the smoothness will be enhanced marginally.

Since current algorithm works segment by segment, any violation for minimum radius of curvature (dog length severity) for a segment can be compensated by modifying coefficients. Even after modification the problem does not get rectified then it will be reported to the user. Since the coefficients are not so sensitive to radius of curvature so there may be a situation to change the input parameters (points, tangents, pre hold length and post hold length). Current observation also matches with Vassilev[9] and Park *et al* [10]. In these work [9,10] the third energy coefficient has not been considered. They have used global interpolation technique, which may not be relevant for well path application. In this work there are two straight elements in each segment with zero energy coefficient values, which help in maintaining linearity. Fang and Gossard[12] tried his algorithm for Z-shaped data points to get optimal number of elements. They have found that if the number of element increases it does not give smoother curvature. Less number of elements may give better smoothness. In this work, Hermite interpolation function has been used. Results shown are for 2D data points. In my opinion the iterative technique suggested by Fang and Gossard[12] is better to limit number of elements. I have obtained satisfactory results with three elements for each segment of the well. Curvature plots shows this fact (see Figures 4, 6 and 8). All the results have been implemented for 3D data points.

In the present work, three energy terms have been used which is giving smoother curvature than Vassilev[9] and Park *et al*[10]. In references [9,10] only two energy terms have been used. In many applications it is needed to have a smoother curvature. Drilling industry can directly use results of this work. Output of this formulation gives a list of survey points. Observation of the results shows the importance of energy and fitting coefficients in smoothness of curve. When all four coefficients are present then effect of individuals is not so sensitive. All the results for well are drawn from the survey data points, which are final output for a drilling engineer.

5.   CONCLUSION

Local minimum-energy, finite-element based formulation for well-path design has been performed in this work. The effect of energy and fitting coefficients on minimum energy well has been studied in this work. This work helps in iterative well design. The segment which is not satisfying drilling criteria can be interactively redesigned. Even making the energy coefficients zero for straight elements there are fitting errors coming due to the energy and fitting coefficients given for the curved elements. Magnitude of error is small and that can be controlled by modifying the coefficients. One reason for smaller magnitude of the error is that the input points are in two straight lines for each segment.

The cost analysis of the well for different length can be considered as the extension of the work. This work can be used in energy based surface fitting. Getting optimal coefficients are also an open issue for further research.

References

(1)   Joshi, S D, 1991, "Horizontal Well Technology ", 1991, PennWell publication

(2)   Helmy, M W, Khalaf, F and Darwish, T A, 1997.
         " Well Design using a Computer Model ", paper (SPE 37708), 42-47.

(3)   Piegl, L and Tiller, W, 1995 , "The NURBS book" New York, Springer

(4)   Kjellander, J A P, 1983, "Smoothing of cubic parametric splines", Computer Aided Design, 15(3), pp-175-179.

(5)   Kallay, M, 1987, "Method to approximate the space curve of least energy and prescribed length", Computer Aided Design, 19(2), 73-76.

(6)   Lee, E T Y, 1990, "Energy, fairness and a counter example", Computer Aided Design, 22(1), pp37-40.

(7)   Cheng, F and Barsky, B A, 1991, "Interproximation: interpolation and approximation using cubic spline curves", Computer Aided Design, 23(10), 700-706.

(8)   Wang, X, Cheng, F, Barsky, B A, 1997, "Energy and B-spline interproximation", 29(7), pp485-496.

(9)   Vassilev, T I, 1996, "Fair interpolation and approximation of B-Splines by energy minimization and point insertion", Computer Aided Design, 28(9), 753-760.

(10) Park, H, Kim, K and Lee, Sang-Chang , 2000, ``A method for approximate NURBS curve compatibility based on multiple curve refitting", Computer Aided Design 32(20) 237-252

(11) Celniker, G, and Gossard, D C, 1991, ``Deformable Curve and surface finite elements for free-form shape design", Comput Graph 1991, 26(2), PP 157-66.

(12) Fang, L and Gossard, D C, 1995, ``multidimensional curve fitting to unorganized data points by non-linear minimization.", Computer Aided Design, 27(1), 45-58.

	$\gamma 1$	α	β	γ	Total Length	Max curvature	Max Error
Test Example Shown In Figure-7	$1.0e^6$	0.0	0.0	0.0455	817.54	.247607	.0442646
				0.0855	817.357	.247289	.0469721
				0.155	817.039	.246728	.0517032
				0.455	815.678	.244214	.0722978
				0.855	813.899	.240641	.0997441
				1.055	813.024	.238768	.113371
				1.455	811.302	.234873	.140366
				2.055	808.789	.228713	.184756
				2.455	807.158	.224438	.221493
				2.955	805.167	.218955	.266617
		1.0	0.2	0.0455	817.283	.247064	.0481637
				0.0855	817.1	.246742	.0508914
				0.155	816.782	.246174	.0556486
				0.455	815.426	.243633	.0762861
				0.855	813.651	.240027	.103722
				1.055	812.778	.23814	.117333
				1.455	811.061	.234219	.144288
				2.055	808.555	.228029	.190092
				2.455	806.928	.223739	.226758
				2.955	804.943	.218242	.271794

Table-1 (effect of γ on well path)

Test Example	α	β	γ	γ1	Total Length	Max curvature	Max Error
1	0.0	0.0	0.0	$1.0e^6$	3327.3	0.640471	.0410935
1	0.0	0.0	0.0455	$1.0e^6$	2048.01	0.074923	0.083902
1	1.0	0.2	0.0	$1.0e^6$	1857.69	0.225023	.108946
1	0.2	1.0	0.0	$1.0e^6$	1911.97	.252363	.0987742
2	1.0	0.2	0.0	$1.0e^4$	794.767	.203723	.51956
2	1.0	.20	0.0	$1.0e^6$	817.493	.247427	.0450737
3	0.0	0.0	0.0	$1.0e^6$	2002.27	.0851196	$1.78e^{-12}$
3	0.0	0.02	0.0	$1.0e^6$	1996.95	.06137	.0001570

Table-2 (effect of α, β, γ and γ1 on well path)

From digital product development to an extended digital Enterprise

Uwe F. Baake and Osmir R. Torres
debis humaita IT Services L.A., Business Unit Engineering and Manufacturing, Al. Campinas, 1070 - Jd. Paulista, 01404-002 Sao Paulo / SP – Brasil, phone: +55-11-3886-2779· facsimile: +55-11-3886-2801, e-mail: {uwe.baake,osmir.torres}@debis.com.br; Internet: http://www.debis-humaita.com.br

Abstract: The modern world-wide (so-called global) economy has a deep influence on the way people and companies think and behave, in such way as to streamline the whole development and manufacturing process, now with a distributed, flexible and integrated structure. Collaborative software is changing the way companies do business. Applications that facilitate communication early in the design cycle and throughout the manufacturing process save manufacturers time and money.

Key words: Digital Product development, Virtual enterprise, Life-cycle engineering, collaborative Product Definition management (cPDm), Web Portals, Collaborative technologies

1. INTRODUCTION

The challenge of providing a global product line as a response to an utmost need to comply with an increasingly demanding and competitive market involves changes to the political, technological, financial, cultural and social sectors. The trends of this evolution indicate that the future will take us to a world that is more and more global, "small" and connected through a world-wide communications network, which is increasingly more integrated, powerful and available to everyone. Simultaneous and decentralised development make the required document management control more and more severe and accurate throughout the several phases of the product life cycle, starting from the first idea until reuse and recycling. Because of the major trend towards business decentralisation, collaboration technologies become an important component of those changes as operations are outsourced through supply chains, new strategic partnerships are formed in extended enterprises, and facilities are dispersed around the world. Figure 1 illustrates the evolution from the function oriented enterprises in the past to a new form of networking enterprises today. The vision is to implement company Units with focus on networks that will be virtually defined and organised. In a partly virtual company all participants of the supply chain will be integrated in an IT-based, sensible innovation, order fulfilment and sales process. This integration includes the complete product and factory life cycle.

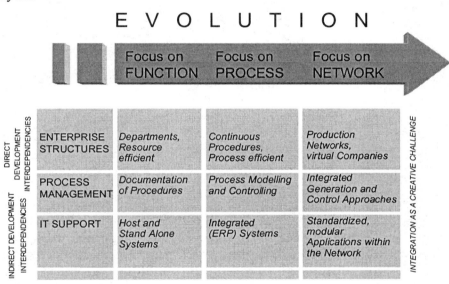

Figure 1. From Traditional Business to an Extended Enterprise

Information Technology (IT) offers an efficient way to build these new evolution paths (Jacucci, Olling, Preiss, Wozny, 1998). Also, as a major responsible for the availability of resources provided to the research and development of new technologies, Information Technology keeps on contributing in a large scale towards product development processes to be supported by digital solutions, keeping customers, suppliers, collaborators and partners connected on a world-wide basis, and focussed on the development of a same product as outlined in Figure 2.

With these new technological resources, Collaborative Engineering becomes an effective global reality. Product Data Management (PDM) environments evolve to a concept that is closer to the collaborative Product

Definition management (cPDm), in which the spirit of shared development with creativity, flexibility, speed in communication, for faster decision making, and synchronism of activities will generate products in a better, cost-effective and faster way (Miller, 2000).

... purchasing, marketing, ...

Figure 2. Networking enterprise

Beside the technological challenges cultural and organisational issues present the greatest obstacle to overcome (Baake, Haussmann, 1998). For exchanges to be of value, organisations have to determine how overall processes must be configured, where information should be route, and who should handle these tasks. In many cases, long established procedures will be affected, and the resulting trauma may not be fully understood. But industry gains additional experiences with the use of exchanges, this experience will help all companies make better use of them and achieve higher value as a result (Miller, 2001).

2.   GLOBAL DATA MANAGEMENT

Companies throughout the manufacturing industry recognise that global data management is becoming a competitive necessity. Information must be structured and available to the several teams involved (designers´ view, production view, marketing view etc.) so that safe and accurate information can be timely shared by everyone. As a data integration and management technology, PDM (Product Data Management) manages product data throughout the extended enterprise as depicted in Figure 3, ensuring that the right information is available for the right person at the right time and in the right form. In this way, PDM improves communication and co-operation between diverse groups, and forms the basis for organisations to restructure their product development processes and institute initiatives such as concurrent engineering and collaborative product development. The result is faster work, fewer errors, less redundancy, greater co-operation, and smoother workflow. This translates into bottom-line money savings and time reductions (Smith, Reinertsen, 1998).

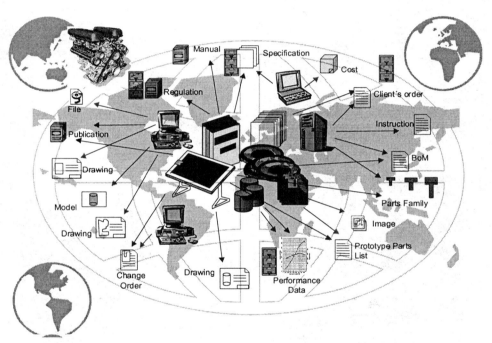

Figure 3. Global Data management

With PDM becoming more pervasive throughout a broader range of companies, the use of the technology is expanding beyond design activities to include information management throughout the product lifecycle. In this way, PDM works closely with ERP in managing product definition, product production, and business operations support from the early conceptual stages of product development, through manufacturing, and finally to product support. (Miller, 1999). A major Brazilian example is the implementation of a PDM solution at Mercedes-Benz Brazil in compliance with a world-wide decision for the whole DaimlerChrysler group.

Advanced data management environments, such as the one developed by DaimlerChrysler group provides a Client/Server architecture with world-wide range, what enables an independent co-operation development. These environments include the introduction of Digital Mock-Up, numerical simulation, management of digital prototypes, and consistent product documentation (Baake, Haussmann, Stratil, 1999). One of the most important tasks in the implementation of global PDM solutions is the definition and installation of IT and process technologies, taking into account the issues concerning the development of a global and multicultural project.

3.   VIRTUAL PRODUCT DEVELOPMENT

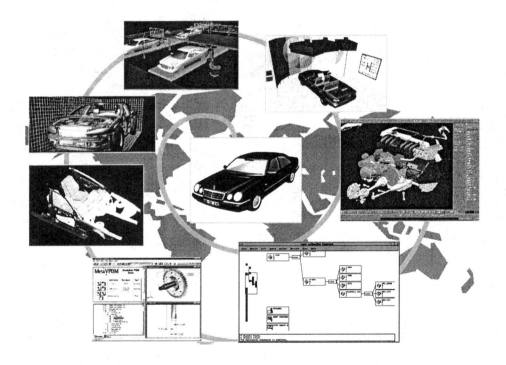

Figure 4. Virtual Product Development

The generation of virtual prototypes as outlined in Figure 4 is becoming every day a closer reality due to the new available technologies, and it is increasingly consolidated as an essential requirement in accomplishing new projects. Virtual manufacturing technology can be applied when new models have been designed using digital engineering tools and the production process is about to be developed.

3.1   Digital engineering

Virtual prototypes distinguish themselves within the product development cycle by making available simulation phases, providing accurate information on physical and mechanical functions of certain systems groups of a product. Within a virtual environment we are able to analyse the application of parts from different projects in order to reduce final production costs without the need of building physical prototypes.

The procedures for the construction and checking of physical prototypes are very expensive and time-consuming, and can also cause delays or cost increase during the development process. The major reason for these additional costs is the non-inclusion of changes occurred in the development during the construction of these physical prototypes. At the end of these prototypes' construction, assemblies and subassemblies may not represent the project's reality anymore. The digital prototyping techniques (e.g. Digital Mock-Up – DMU), embedded in the PDM environment, minimises this possibility, as the physical elements are not incorporated into the process. In addition, the digital analysis is always based on most recent/updated information or on a severe release control. Major benefits of the digital prototyping technology application are:

– Work developed on a shared virtual environment;
– Evaluation of maximum and minimum distances;
– Fast responses to the development teams;
– Change forecast and management;
– Analysis and Simulations of assemblies/disassemblies;
– Optimisation of physical prototype manufacturing, when necessary;
– Interference/collision analysis and control;
– Provision of managerial information throughout the development;
– Possibility of detailed evaluation of digital prototypes by work teams located at other plants.

3.2   Virtual manufacturing, the next step

When new models have been designed and the production process is about to be developed, virtual manufacturing technology can be used to increase the automation level as shown in Figure 5. Not only during the engineering phase the automation level is quite high due to different virtual product development technologies mentioned above, but also digital manufacturing, over the last years, has moved out of the laboratory and is now part of the business process.

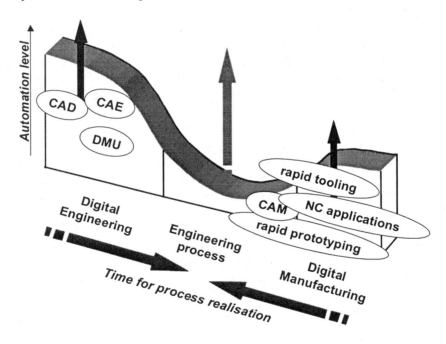

Figure 5. Different automation levels

This technology allows users to do design both for assembly and manufacturing before the design is frozen. One sure way to improve the manufacturing process is to make sure the model is correct when reaching the shop floor (Reinertsen, 1997). Computer-aided design packages are equipped to allow collaboration so that everyone involved in the design process is working on the same model version using a common database as outlined in Figure 6. Thus, virtual manufacturing tools allow engineers to share models with those who handle product design, manufacturing engineering design, and industrial engineering design for layout and facilities.

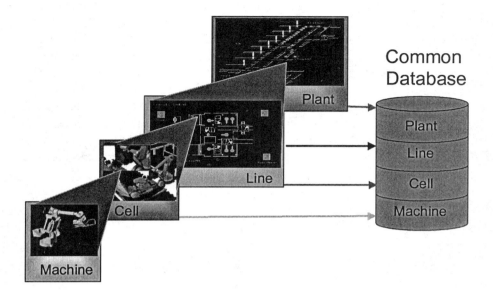

Figure 6. Digital Manufacturing

4. INTERNET TECHNOLOGY IN COLLABORATIVE DEVELOPMENT PROCESSES

Collaboration supports decentralisation of engineering and manufacturing operations by providing a framework for distributed working across the extended enterprise, supporting processes with integrated tools and production information, providing methods to help users visualise and access information supporting data consistency and integrity in a shared environment, and allowing geographically distributed product development through dispersed teams. Thanks to the power of the Internet and the proliferation of Web-based software, collaboration can be accomplished in a cost-effective manner. The provided tools and services bring to the same global information network all the fundamental elements for the development of products, integrating clients, partners and suppliers through Engineering, product design, product documentation, product simulation, change management, process and factory planning, sales, purchasing etc. The most far-reaching impact of collaboration technology is not to speed up current methods of operation but rather change the way that companies operate. Specifically, Web-based collaborative tools are facilitating major shifts in organisation, workflow, geographic distribution, and relationships between companies. Undoubtedly, manufacturers that encourage employees to collaborate with partners, suppliers, customers, and each other will surely save time, reduce costs, and bring products to market faster than the competition.

For the integration of all the processes, technologies, products and services involved, controlled and safe access portals have been developed under the Web environment. These portals provide access to several databases as depicted in Figure 7, with information made available all over the world for the collaborative product development.

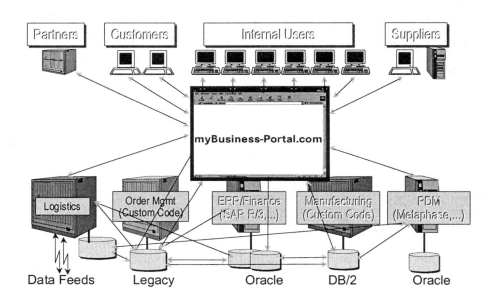

Figure 7. Business Portal concept

Thus, the product life cycle can be managed, and the knowledge acquired throughout this period (bad and good experiences) can be applied to new developments.

5. cPDm – cOLLABORATIVE PRODUCT DEFINITION mANAGEMENT

Nowadays, with the availability of a broad range of more sophisticated technological resources, mainly in the e-business world, by means of data base accesses through specific portals as shown in Figure 8, the product development cycle becomes much more comprehensive. With these technologies the world of Product Data Management and the world of Enterprise Resource Planning (ERP) are getting closer and more integrated. By implementing e-business technology together with PDM concepts we have the so-called collaborative Product Definition management with a tremendous impact on the ways companies operate. A major change in enterprise computing is on the way as companies apply best-practice processes in combination with a wide range of technologies including data management, collaboration, visualisation, collaborative product commerce, component supplier management, and others.

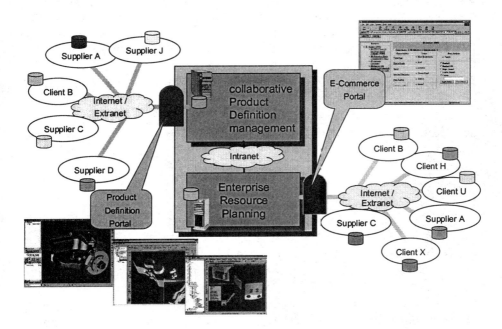

Figure 8. PDM Evolution: cPDm by using Portal technology

The vision of cPDm requires companies to expand their view beyond CAD management to management of the entire product definition, from the design phase of product development to the entire product lifecycle, and from the engineering department to the extended enterprise. The broadened vision reflected by cPDm is changing the industry in two ways: first, users and suppliers are shifting their focus from technology to solutions targeted at specific business problems; and second, the focus is shifting from data management to business processes. Companies are concerned with the best way to change their overall business processes and leverage technologies on an enterprise-wide basis.

In the past years, companies would implement product data management technology and then try to adjust their practices to fit the new system. In contrast, organisations are presently turning this order around by first assessing and re-engineering their processes towards strategic corporate goals. Then, they select and implement technologies that can best facilitate these corporate processes. This shift in focus towards process-based solutions is an evolutionary step in corporate computing. But some newer technologies are having a revolutionary impact by providing a quantum jump in the speed with which new processes now can be implemented. Web-based systems can broadly disseminate information and support widely distributed operations. Collaborative tools allow dispersed teams and groups to review and discuss designs of parts and assemblies, and work together in jointly developing products. Web technology and collaborative tools are becoming powerful drivers for the industry and are increasingly being focused on applying e-business fundamentals to the product definition process.

The pace of improvements in cPDm-related technologies continues to accelerate, and companies use these tools to implement process change faster than ever.

6.   CONCLUSION

To introduce a digital enterprise means to radically redesign/streamline the processes. Fundamental changes to the business processes interfere with all organisation areas. When a process undergoes a reengineering process, the activities involved suffer multidimensional changes, the working units change from departmental functions to process teams, and the staffs' working rules change from controlled approach to a policy that is totally aimed at the commitment, involvement, initiative, creativity and responsibility (empowerment).

7. REFERENCES

Baake, U.; Haussmann, D. and Stratil, P. (1999), Optimization and Management of Concurrent Product Development, International Journal on Concurrent Engineering: Research and Applications, Technomic Publishing Co. Inc.

Baake, U. and Haussmann D. (1998), Managing Technology and Process Changes within a multicultural PDM Implementation (eds. U. Baake and R. Zobel), SCS Publication.

Dimancescu D. and Dwenger K. (1996), World-class new product development: benchmarking best practices of agile manufactures, amacom American Management Association.

Gianni Jacucci, Gustav J. Olling, Kenneth Preiss, Michael Wozny (1998), Globalization of Manufacturing in the Digital Communications Era of the 21st Century: Innovation, Agility, and the Virtual Enterprise, Kluwer Academic Publishers.

Miller Ed. (1999) Managing Enterprise Information with PDM, CIMdata Publication 1999.

Miller Ed. (2000) Collaborative Product Definition Management for the 21st Century, Computer-Aided Engineering Magazine, March 2000.

Miller Ed. (2001) Exchanges: More than Marketplaces, Computer-Aided Engineering Magazine, January 2001.

Reinertsen, D. (1997), Managing the Design Factory – A Product Developers Toolkit, The Free Press

Smith P. and Reinertsen, D. (1998), Developing Products in Half the Time, John Wiley and Sons. Inc.

8. BIOGRAPHY

Uwe Baake received the Dipl.-Ing. degree in electrical engineering from the University of Siegen in 1990, and the Dr.-Ing. degree in computer science from Darmstadt University of Technology in 1995, Germany. In 1990 he joined the Computer Science Department of Darmstadt University of Technology, working in the Computer Aided Design Area. From 1995 to 1998, Dr. Baake worked at the Corporate Research and the headquarters of the Commercial Vehicles division of Daimler-Benz AG. From February 1998 to December 2000 he was responsible for the Process and Technology Department of debis humaitá IT Services L.A. and since 2001 he has been head of the Business unit *Engineering and Manufacturing* of debis humaitá.

Osmir Torres received the Bachelor degree in Mechanical Engineering from the University of Sao Paulo in. He graduated in Computer Science by the Fundação Armando Alvares Penteado in. He started his professional activities at General Motors in 1972 in the implementation of new technologies. He developed his activities at DaimlerChrysler Brazil in the implementation of CATIA CAD/CAE/CAM environments. He has been working since 1995 at debis Humaita IT Services as a Consulting Manager for the implementation of technologies applied to Digital Product Development.

Sustaining the introduction of innovation in digital enterprises
Challenges and demands from Small Medium Enterprises in the Italian Northeast region

Maurizio Marchese and Gianni Jacucci
Department of Computer and Management Sciences, University of Trento, Italy

Abstract: Small Medium Enterprises (SMEs) face specific challenges as the pace of technological advancement quickens. In order to keep pace with market demands, these SMEs need, among others, new tools to manage information content and new problem solving skills. They also need personnel capable to analyse the new tools and methodologies, select the most appropriate, deploy the solutions and train their workforce. Larger corporations can relay on the presence of various skills within the company to create successful teams; they can afford external consultants to assist the change management, and broadly based training programs, while SMEs cannot.

In this paper we analyse and discuss specific challenges and demands from Italian North-East SMEs as they have emerged from our experience in research and internship projects. We present some interesting case history data and discuss some general trends emerging from them. In particular we see that the type of problem in SMEs are not very different to the one encountered in large companies: they are of the same class, and the challenge here is to face efficiently these problems with limited resources.

Key words: Technological Innovation, Information Systems to support Innovation, Enterprise Knowledge Management, E-Business models

1.   SMES CHALLENGES AND DEMANDS

Globalisation of the markets together with the transition towards a knowledge-base economy is changing drastically small medium enterprises business environments.   Both trends rise specific challenges for SMEs. These problems are strictly correlated with the shortage of "intangible" investments like an efficient access to and use of technology, suitable managerial skills, effective training for the work force, high organizational quality and customer relationship management skills as well as new sales skills.

The main difficulties that SMES meet when they try to face the challenges of globalisation are often related to their tendency to remain tied to their "domestic market". In quantitative terms, some studies [1] indicate that less then 10% of Italian SMEs are currently present in other countries and/or continents and therefore cannot be considered ready to become global market actors.

Data collected in Northern Italy [1] indicate that approximately only 25% of manufacturing SMEs are competitive on an international level. Those that will not adapt to the global market challenges - between 25-50% of the total number of SMEs - will likely disappear. To survive, SMEs need to modify their central business objectives: not exclusively product quality, but also fast product availability, optimal logistics and improved customer services.

The business context for SMEs is in fact radically changed: in the classic model of mass production the enterprise produces a certain constant product, with limited requests for modifications of the production processes for a considerable period of time. Both products and production processes remain relatively constant. In this kind of environment, typical for SMEs, it is not crucial to transfer information between one process to the other in the value added chain. The product itself connects the different manufacturing phases and the warehouse acts as a buffer. In this old scenario, the main goal for the enterprise was the product and the main aim was to produce high quality products.

Now the context is changed from static production processes to a dynamic system of coupled processes. In this new scenario, the main goal is to correlate effectively all production processes. Naturally the product is still the final result, but the reason to choose one specific enterprise relies more in its production processes than in its products. The new aim is to improve the level and the quality of the link between internal production processes and suppliers and customers processes. To maintain a competitive advantage enterprises do not seek only to be selected as suppliers of high-quality/competitive-prices products. They must also became an integral part of the supplier-manufacturer-customer supply chain [2].

1.1   What kind of challenges for SMEs ?

The new business models requires to face the change in three fundamental dimensions of the production activity, both from SMEs and large organization:

- <u>new procedures for the cooperation and the interaction with the suppliers</u>: as enterprises business relations evolve to include all information fluxes between customers and suppliers, new information exchange procedures emerge and need to be implemented. They range from suitable B2B applications to integrate the different production activities to new organizational methods to enrich customer-supplier relationship, like innovative methods to gather more richer information from customers (feed-backs, suggestions, etc.) and involve business customers in both risk and profit investment.
- <u>business processes reengineering</u> for the enterprise to support the transition from the first phase of computerization (not-interconnected information systems) to the second phase of a distributed enterprise information system capable to assist communication and information exchange. The basic skills required here include analysis capabilities and tools, design and management of communication infrastructures, of knowledge sharing tools, and of enterprise procedure optimisation and automation tools. In this respect, the new business paradigm underlines efficient work organization, fast and appropriate introduction of innovation and effective training of the work force as key competencies for the successful enterprise.
- <u>new procedures for the interaction with the customers</u>: the new industrial activities are focused on the creation of profitable and long-lasting relation with its own customers. In order to pursue efficiently this aim, companies must - collect and use customers data - increase the quality of the product, investing in tools, methods of production planning; - increase the offer of value added services to their customers.

1.2   What type of demands from SMEs?

In order to keep pace with these new changes, enterprises need ,among others, new problem solving and knowledge management techniques [3]. They need to analyse the new methodologies and tools, select the most appropriate, deploy the solutions and train their workforce.

By collecting and monitoring proposal for research and internships projects from SMEs in the Italian North-East region, we have been able to identify some classes of demands. The majority of the proposed projects involved the deployment of a particular innovation (technological or organizational) that the middle management of the enterprises believed to provide a competitive edge. Most of the times the innovation required massive entrepreneurial challenges: radical re-invention of internal or external processes, managing alliances and networking strategies, issues related to organizational memory and knowledge management problems.

In order to identify specific responses to the demands, we have grouped them into the following five categories:

- new management methodologies to examine competitive advantages in networks of alliances: managers demands tools for and training in the decision making process. In particular they seek methodologies and tools to evaluate risks and benefits in alliances, merging and outsourcing activities. This group of demands respond to the challenge for the management of mastering new procedures for the cooperation and the interaction with the other actors of the particular market and is closely related to the following point.

- new form of work organization and supporting tools: here the demands from the management are concentrated on the development of a company culture for teamwork, open to a flexible and modular reconfiguration of roles and activities. In this respect the introduction of effective teamwork methodologies and specific groupware tools need to be considered on a case to case basis in order to assess its usefulness together with related phenomena as innovation drift, care and tolerance [4].

- fast, appropriate and integrated introduction of innovation in the production and management processes. The emphasis here is in a demand of qualified human resources capable to grasp the inherent interdisciplinary approach to an innovative solution. The radical restructuring in enterprises has in fact created a new class of problems requiring more than ever a synthesis of organizational and technological skills [5]. This class of problems is of central importance to all enterprises today and its solution is critical to the success of innovation for SMEs.

- the effective use of the new information channels offered by telecommunication developments. The rapid diffusion of enterprises Intranets and Extranets together with a widespread diffusion of Internet access pave the road to a new series of application and services for SMEs to support business activities. In this setting E-Business is a used term in order to indicate all automated procedures aimed at the management of all aspects business operations. Typical demands are B2B solution for supply chain management and tools for customer relationship management.

- **new methodologies and tools for enterprise knowledge management**: here the emphasis is in the research on the conditions that promote enterprise learning and on methodologies and tools that assist in organizational memory processes.

2.   CASE STUDIES AND ANALYSIS

2.1   General data on internship projects

The Laboratory for Informatics Engineering (LII) at the University of Trento is involved in a multi-year program of research, technology development, technology transfer, and professional training on the use of emerging information and communication technologies for supporting enterprise innovation. The main body of experiences and data we have gather from our work with local enterprises has been focused on the segment of Small and Medium Enterprises (SMEs) in the private sector and with relatively large enterprises both in the private and public public sector.

Our experiences on the problems related to the introduction of innovation have been collected during the years through three main channels:
- applied research projects with local enterprises SMEs,
- internships projects in connection with a three year degree course in Engineering Informatics of the faculty of Engineering, and
- internships projects in connection with higher education training courses both for unemployed people and for life-long training of enterprises personnel.

The different types of training courses we have running have involved so far about 420 students (ca. 240 degree students, 110 post-diploma students and 70 post-degree graduate). In the course of our training activities we have therefore organized and run about 420 internship projects. Approximately half of them involved SMEs. The remaining projects involved larger corporation, like public administration, large research institution and large enterprises. We report in the following data from the SMEs internship projects collected and analysed at different levels.

Table I summarize total internship data, covering a five-year period from 1996 to 2000 for a total of 258 projects, according to the previous proposed categories of SMEs demands.

Table I: Categories of Internship Projects in SMEs

Category	
Alliances / Outsourcing / DSS	13 %
Teamwork / Groupware	26 %
Business Process Re-engineering	37 %
E-Business	14 %
Knowledge Management	10 %

All categories are significantly populated. The main two project classes involve BPR and re-organization of work activities through teamwork methodologies supported by groupware tools. In our activities we have monitor the evolution of the types of project proposed by local SMEs in the past five years. Table II presents this evolution in time.

Table II: Evolution of Categories of Internship Projects

Category	1996	1997	1998	1999	2000
Alliances / Outsourcing / DSS	11%	12%	14%	15%	12%
Teamwork / Groupware	27%	25%	26%	26%	25%
Business Process Re-engineering	48%	44%	44%	34%	16%
E-Business	12%	7%	11%	13%	29%
Knowledge Management	2%	12%	5%	13%	18%
No. of projects	*41*	*52*	*58*	*56*	*51*

The growth of demands in the area of E-business and Knowledge Management is important and correspond to a clear trend in the local economy towards an extensive use of the new information channels that are mature for use. This can take place and advantage from the continuous effort of reorganization of the business processes that has started in the past and is continuing in the present year.

2.2   Presentation of case studies

To extract more detailed information from our data, in the following we present three interesting case studies and take the opportunity to discuss some general trends emerging from them and supported also by the overall collected data.

2.2.1   Assessing the deployment of groupware tools in a network of local banking institutes

This project has dealt with the monitoring and assessment of the deployment of groupware tools, based on Lotus Notes, in a network of local banking institutes of the Federazione delle Cooperative Trentine. The banking network includes 69 independent institutes (Casse Rurali) for a total of 330 front offices. This environment covers very different realities ranging from small offices with only four workers to medium sized offices with more than 100 workers.

The project has involved several skills and operation like:

- integration of the project team (2 students) in an existing monitoring committee
- the preparation and distribution of digital questionnaires
- the preparation of interviews with final users of the systems: top managers, middle managers and operator
- statistical and sociological analysis of the collected data

During the project the management of the consortium of banking institutes has emphasized four area of interest, namely: monitoring the use of the software, analysing its organizational impact and analysing peoples resistance to the new tool. After a first period devoted to get in touch with the company reality, the first phase of the project has been devoted to the preparation, distribution and analysis of a digital questionnaire to all people in the network. The work of analysis here has enable a first clustering of institutes on the basis of their kind of use of the tool. The second phase has concentrated in the preparation of interviews with selected final users of the system. The final outcome of the internship has been a report and a presentation of the findings to the top management of the banking network.

Major findings from the present project include:

- wide differences in the attitude towards the particular innovation are present in the work-force: top management being more prone to adopt innovation, while middle management resist the change;
- the majority of the workers feels the need for special training courses to improve their use of the tools;
- groupware can serve as a set of tools for facilitating closer interaction and better communication between individuals, especially if they are geographically dispersed; however, the use of technology will not bring about improvements per se. It must be combined with the creation of an information-sharing culture and adequate work-processes that support it.

2.2.2 Definition of a model for the knowledge management in a testing laboratory of a manufacturing enterprise

The initial proposed project dealt with the development of a database to assist the management of the material-testing laboratory in Novurania, a manufacturing medium enterprise of plastic fabrics for nautical applications. During the course of the work it became clear that the collection of internal data for the creation of databases involved harvesting of internal non-formalized procedures.

In a second phase, after discussion with the company managers, the objectives of the project were broaden to include:

- an exhaustive analysis of the enterprise testing procedures in the various departments;
- an analysis and selection of enterprise best-practice to be capture in a proposed knowledge database
- the development of a prototype knowledge database
- the use of the knowledge database to propose improvements to enterprise procedures;

Different skills, methodologies and tools were useful in the project. In the first part the work focused in the development of Data Flow Diagram to capture company procedures (see Figure 1). To gather the necessary information a number of interviews with the middle managers involved in the testing procedure were prepared and carried out by the student. The Data Flow Diagrams were used to capture the actual information flow between enterprise departments and to underline criticalities (redundancies, possible coordination problems ..).

The second phase of the project was devoted to the design of the Entity Relation Diagram for the proposed knowledge database. The developed DB tool has been extended to include decision support features, like the automatic proposition of categories of testing to execute to new materials based on stored company procedures. The developed tool is actually deployed in the different departments and our research group monitors its use after the end of the internship project.

Naturally the deployment of the tool has impacted with existing working practice and organization. The success of the introduction of the innovation has been strongly correlated by the human and organizational factors that were discovered and analysed in the design phase.

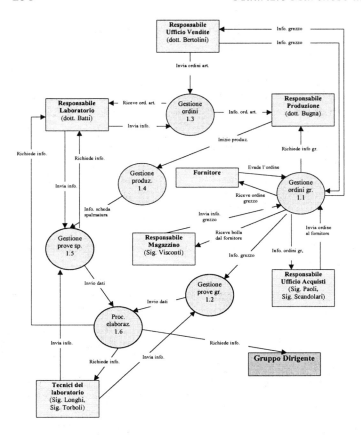

Figure 1: Example of Data Flow Diagram for overall testing procedure

2.2.3   E-Business Models for a SME: Dr. Schär case-study

Dr. Schär is a SMEs (ca. 100 employes) situated in Burgstall, Italy, a small village about 20 km north-west from Bolzano, the provincial capital of South Tyrol. Dr.Schär has earned an excellent reputation as leader and specialist in the European market for high-quality dietetic products. Starting in 1997, Dr. Schär management decides to explore the opportunities offered by the new telecommunication developments. The company goes on Internet under www.schaer.com and at the same time participates to a local project named Telepraktikum [6] which involve the local Business Innovation Center and some Italian and Austrian Universities.

The Telepraktikum project focuses on the introduction of e-business models in local enterprises with particular attention to the technologies supporting telework.

Within this framework the internship project has dealt with the requirements analysis of Dr. Schär. The proposed technological tools to improve the company e-business profile included:

- Unified Messaging: integrated information system to support e-mail-telefax and phone enterprises messages; the requirement analysis focused on the groupware activities that could be implemented by the use of such a system;
- use of VideoConferencing to support communication with suppliers/resellers located in Europe;
- development of a Call Center (see Figure 2) to manage all Consumer contacts with the enterprise

Through the implementation of these tools, Dr.Schär desires to innovate its enterprise information system and in particular:

- to improve its control on the information fluxes;
- to update its internal and external communication system
- to explore and test new emerging business models and include them in

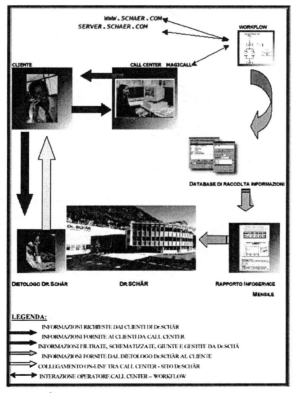

the company culture.

Figure 2: Information fluxes in Dr.Schär Call Center

3.   DISCUSSION OF CASE STUDIES

Some general aspects emerges from these reports, and are supported by
other internship data:
- the problems and demands we have found in our local SMEs do not
  differ from the ones encountered in large enterprises. They belong to the
  same class of problems, inherently involving organizational as well as
  technical skills acquisition and practice. Larger corporations can relay on
  the presence of various skills within the company to create successful
  teams, can afford external consultants to assist the change management,
  and broadly based training programs, while SMEs cannot. The typical
  result from this lack of resources in SMEs is the focusing of their
  interest on particular one-dimensional problems, connected to a single
  aspect only (either technological or organizational). But in order to
  actually solve the problem the overall issues must be successfully
  addressed;
- a different attitudes is present between top management and middle
  management in regard to the introduction of innovation. Top
  management is prepared to face the challenge of an integrated approach
  to innovation. They are very sensitive and open to the idea of a full-
  spectrum analysis of particular topics and they seek contacts with
  possible partners to face the challenges (university, consultant etc.)
  Middle management are resisting these ideas and are more interested in
  day by day routine innovation problems, more easily separated in the
  two macro-area (either technological or organizational);
- a gradual approach in the introduction of the innovation is often desired
  and expected by the workers and supporting training programmes can
  assist in the change management aspects of the issue;
- another general aspect is the separation of the projects proposed by
  enterprises, in regard to innovation, into three main groups: analysis
  prior to innovation, deployment of innovation, monitoring the process
  after innovation.
    - ✓ analysis prior to innovation: these projects are more likely to
      present integrated problems and are the ones at the top of the
      interest for the top management;
    - ✓ deployment of innovation: essentially operative problems, most
      often of a technological type that requires operative skills
      (programming, customisation, scheduling of activities etc.);
    - ✓ monitoring after innovation: again the need for integrated
      knowledge and culture is high. The collection of data, the
      identification of critical points, and the final analysis requires

different kind of skills like contextual analysis, data gathering, knowledge of technology solutions.

4. REFERENCES

[1] Rapporto sulla Situazione Economico e Sociale della Regione Trentino-Alto Adige, Agenzia del Lavoro, Trento, Italy, March 2000.

[2] J.A. Sharp, R. Beach, A.P. Muhlemann, A. Paterson and D.H.R. Price, "Towards Globalisation and the Virtual Enterprise", Proceedings of 10th IFIP WG5.2/5.3 International PROLAMAT Conference, p737, 1998, Trento, Italy.

[3] J. David; P.V. Laure, "Research note: Attitudes towards graduate employment in the SME sector", International Small Business Journal., vol.11 n.4, pp. 65-70, Jul-Sep 1993;

[4] C. Ciborra, "Groupware and teamwork: invisible aid or technical hindrance ?", John Wiley & Sons, London, UK, July 1997.

[5] M. Marchese, C. Cattani , V. D'Andrea and G. Jacucci, "Training Students to Intervene in Information Systems Inherently Involves Organizational and Technology Skill Acquisition", in Proceedings of 8th European Conference on Information Systems – ECIS 2000, July 3-5, 2000, Wien, Austria.

[6] for more information on the Telepraktikum project see: http://www.teleworker.org

5. CURRICULUM VITAE

Maurizio Marchese is a Assistant Professor at the University of Trento. Main research interests are in the introduction of Information Systems innovation in enterprises and knowledge management models and tools. E-mail: marchese@science.unitn.it.

Gianni Jacucci is Professor of Informatics at the University of Trento. He leads activities of service and training for innovation in small enterprises and economic development of local communities, including virtual enterprises and telematics for tourism. He also coordinates the activities of a University Diploma Course in Informatics Engineering. E-mail: gianni.jacucci@lii.unitn.it

A Planning and Management Infrastructure for Complex Distributed Organizations

George L. Kovács
Computer and Automation Research Institute (www.sztaki.hu)
Kende u. 13, H-1111 Budapest, Hungary, gyorgy.kovacs@sztaki.hu

Paolo Paganelli
Gruppo Formula Spa. (www.gformula.com), Via Matteotti, 5. I - 40055
Villanova di Castenaso (BO), Italy, paolo.paganelli@formula.it

Abstract: The objective of our work is to provide a web-based, software supported planning and management infrastructure for complex, distributed organizations working on large scale engineering projects. Such projects are characterized by huge investments in both materials and human resources and by concurrent, disparate activities – manufacturing, design and services as well. These type of projects are rarely carried out within the scope of a single organization, or at least several, distributed parts of the organization and/or subcontractor small and medium enterprises (SME) are often involved, too.

Key words: management, complex enterprises, web-based, infrastructure, engineering

1   INTRODUCTION

In the following the main ideas (problems and suggested solutions) of a European joint project are going to be detailed. The WHALES (**Web**-linking **H**eterogeneous **A**pplications for **L**arge-scale **E**ngineering and **S**ervices) project started in March, 2000 and it has Italian, German, Portuguese and Hungarian participants. In the following to distinguish between the *project itself* and *projects managed by WHALES*, the word **project** will be used for the managed projects, and the running project itself will be called the WHALES **system,** or simply the system.

The objective of our system (under development) is to provide a planning and management infrastructure for complex distributed organizations working on large scale engineering projects, characterized by huge investments in both materials and human resources and by concurrent, disparate activities – manufacturing, design and services as well. Managing projects of this kind means dealing with several problems at the same time, as:

- Complexity of scope, in terms of time and resources employed, and variety of activities to be planned, synchronized and monitored;
- Distributed organization, spanning through several companies and involving a multiplicity of actors and competencies;
- One-of-a-kind design, increasing planning complexity, hard to apply product and process standardization;
- Geographic distribution of project activities, sometimes in unprepared or hostile environments;
- Strict time constraints, with complex milestones and dangerous critical-path dependencies;
- Contingency risks, due to the high planning uncertainty and difficult re-alignment of activities;
- Revenue-loss risks, due to difficulties in budgeting and high contingency costs.

Such projects are rarely carried out within the scope of a single organization. More often the prime contractor, typically a large company with adequate know-how, references and financial resources to sustain the project, outsources specific components and services to smaller firms through several forms of sub-contracting. This way SMEs are often involved.

For the prime contractor and its major partners, winning a project represents a demanding and risky activity in itself. The best available technical and management skills are required to present competitive offers in a world-scale market where ability to perform, rapid implementation and acceptable cost are at least as important as background expertise and technical quality. The goal of this system is to try to answer the above challenges even if there are the following problems, too:

- High direct and indirect costs of basic resources;
- Complex and hierarchical organizations grown up in better times of unchallenged and stable demands (e.g., markets protected by local governments);
- Low operative margins, putting short-term activities and contingency management ahead of technology and business process improvements;
- Cultural and organizational obstacles to apply "virtual enterprise" partnership models;
- Low flexibility that, burdened with complexity, makes it almost impossible to prepare reliable plans and project budgets.

2   EXPECTED RESULTS – BENEFITS

As an alternative to massive restructuring, which would entail significant losses in terms of employment, technical potential and historical background, we intend to enhance the project planning and deployment capabilities of the involved firms, thanks to a software infrastructure producing the following measurable results on the end-users business:

2.1   Improved planning and budgeting

These will be measured in percentage of successful bids, and in planned vs. actual costs/duration, etc. and will be achieved by means of designing and implementing several subsystems, and taking into consideration several important factors at the same time, as:

- A planning and financial analysis system for bid preparation (tendering): projects managed by our system have to go over a hard preliminary competition, commonly called "tendering". The price, the deadlines, the technical specifications have to be given in a bid, so a poorly calculated bid can result winning the order but failing the profit and the delivery terms;
- Considering the whole scope of project activities (including sub-contracted ones, as well as needed materials, machinery and human resources at all project sites) when taking project timing and allocation decisions;
- All decisions will be based on updated and consistent information about running activities and availability of resources, on aggregating and normalizing data from heterogeneous applications and different functional domains both inside the company and across the consortium network – using appropriate DSS (Decision Support Systems);
- Analyzing and comparing alternative scenarios, generated through alternatives on a full scale project model, and evaluated and compared by means of advanced on-line analytical processing tools – using appropriate simulation and DSS tools;

2.2   Improved monitoring, cost and risk assessment

These will be measured in manpower/assets utilization, in reduction of "idle" time (waiting for unfinished activities), etc. and they will be achieved by means of:

- On-line access to the current status of project activities (the resources consumed at all project sites, the quantifiable results and costs borne, and any other indicators relevant to project progress evaluation and current risk assessment);

- Real-time notification of events and conditions constituting potential failure sources to the appropriate actors in the project network, according to rules and criteria set case by case (e.g., maximum delay on a critical path activity);
- Automatic update of the project plan, highlighting deviations between actual and planned activities, their impact on related tasks and milestones, and corrections required to meet the project objectives.

2.3   Effective contingency management

This is measurable by the percentage of "perfect orders", i.e. orders delivered according to original request, and by the reduction in number of cancelled or re-negotiated orders, - and will be gained by:

- Pro-active risk analysis in the project planning phase, where alternative solutions are compared considering both internal factors (e.g. strict time-dependencies) and external factors (e.g. casual distribution of machine failure or manpower shortage at project sites);
- Re-planning in the project deployment phase in reaction to alerts and deviations notified by monitoring functions, selecting backup options on the basis of cost, perturbations on running activities, need for re-negotiation of already set plans, etc.;
- Alignment of the whole project plan to changes on re-planned activities, promptly updating all project sites and management levels, and tracking of project revisions for costs evaluation and statistical risk analysis.

2.4   Higher flexibility and efficiency

The best measures for them are the increase in bids processed by the same organization, the reduction of bid preparation cycle time and the profit margins by order and project unit. These can be approached by:

- Ability to respond quickly to customers' requests for proposals and requests for changes by involving the appropriate technical and management skills at all project sites;
- Better exploitation of the network resources, thanks to a decision-support environment which is aware of co-operation possibilities (e.g., roles to be fulfilled in a project under planning) and available partners' skills and capacities;
- Prompt negotiation of planning and re-planning options, by means of a communication infrastructure that circulates decisions and events between the appropriate actors, crossing companies and organizational units boundaries;

3   SYSTEM GOALS - TASKS TO BE SOLVED

To achieve the above improvements requires dealing with different enterprise functions and information sources, supported by heterogeneous and poorly integrated software applications, as:

- Enterprise Resources Planning systems (ERP) (as SAP, Baan, etc.) represent the companies' administration backbone, and provide basic transactions for bids and contracts management, job order stages and costs reporting, billing and invoicing;
- Production Planning and Control (PPC) and Warehousing systems, often sold as ERP components, support materials management and long- to short-term production planning;
- Project planning tools provide graphical editing of GANNT and PERT project diagrams, along with on-line display of resources workload and activities timing;
- Human Resources (HR) packages support company organization management, identifying key project roles, skills and positions, as well as project personnel costs and time-tables.

None of these systems alone covers the full spectrum of project management requirements, that in complex organizations range from financial planning and cost analysis to human resources recruiting and assignment, to procurement and allocation of manufacturing resources and materials. Moreover, none of these systems provides a data and communication infrastructure for the whole project network, i.e., the multi-site, multi-company organization created to fulfil specific project objectives. As a temporary and goal-oriented structure, although it can last years and absorb large turnover shares, the project network presents typical "virtual enterprise" properties that make it impossible to map it on traditional, enterprise-centric information systems.

In response to these requirements, the WHALES system will pursue two main objectives, which should be solved by the system software:

- To design and develop a set of software components supporting integrated planning, deployment and monitoring of large projects in multi-site, multi-enterprise organizations;
- To demonstrate the applicability and benefits of the developed software components through analysis, implementation and experimental usage on different type of pilot business cases presented by different users in different countries spread in Europe.

As an innovative system for project management in complex and distributed organizations, the system shall implement the following general features:

3.1 **Provide a unified and generalized representation** of project activities and related artifacts, comprising all material and immaterial work items (e.g., products, knowledge, design documents in different stages) that need to be organized in complex projects (in our test cases: shipbuilding, engineering industry, plant repair and maintenance services).

3.2 **Support distributed organization models**, crossing hierarchies and company boundaries; to be general and commercially exploitable, the system shall not rely on any pre-defined organization schema, but will support a case-by-case definition of links between companies, organizational units and employees involved in each project.

3.3 **Provide a scalable and flexible co-operation environment**. The system will provide a project network infrastructure accessible to every node (company or organization unit) independently of its size and information system. It will support nodes and individuals in readjusting their role and interface toward the network (for example to reflect changes in the node internal organization, or to make new resources available to any project).

3.4 **Integrate and distribute relevant information** across the project network. Data maintained by each node and related to a specific project will be given a generalized representation and shared with the other project participants through a web-based environment according to visibility and consistency rules mirroring the project organization model and management responsibilities.

3.5 **Support decision-making** in the project ideation, definition and deployment phases. This means to select potential partners on the basis of their past performance, cost and capabilities, to generate detailed plans considering both activities' timing, equipment and materials availability, and to find substitute resources for a running activity, etc.

3.6 **Manage and synchronize** the flow of decisions and events in the project network. The system will manage the distributed workflow associated to a project e.g. circulating planning proposals among the partners, integrating multiple decision threads in a consistent and transparent fashion, and dispatching monitored exceptions to the responsible actor(s) for contingency management.

3.7 **Integrate** with local management and planning systems. It means to safeguard the nodes' autonomy and IT investments. The system shall not interfere with node internal procedures and management tools, as ERP, PPC, Human Resources, stand-alone Project Planning and Budgeting packages. Instead, proper interfaces shall be designed for real-time information exchange between these systems and our system network infrastructure.

4   INNOVATIVE SYSTEM FEATURES

The most innovative aspects of our system solution lie in its *distributed architecture design*, that provides an *integrated data and process infrastructure* for different companies and actors participating in large projects' planning and execution, at the same time safeguarding each node's autonomy as regards local operations management and information system. These features match key requirements of the so called "virtual enterprise" organizations working on large one-of-a-kind industrial projects, recently highlighted by a survey on European large scale engineering companies carried out in the IV. Framework Programme ([1]).

The main findings of the survey are: lack of data models, communications and workflow infrastructures for project teams "extended" to suppliers and sub-contractors; lack of lifecycle planning, costing and risk assessment tools for complex distributed projects. State-of-the-art software applications offer only partial responses to the above needs, being still too much dependent on specific industrial sectors, organization models or ERP platforms, and approaching project management with a solution- rather than a problem-oriented approach, focused on a specific tool or technology application. Their goal is to optimize a single aspect of project life-cycle management, as:

- *ERP packages' Management extensions*. ERP systems are adopted as the enterprise backbone for execution functions. World-class packages (e.g., SAP, JDEdwards, Baan) provide project management modules capable to integrate typical ERP functions like job orders management, accounting and purchase, with higher level features like Work Breakdown Structure (WBS) or project profitability analysis.
- *Project Management applications* include a wide range of software products. Professional project planning and project accounting suites (e.g., SAS, Solomon Software) provide advanced decision support, scheduling and on-line analysis features. Office project applications (e.g., MS Project) provide graph- and table-based editing for manually planned projects, easy to use and integrated with common office tools. Dedicated packages provide a broad range of project management features for specific industrial sectors (e.g., ABT for software development projects, others in many sectors).
- *Data Interchange and Workflow infrastructures* recently emerged as means for improving efficiency and standardize operations of complex distributed organizations, including engineering networks. On the one side, standards for data and documents interchange (e.g., STEP, EDI) provide the foundations for knowledge sharing and communication of

engineering and commercial information. On the other side, communication and workflow technologies provide process automation features for real-time electronic business interactions.

WHALES will introduce a significant advance on project management practices supported by state-of-the-art applications, thanks to a flexible architecture integrating project-related data from heterogeneous applications, workflow automation, and decision support functions into a *web-based environment*. The resulting system is expected to accommodate the needs of project networks independently of the industrial sector, thanks to its general and adaptable design, that comes from features like:

- Distributed project management environment,
- Decentralized architecture and accountability structure,
- Powerful project and network data model,
- Flexible decision-support tools.

The decentralized and flexible system model will safeguard the autonomy and visibility of each network node, independently of its size. This will prevent the constitution of hierarchical project networks, actually dominated by a single, large contractor. Typically this happens when planning and logistics departments of large firms, faced with tasks surpassing their traditional responsibilities, tend to pass part of this complexity onto their suppliers. These are often SMEs, whose resources and commercial strength are insufficient to deal with such demanding scenarios, with consequent problems in terms of competitiveness, losses and high risk on investments.

Concerning the research state of the art, we can identify two main directions pursued in the last years by many projects:

- On the one side, standards and systems are sought for product and process data modeling and interchange, and to support distributed design in concurrent and co-operative engineering environments. This category of projects focuses on the "what", i.e., on the contents specifications for a product or project, rather than on the "how" and "when" that are typical project management concerns. References to some of these projects are in [2 and 3].

- On the other side, virtual enterprises are studied as evolving organisms, investigating environmental, legal and socio-economic conditions for the creation of enterprise networks.

- Considerably less effort has been directed to the analysis of the planning and monitoring problems characterizing such networks, and to how co-operation can be sustained and managed on a daily basis. References to some of these projects are in [4 and 5].

5   SYSTEM TEST CASES AND ARCHITECTURE

The WHALES results will be demonstrated by four different type of pilot cases in four different European countries:

- Lisnave, an important *Portuguese* company with a long tradition in shipbuilding and ship-repairing services. It is presently suffering aggressive competition from the far East;
- FATA, a large *Italian* engineering company, it represents a complex and hard-to-manage business with respect to leaner and faster manufacturing SMEs in the North-East;
- MTS, from *Hungary* is facing foreign competition with a business organization typical of state-owned companies operating on local, protected markets;
- METZ, a German medium-sized firm delivering customized vehicles and vehicle equipment services. It has to synchronize engineering, manufacturing and procurement activities in a typical one-of-a-kind environment.

The system architecture has been designed to match the project wide application scope, the complexity of technical objectives, the variety and extent of business cases to be analyzed and implemented at pilot users' sites. Each of these topics raise different categories of problems, requiring specific competencies and additional co-ordination along with conventional project management and software development activities. The work to be undertaken has been subdivided into two basic thematic areas. See Fig. 1. for all important relationships.

The two basic thematic areas (1. and 2. in Fig. 1.) in the system development are the following:

5.1   Network Architecture & Software Components

The project main body consists of five technical parts devoted to the study and development of the ICT architecture and software components that are going to support the system network organization model. Each part includes the fundamental activities of a quality-based software development process: *requirements, analysis and design, implementation, test and deployment*. A sixth part provides a *common development infrastructure* for the teams dealing with: methodology and tools to be used, selection of existing re-usable components, co-ordination of joint developments, maintenance of a technical data repository, configuration and change management.

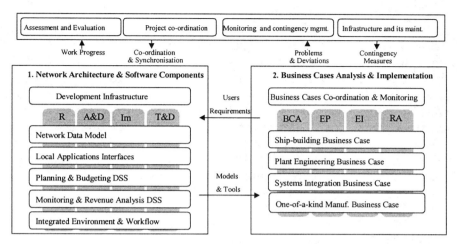

R: Requirements - A&D: Analysis and Design - Im: Implementation, T&D: Test and Deployment,
BCA: Business Cases Analysis - EP: Experiment Preparation EI: Experiment Implementation,
RA: Results'Assessment.

Figure 1: system architecture

5.2   Business Cases Analysis & Implementation

In parallel with technical developments, the proposed organization
model and tools are introduced and applied on four business cases proposed
by users in different industrial sectors in different countries. Each business
case consists of the fundamental activities of *business case analysis,
experiment selection and preparation, experiment implementation and
results assessment*. All business cases will apply common methodologies,
metrics and best practices to ensure uniformity, comparison and joint
evaluation of all results produced.

In addition due to the complexity of the tasks a solid *Project
Management* is necessary to take care of monitoring and accounting of
project activities, risk assessment and contingency management and project
infrastructure maintenance.

The main relationships of the thematic areas (and of all other important
parts) are represented by arrows in Fig. 1:

1) *User requirements* produced in business cases analysis are a
   necessary input to software analysis and design specifications.
2) *Models* specifications and *software tools* produced in the technical
   segment are necessary for experiments preparation and
   implementation of all business cases.

6   CONCLUSIONS

The WHALES system is designed and implemented by a European team of four countries. The work started with defining common tools and principles to avoid any misunderstanding. A well organized management takes care of all harmonization problems, and four test cases will prove the applicability of the results. The results will provide a web-based planning and management infrastructure for complex distributed organizations working on large scale engineering projects, characterized by huge investments in both materials and human resources and by concurrent, disparate activities – manufacturing, design and services as well.

7   REFERENCES

1. EP 20876 ELSEWISE (www.cordis.lu)
2. EP 20377 OPAL (www.cordis.lu)
3. EP 20408 VEGA (www.cordis.lu)
4. IMS GLOBEMAN21 (www.ims.org)
5. EP 26854 VIVE (www.cordis.lu)

8   BIOGRAPHIES

George L. Kovács got his Dr.Techn. degree (Ph.D.) at the Technical University of Budapest,. 1976. and he is Prof. at the same university since 1995. In 1997 he got the Dr. of the Academy degree from the Hungarian Academy of Sciences (HAS). He is with the Computer and Automation Research Institute of HAS since 1966, recently head of the CIM Research Laboratory. Visiting researcher in the USA, Soviet Union, West Germany. Visiting professor in Mexico and at the University of Trento, Italy. Author of more than 250 scientific publications. Member of several Hungarian and international scientific organizations, including IEEE, IFAC and IFIP. Project manager of several Hungarian and international R&D projects.

Paolo Paganelli graduated in Electronic Engineering at the University of Bologna in 1991. Research assistant at the University of Modena (1992-1994). From 1994 to 1997 he worked at Democenter, in the R&D department as technical manager in different ESPRIT projects. In 1997 joined Gruppo Formula S.p.A., as senior analyst involved in the design of Formula's next generation ERP software. He is now Formula's Product Manager for Supply Chain Management solutions. He is the initiator and project manager of the ESPRIT projects FLUENT (Flow-oriented Logistics Upgrade for Enterprise Networks) and WHALES. He published around 30 papers, including presentations at international conferences.

MERCI: Standards based exchange of component information to support e-business applications

Wolfgang Wilkes
University of Hagen, Germany
wolfgang.wilkes@fernuni-hagen.de

Abstract: Product developers need high quality information about the components which they use in their product development. The exchange of this component information is today often still in an "archaic" stage: component data hardly exists in computer sensible form, often it is distributed only in form of textual data sheets in pdf format. This paper presents the MERCI system which exploits the PLIB standard for the exchange of component data between the processing systems of manufacturers and users.

Key words: supply chain management, component information exchange, PLIB

1. INTRODUCTION

Today, the design and manufacturing of products cannot be done by a single company. Almost all products integrate components which are produced by other companies. The relationship between product manufacturers and their suppliers, the component manufacturers, varies from a close relationship where component manufacturers are integrated in the product development process (they produce specific components for specific products) to a loosely relationship where independent component manufacturers offer their components on the market to potential customers. The latter case can be found quite often in the area of electronic components (e.g. memories), whereas the former one is found often in the automotive industries.

In both cases, the product manufacturer is dependent on information about the components: This information is needed in the product development and design process. In case of close relationships, often the customer (the product developer) defines the way in which the supplier (the component manufacturer) has to deliver the data for the specific project. If, on the other hand, components are offered for the market, the component manufacturer decides himself in which format to describe his components and how to characterise them. Unfortunately, even today it is quite common that the important characteristics of electronic components are described mainly in textual form in the so called data sheets. As a result, component users (product manufacturers) have to capture this information again to use it in their design tools and corporate databases.

In this paper, the situation in the domain of electronic components is described. Here components are normally offered on the free market, and all the mentioned problems are the daily burden of engineers. But recently also the manufacturers have begun to notice that a better data organisation is essential for streamlining their own documentation and publishing processes.

The remainder of this paper is organised as follows: Section 2 gives a short description of the overall information flows for electronic component distribution. A key aspect to overcome the current problems are standards for the representation of the component information and dictionaries for the description of the semantics of data. This is the topic of section 3. Finally, section 4 presents how the IST project MERCI addresses these areas. It is illustrated how standards and dictionaries are used as the foundation of the MERCI system architecture with the goal to exchange component information between the processing systems of component manufacturers and their customers.

2.   EXCHANGE OF COMPONENT INFORMATION

2.1   Electronic component information supply chain

In the supply chain of electronic components, three different actors play a role:
- Manufacturers of semiconductors normally sell their components (e.g. memory chips, processors, micro controllers) only to the biggest customers, their key accounts.
- Distributors supply components to the other customers (which still may be fairly big). Brokers provide further services, e.g. the timely provision of all the different components which are used for the production of a specific product. They also check for the availability of alternative components from other manufacturers and the prices of distributors.

– Component users buy their components from manufacturers or distributors. The design engineers use only qualified parts which have been approved by their component experts. Qualifying may also include the selection of specific manufacturers and/or distributors, e.g. on the basis of prices, availability, quality guarantees and internal rating of the manufacturer/distributor.

2.2   Information exchange along the supply chain

The component description is produced by manufacturers and normally provided in form of data sheets in pdf format, i.e. as text. Unfortunately, very often this is the only usable information which the manufacturer can provide to his customers. Many manufacturers do not have a well organised database containing the relevant data about their components with the possibility to generate the paper and Web documentation. Often it is the other way round: Manufacturers notice that there is a lack of data management and start to extract data from their own data sheets.

Distributors only refer to the Web page of the manufacturer, where the datasheet can be downloaded but they do not provide further component information. Often they provide some search databases to select the parts to be purchased. These kind of databases are also provided from independent parties with connection to several distributors.

At the end users side, component information has to be transferred into the design libraries which are used by the design engineers. This is normally a manually process which includes (dependent on the company and the domain) more or less tests and evaluation steps. Thus, very often data is recaptured again by qualified engineers. To have full control about the use of components, it is important to also have access to the commercial information about the component (normally stored in ERP systems), to the use of components in product assemblies (bill of materials), etc. That means, the component information used in the various systems throughout the company has to be linked.

An improvement of this situation has to start with the data organisation of the component manufacturers. This requires the integration of all publishing processes of the manufacturer on the basis of a single database containing all information about components. The source of component data has to become a database from which documents can be generated.

A second requirement is the use of standards: Only if the data exchange can be based on well accepted standards, data provided by several manufacturers can be understood and used by the component users. This will increase the possibilities of end users to easily select between the components from different manufacturers. This comparability also requires a clear description of the data semantics: Only if the customer can easily understand the meaning of a parameter, he can correctly select the right component without investing big research into the selection process.

Finally, for the component users it is of importance to have tools which allow the integration of component data into their design libraries and other databases and which keep the connection between the technical data and the business data (prices, availability, etc.).

3.   STANDARDS AND DICTIONARIES

3.1   Overview about PLIB

For the description of electronic components a lot of information is required. Dependent on the process which is fed, different pieces of information are needed:
- For the selection of the component, parametric data describing the external properties of the component are required.
- The integration into a CAD system is supported by the provision of a logical symbol which describes the logical connection points of the chip.
- For the physical design, the description of the physical pins and the mapping between the logical and physical pins are needed.
- For the manufacturing process, the physical description of the pins is required, for simulation of the product, specific simulation models are needed, etc.

Thus, a suite of different pieces of information is required, and a number of different standards have been developed for the various kinds of information.

As a basis for the representation of component information, MERCI uses the PLIB standard (ISO 13584). PLIB provides a dictionary concept for the description of the data semantics and it allows to represent the parametric data associated with components in various ways:
- In the *explicit representation*, each component is described by all its properties (similar to a table where each component is represented in a single row).
- In the *implicit representation*, all the components of a family are described by a set of mechanisms: Some tables describe the independent properties of the components, other tables represent dependent properties, some dependent properties are described by functions, etc. By using these mechanisms, it is possible, for instance, to factor out the common property values of a family and to specify them only once, or to specify

that the values of some properties are computed from the values of other properties. As such, it allows to model the organisation of data sheets which very often use this kind of implicit description of data.

An example illustrating these two ways of component modelling is given in figure 2: In (a) a family of screws is described. This family is defined by a table which correlates the possible diameter and length properties, a function which computes the width of the screw from its diameter, and a rule which indicates the relationship between the load and the diameter. On the other hand, for more complex components which do not exist in such a big number of variants, it is much easier to represent each component in a single row as illustrated in (b).

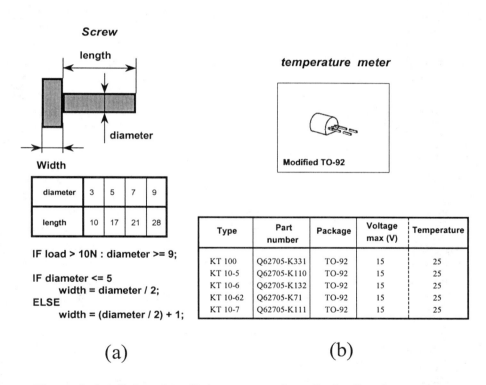

diameter	3	5	7	9
length	10	17	21	28

IF load > 10N : diameter >= 9;

IF diameter <= 5
     width = diameter / 2;
ELSE
     width = (diameter / 2) + 1;

Type	Part number	Package	Voltage max (V)	Temperature
KT 100	Q62705-K331	TO-92	15	25
KT 10-5	Q62705-K110	TO-92	15	25
KT 10-6	Q62705-K132	TO-92	15	25
KT 10-62	Q62705-K71	TO-92	15	25
KT 10-7	Q62705-K111	TO-92	15	25

(a)   (b)

Figure 1. Implicit and explicit representation of a family of components

Originally, PLIB provided mainly the implicit representation. Due to the requirements from MERCI, the explicit representation has been much more strengthened in the standard, and it has been fully implemented in the IS version of PLIB [ISO 13584/24].

In addition to the parametric data which are modelled directly by PLIB, there exist hooks to connect the supplemental information around a component in a well defined way. Thus, it is possible to use other standards in combination with the PLIB kernel and to provide this design data to the customers in the most appropriate format.

The PLIB standard has been defined formally by means of an EXPRESS information model [ISO 10303/11], which has been developed and standardised in the context of the STEP project (Standard for the exchange of product model data, (ISO10303)). This allows the automatic check of data with respect to the model and with respect to the dictionary.

3.2   Use of dictionaries

One central element of the PLIB standard is the use of dictionaries to clearly identify the elements used in the database. In general, a dictionary is a hierarchy of component classes. Associated to a class are the properties by which the components in this class are described, and properties are inherited along the hierarchy. Both for classes and properties, several kinds of information are provided to uniquely identify the meaning of the elements and its representation in databases (e.g. by the data type of properties). In a database or exchange file, the single components and their property values always refer to dictionary entries which describe their meaning and interpretation of codings.

For the domain of electronic components, a reference dictionary based on these concepts has been defined by the IEC [IEC61360/4]. This dictionary was supposed to be widely used, but unfortunately, the IEC claimed an intellectual property right on the dictionary without defining clearly the way how it can be used. As a result, various groups started to define and use their own dictionaries (e.g. ECALS, RosettaNet). In addition, each manufacturer uses internally his own dictionary which has been developed over a couple of years. Thus, we have to deal with a number of dictionaries.

There are two possible ways to deal with different dictionaries in a database:

1. Dictionaries can be extensions of each other. For instance, a manufacturer may choose to extend a reference dictionary (e.g. the IEC dictionary) by its own extensions.
2. Dictionaries can be independent from each other. In this case, a mapping has to be defined between the different dictionaries to make them comparable.

To handle these different situations, PLIB provides two mechanisms to combine dictionaries:

− A reference mechanism allows to connect one dictionary to another dictionary: by the "is-case-of" relationship, a class in a dictionary A can be linked to a class in a dictionary B, and it may inherit specific properties from the A-class.

– A mapping mechanism allows to map dictionaries: By the "a-posterio-case-of" relationship, classes in different dictionaries can be associated to each other and a mapping of the respective properties of these classes can be defined.

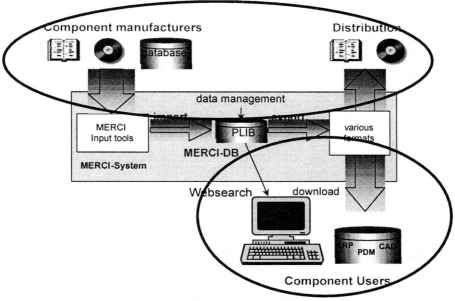

*Figure 2.*Overall vision of MERCI

4. MERCI: DEALING WITH COMPONENT DATA

4.1 Overview and goals

The goal of MERCI is to provide solutions and services to close the information gap between component manufactuers and component users and to support the direct flow of information between their processing systems. Therefore, MERCI provides tools to import data into the standard PLIB format and to generate data and documents in various formats from the standard PLIB format. This is illustrated in figure 2: Manufacturers provide their data in various forms, e.g. in databases, in textual data sheets, on CD, etc. MERCI transfers this data into the neutral PLIB format and stores the

results in a database. This database can be used both by the component
manufacturer and the component user: The manufacturer can generate
various document formats from the database and thus base his publishing
process on the database, and the component user can select components via
the Web, download the information in various formats and document types
and then integrate the downloaded data into his CAD libraries and corporate
databases. Thus, this data is available for his further product development
processes.

4.2   Use of dictionaries in MERCI

A central element of the MERCI system is the MERCI-DB which
contains the component data and the dictionary information. In the MERCI-
DB, dictionaries are used for several purposes:
- for the definition of the semantics of elements of the database
- for a hierarchical search in the component classes and for providing class
  specific query forms,
- for providing manufacturer independent and manufacturer specific
  queries
- for the database organisation of the component descriptions

For users, it is important to have the capability to search across several
component manufacturers. This could be best realised by providing a
common dictionary for all manufactures. On the other hand, different
manufacturers like to provide different properties for their components and
to use this as a means to differentiate themselves from their competitors. To
fulfil both requirements, MERCI deals with dictionaries in the following
way:

The central roles plays the reference dictionaries defined by the IEC.
However, each manufacturer has his own dictionary with a specific entry
point. But the manufacturer dictionaries are connected to the reference
dictionary, i.e. they inherit most of their properties from the IEC dictionary.
Only those properties are defined in the manufacturer specific part of their
dictionaries which are not already part of the reference dictionary. All
components are associated to a manufacturer dictionary (see figure 3).

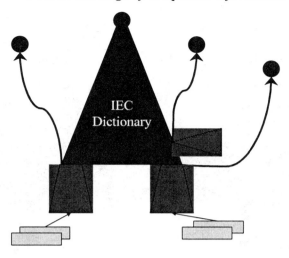

Figure 3. Dealing with dictionaries in the MERCI system

Users can use the reference dictionary for their searches across manufacturers. As long as they restrict their hierarchical search to the IEC dictionary, their query results will contain components from all available manufacturers. If they are interested to use the manufacturer specific properties for their search, then these properties are defined in the manufacturer dictionaries and are thus usable for manufacturer specific queries.

To make the dictionaries comparable, it is necessary to map their common properties to the IEC dictionary. Thus, with the import of data into the database, a mapping process needs to be initiated. The same is true for the export: Most component users also use their specific dictionaries and have to map data which they download accordingly. Thus, both import and export is accompanied by mapping processes, as illustrated in figure 4. But again, dictionaries help: The mapping is defined on the level of the dictionary elements, and the mapping programs on the data level are then generated automatically.

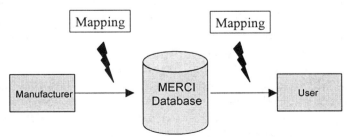

Figure 4. Import and export with mapping to dictionaries

4.3   Architecture of the MERCI system

The overall architecture of the MERCI system is sketched in figure 5.

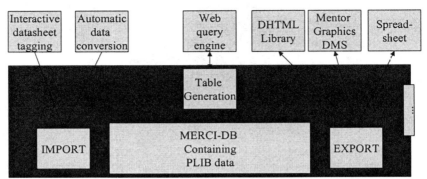

Figure 5. Overall architecture for the MERCI system

Around the PLIB compliant database, the system provides a middleware layer which contains generic mechanisms for dealing with the database. This layer is used by the tools and applications in the application layer to implement their services.

On the input side, component data can be captured. The goal is to get as much data as possible from existing sources. If these sources are structured, then it is possible to implement automatic conversion tools. But unfortunately, many of the data is kept in textual data sheets. This data can only be extracted from the data sheets in a semiautomatic process. For this purpose, a tagging tool has been implemented by MERCI which allows to associate elements of the text (like numbers) with meaning (i.e. the property which this number describes). As a result, a file is generated which contains this information in an explicit form. This serves as input for the MERCI DB.

The component data in the database can be queried via a powerful search engine through the Web. Information about selected components can be downloaded in various formats:

- Step File: This format has been defined by ISO in the context of STEP [ISO 10303/21], and it allows the exchange of data which are structured according to an EXPRESS model. This format is mainly used for the downstream import of data into a corporate database of the end user's system environment. For example, for MERCI an interface to the Mentor Graphics DMS system has been implemented. This allows the import of the downloaded STEP File into DMS from where the data can be imported into various CAD systems and a connection to PDM, ERP and BOM systems can be established.

- EXCEL file: The tabular information about the set of components can be directly imported into EXCEL spreadsheets.
- XML can be generated according to the XML binding of STEP [ISO10303/28], according to the SimPLIB DTD [SimPLIB] and according to the RosettaNet standard.
- DHTML library format: This set of dynamic HTML files contains a full component library which comprises the data about the selected components combined with the appropriate part of the dictionary. The DHTML library provides search capabilities for selecting components, it can print order forms or it can be directly connected to a distributor channel for online ordering of components. Thus, it provides similar functionality on a local basis as the MERCI database on a global basis. It is meant as a low cost means for small and medium companies to work with the data downloaded from the MERCI DB: They get a full working library system, but the only software they need is an Internet browser.

5. CONCLUSION AND OUTLOOOK

The paper has shown how the MERCI system exploits the PLIB standard and associated dictionaries for developing a system which allows a more efficient exchange of component data between component manufacturers and component users. The system will support the manufacturers to switch from documents as their source of information to data as the primary source – documents can then be regarded just as a representation forms of data. Based on this philosophy, both the publishing processes can be streamlined and better organised and the information exchange with customers can be improved.

The MERCI project will finish in the first half of 2002, and it is planned to install afterwards a commercial service for manufacturers and users of components providing tools and services for data organisation, document and Web publishing and data exchange. The current prototype and demonstrator can be visited on the Web under www.merci-project.net.

6. ACKNOWLEDGEMENTS

Part of this work has been funded by the EU under the IST project MERCI (IST-1999-12238). The author likes to thank the partners in the MERCI project for their contributions. Partners in the project are Adepa [F], EADS (Aerospatiale Matra Missile) [F], ENSMA/LISI [F], EPM

Technologies [N], Infineon [D], Rosemann & Lauridsen [D], Descon / Mentor Graphics [D], and University of Hagen [D]. The data conversion tool has been implemented by Rosemann & Lauridsen, the tagging tool and the DHTML library generator has been implemented by ENSMA/LISI, and the integration facility into the DMS system has been implemented by Descon / Mentor Graphics. The MERCI database developed at the University of Hagen is based on the EXPRESS Data Manager of EPM.

7.   REFERENCES:

[IEC61360/4]   IEC 61360-4: Standard data element types with associated classification scheme for electric components - Part 4: IEC reference collection of standard data element types, component classes and terms, 1997

[ISO13584/24]   ISO/IS 13584-24: Parts library : Logical Resources : Logical model of supplier library. 2001.

[ISO13584/42]   ISO/IS 13584-42: Parts library: Description methodology: Methodology for structuring parts families. 1997.

[ISO10303/11]   ISO IS 10303-11: EXPRESS Language Reference Manual. 1994

[ISO10303/21]   ISO IS 10303-21: Clear Text encoding of the exchange structure. 1994

[ISO10303/28]   ISO CD/TS 10303/28: XML representation for EXPRESS driver data. 2001

[SimBLIB]   Pierra, G.; Potier, J.-C.; Sardet, E.: From digital libraries to electronic catalogues for engineering and manufacturing. International Journal of Computer Application in Technology, July 2000.

VIRTUAL COMPOSITE – AN EFFECTIVE TOOL OF RESEARCH

Miloslav Kosek
Technical University of Liberec, Halkova 6, 46117 Liberec 1, Czech Republic,
e_mail: miloslav.kosek@vslib.cz

Abstract: The paper deals with analysis of the structure of textile composite. The quasi-periodic yarn shape, as basic structure unit, is modelled by the Fourier series that allows high flexibility, for example very effective approximation. Amplitudes and phase constants in Fourier series exhibit the distribution very similar to normal one. By the use of random number generators, virtual yarns can be generated and virtual composite realised from them. The structures are effectively visualised by a virtual reality system. The method of visualisation allows the immediately use of finite element method to predict the properties of virtual composite. Efficient model of virtual composite saves a lot of expensive experimental work.

Key words: Textile composite, composite structure, image processing, Fourier series, discrete Fourier transform, 3D imaging, virtual reality, statistics distributions, finite element method

1.   INTRODUCTION

Composites are new perspective materials because of their unusual mechanical properties. An important group of them are composites reinforced by a two-dimensional fabrics – textile composites. They are produced by a relatively simple technology. Several layers of woven cloth are impregnated with a suitable resin, placed one another, pressed and heated in a specified regime. The cloths, or yarns creating the cloths, are termed reinforcement and the resin is known as a matrix in composite terminology.

The basic composite feature, irrespective of its type and realisation, is that composite properties are not a sum or another analytical function of individual component properties, reinforcement and matrix, for example. Also the composite structure has a strong effect on its properties. This is a reason of extensive study of the composite structure. If the structure is known with a satisfactory accuracy, the composite can be analysed, for example by the finite element method (FEM), and its properties can be predicted from suitable models.

The three-dimensional (3D) structure of textile composite is determined by yarns forming the reinforcement. This structure during composite technological processing changes from almost regular ideal to quasi-regular real structure. A very convenient way of composite structure study are images of cross-sectional cuts made by standard metallographic technique. The yarn shape can be derived from the image. Among many ways of real yarn shape description, the most suitable appears the Fourier series.

By such an approach we get only random specific samples of quasi-regular real structure. A high degree of uncertainty is typical for this method. Therefore the information allows only to create a model of real composite structure, which is termed a virtual composite. Statistical methods should be a base for such model. The paper presents preliminary results necessary to realisation of the virtual composite.

2.   YARN SHAPE

Yarn is a basic unit of the textile composite, therefore composite modelling should start by yarn analysis. The first step of such analysis is yarn visualisation. As the cross-sectional cuts were long and narrow, their parts were scanned by LUCIA imaging systems and then the whole image was composed form partial ones. Our original HW and SW was used [1]. An example of typical image is in Fig. 1. Yarns, making the reinforcement, are visible by human eye, but not by a computer, therefore the yarn axis co-ordinate reading was made manually.

Figure 1. Image of composite cut

The second step is yarn analytic description. There is a lot of possibilities how to approximate yarn axis: polynomial [2], splines [3], Fourier series [4]. We preferred the use of Fourier series, as all its parameters have well defined interpretation and different level of approximation can be made by selecting lower or higher number of parameters. If the yarn axis is given as the function $y(x)$ defined on the cut of length L, the most suitable Fourier series has the form

$$y(x) = \frac{A_0}{2} + \sum_{n=1}^{N} A_n \sin\left(2\pi n \frac{x}{L} + \varphi_n\right)$$ (1)

where A_n and φ_n are the amplitude and phase constant of the nth harmonics, respectively. They were obtained by the discrete Fourier transform (DFT) of the axis points get manually in the previous step.

The use of Fourier series has other advantage – the yarn axis can be described ether in the co-ordinate or wave domain as it is typical in electrical engineering. The co-ordinate domain is the usual graphical description while the wave domain consists of amplitude and phase constant spectrum. Both the forms are demonstrated in Fig. 2 and 3. Only amplitude spectrum is shown for simplicity in Fig. 3. It must be stressed that both the forms contain the same amount of information, but the information is presented in different ways. Some features are pronounced in co-ordinate domain and others in wave domain.

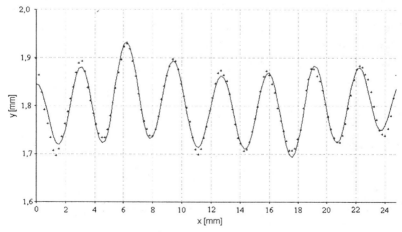

Figure 2. Yarn axis presentation in co-ordinate domain

It is evident from randomly selected yarn shape in Fig. 2 that the ideal sinusoid changes considerably during processing. Fig. 2 also illustrates the possibility of approximation. Only seven highest harmonics of spectra (see Fig. 3) were used and the agreement with experiment is satisfactory. Important harmonics have simple and straightforward interpretation. The highest, for example mth harmonics, determines the period, or wavelength, $\lambda=L/m$. Typically $m=7$. Harmonics near to maximum are responsible for the deviation from an ideal sinusoid shape. Low harmonics determine the distortion of the layer from its initial plane shape, see sinusoid axis in Fig. 2.

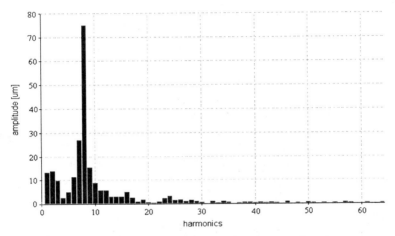

Figure 3. Amplitude spectrum of yarn – wave domain

If each yarn in the cut is approximated, we get an analytic description of the cut, which is probably the maximum information that can be get from the cut image. An example of the analytically described cut is in Fig 4. We see that yarns differ one another.

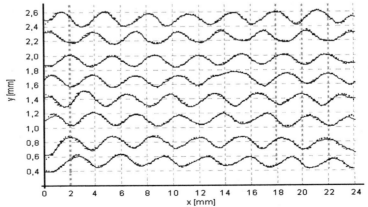

Figure 4. Approximated yarns in composite cut

3.   YARN STATISTICS

The method of random cuts yields yarns that are only a small random representation of all the yarns. Their parameters should be processed statistically. As a result of extended experimental work on plane weave composite, 96 samples of each harmonics were obtained. Since the amount of data is relatively low for statistical processing, non-standard procedures were used in parallel with standard ones.

As it is clear from equation (1), each harmonics is done by the amplitude and phase constant. Therefore the statistic parameters of both the amplitude and phase must be found. Only a few statistic parameters are important for us: the type of distribution and, if the distribution was found, its mean value and standard deviation. As for the type of distribution, we predict, from analysis of composite technology, that amplitudes exhibit normal and the distribution of phase constants is uniform.

Results of standard statistical SW were checked by our own programs and similar conclusions from all the systems were obtained: the phase constant distribution is uniform for all harmonics, as we expected. However, in contradiction to our prediction, normal distribution of amplitudes was found only in few cases. About 90% of harmonics exhibited the Weibull distribution. For completeness, in few cases also a log-normal distribution appeared and no other distribution was detected. All the distributions are subject on any statistics textbook.

As for statistical parameters, typically values of the variation coefficient of amplitudes is about 30%. This means a large dissipation of important amplitudes from yarn to yarn and large variation of structure from sample to sample. The parameters depending on the structure should vary roughly in the same extent.

4.   COMPOSITE MODELING

As the composite exhibits 3D relatively complicated structure, the computer graphics should be used for its modeling. There is a lot of methods of computer graphics which are applicable to the structure visualization. However, since final users should be textile related specialists, simple and efficient solution is necessary. As an ideal approach the Virtual Reality Modeling Language (VRML) was found for this case, since it allows many forms of structure inspection and the structure description is in a form of text file that can be quite complicated, of course.

Basic unit of composite structure is the yarn. Since its cross-section does not change, the yarn can be effectively modeled by a set of small cylinders.

270   *Miloslav Kosek*

The principle is illustrated in Fig. 5. From case a) to d) the number of
cylinders increases, while both their height and the distance between
neighbors decrease. Quite realistic image of the yarn can be get, see part d).

Another and in some cases better choice of model element exists [5]. We
preferred simple cylindrical element, since the geometrical model should be
used in the FEM. Furthermore, cylinders are basic elements of VRML.

Almost all ideal 2D and 3D textile structures can be simply modeled in
this way and this approach is widely used [6]. The top of structure hierarchy
is 3D model of an ideal composite that is shown in Fig. 6.

Figure 5. Principle of the yarn model

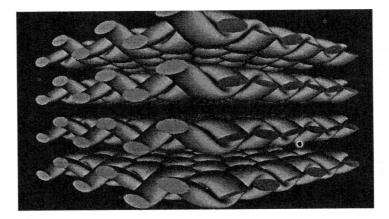

Figure 6.   Ideal composite

5. VIRTUAL COMPOSITE

The only knowledge on composite structure that can be get from the method of cuts, is the statistical distribution of amplitudes and phase constants and parameters of these distributions. It has been already mentioned that the most frequent amplitude distribution was the Weibull one, about 90%. However, we have found that there is little difference between Weibull and normal distribution from the practical point of view in most cases. Therefore, we have used standard generators of uniform distribution to generate phase constants and MATLAB generator of normal distribution to get amplitudes. By the use of Fourier series (1) the virtual yarn can be realized from the generated random amplitudes and phase constants. An example of a virtual cut is in Fig. 7. If we compare the virtual cut with a real cut in Fig. 4, we see that the virtual structure exhibits higher irregularity than the real one.

Figure 7. Cut in virtual composite

In principle the virtual textile layer and set of layers, virtual composite, can be realised by the same way. Practically, the generation of virtual textile layer is extremely difficult, since the virtual yarn intersections must be avoided. By another words, the second perpendicular set of yarns in the layer must respect a lot of constraints. This problem is not present in the case of ideal regular structure.

One of solutions of this problem supposes non-standard application of advanced methods of computer graphics to find intersections and then modify the yarn shape to avoid the intersections. Simpler solution is to find constraints from the first set of yarns and then generate random yarns of the second set until all the constraint condition are solved. The second solution is in progress.

6.   CONCLUSION

The key step of virtual composite realization, inspection and processing is the reliable virtual yarn generation. Results from our simple approach reveal that there are some differences between cuts of real and virtual composite. The reason of the difference is probably the improper use of statistical data. Especially the yarn shape is very sensitive to phase constants in Fourier series. In order to get a better insight into phase constant distribution, phase shifts relative to the most important harmonics (usually the 7th or 8th harmonics) should be treated instead of absolute phase constants. Also comparison of random amplitudes with real ones is necessary. May be, lower limits of random number generators should be applied. On the other hand the preliminary results confirm that the used method is correct, but it requires corrections and refinements.

The generation of virtual composite is a complicated task that requires a lot of non-trivial work. However, the solution of this problem has a high value. It will make possible to generate structures very similar to real ones without expensive experiments. The virtual composite structure can be analysed by standard methods, for example by FEM. Such an approach allows to predict properties of real samples with satisfactory agreement with experiment. Furthermore, limiting cases of structure, practically impossible from experiment, can be generated to find composite behaviour in such seldom and extreme situations.

In parallel to virtual composite realisation an experiment starts that will try to get a detailed structure of a few real samples of composite. The procedure is to cut off very thin layers and make the image of the surfaces. Complete 3D structure of real composite can be composed from the images and compared with the structure of virtual composites. Unfortunately, this very extensive and time consuming experiment will have at its output only few real structures. Its main result will be to verify the method used for virtual composite.

ACKNOWLEDGEMENTS

This study was supported by the Grant Agency of the Czech Republic within the grant project No. 102/00/0696.

REFERENCES

[1] Kosek M., Myslivec M. Simple Graphical System for Determination of Actual Yarn Geometry of Textile Composites. Proc. of Spring Conference on Computer Graphics, Budmerice, Slovakia, April 28th – May 1st, 1999, pp. 25-26.

[2] Yurgartis S.W., Morey K., Jortner J. Measurement of Yarn Shape and Nesting in Plain-Wave Composites. J. Composite Sci & Technol., 1993;1:39-50.

[3] Vopicka S., Koskova B., Glogar, P. Approximation of Actual Yarn Shape in Textile Composites Using Cubic Splines. Proc. of 3rd Internat. Conference "Textile Science'98", Technical University in Liberec, Czech Republic, 1998; vol. 3, pp. 580-583.

[4] Kosek M., Koskova B. Analysis of yarn wavy path periodicity of textile composites using discrete Fourier transform. Proc. International Conference on Composite Engineering, ICCE/6, Orlando, 1999, pp. 427-428.

[5] Groller E., Rau R.T., Strasser W. Modelling and Visualisation of Knitewear, IEEE Trans. on Visualisation and Computer Graphics, 1995;4:302-310.

[6] Lomov S.V., Huyasmas G., Luo Y., Prodromou A., Verpoest I. Textile geometry preprocessor for meso-mechanical and permeability models of composites, Proceedings of 9th European Conference on Composite Materials (ECCM-9), Brighton, UK, June 4-7th 2000. CD-ROM.

A Virtual Glovebox for the Digital Enterprise

Christian Seiler, Ralf Dörner
Fraunhofer Applications Center for Computer Graphics, D 60486 Frankfurt am Main, Germany
{cseiler, doerner}@agc.fhg.de

Abstract: The employment of simulation technology together with the exploration of virtual environments promise to be a unique advantage for digital enterprises in diverse fields like planning, rapid prototyping, marketing and presentations, as well as information dissemination and training. However, there are serious obstacles that inhibit digital enterprises to take a competitive advantage when using 3D for business processes. Most of these obstacles are concerned with the fact that 2D interaction metaphors are used for interacting with 3D content. With the motivation to help a digital enterprise to benefit from the advantages of applying 3D and virtual reality for different purposes and different user groups, this paper introduces an innovative dedicated interaction device for the digital enterprise. It employs the metaphor of a glove box where stereoscopic visualization together with haptic feedback is given in a natural manner. This makes it intuitive to manipulate three-dimensional content and makes it easy for users to navigate and to orient themselves in a virtual 3D world. Beside presenting the Virtual Glove Box device, we will discuss its advantages for new industrial organisations and deal with the question how to integrate it in existing structures and processes of an enterprise.

Key words: Virtual Glove Box, Input and Output Devices, Digital Enterprise, Three-Dimensional Content, Virtual Reality, Haptic Feedback

1. INTRODUCTION

Economic trends like globalisation lead to new organisation structures of enterprises that depend heavily on digitised information and digital data. The employment of simulation technology together with the exploration of virtual environments [4] promise to be a unique advantage for these digital enterprises in diverse fields like planning, rapid prototyping, marketing and presentations, or information dissemination and training. This is specifically true if we focus on 3D which is important since many products are inherently three-dimensional and their spatial layout is important (e.g. assembly of parts, planning in the field of pipeline construction). However, its up to modern digital enterprises to overcome obstacles in the usage of 3D in order to take a competitive advantage when using 3D for business processes. Most of these obstacles are concerned with the fact that 2D interaction metaphors are used for interacting with 3D content. This makes it non-intuitive to manipulate three-dimensional content and is a reason why most people find it difficult to navigate and to orient themselves in a virtual 3D world [3]. In addition, there is no haptic feedback and the 3D world is displayed on a 2D screen although haptic feedback devices as well as stereo systems are more and more available in low priced versions.

With the motivation to help a digital enterprise to benefit from the advantages of applying 3D and virtual reality for different purposes and different user groups, this paper introduces an innovative dedicated interaction device for the digital enterprise. This interaction device overcomes limitations of traditional interaction devices even the ones that are dedicated to 3D. In the paper we will present and describe this interaction device, called *Virtual Glove Box*. Moreover, we will discuss its advantages for new industrial organisations and deal with the question how to integrate it in existing structures and processes of an enterprise.

The paper is organized as follows. In the next section we present how the Virtual Glove Box device is built and used. In section 3 we will discuss how the Virtual Glove Box can be seamlessly introduced and integrated in a digital enterprise. Finally, section 4 gives an example application of the Virtual Glove Box.

2. THE VIRTUAL GLOVE BOX INTERACTION DEVICE

Although there was a constant advance in 3D computer graphics over the recent years, 3D computer graphics had not much impact on the digital enterprise. Most 3D Computer graphics applications stem from CAD products and are often not interactive. Some of the main reason for the lack of 3D computer graphics were its demand for dedicated hardware, the heterogeneity of the hardware used in the enterprise and the lack of adequate I/O devices to interact with 3D graphics.

With the dramatic increase of 3D performance of low cost graphics hardware the first obstacle for 3D graphics in the digital enterprise is about to disappear. The problem of heterogeneity an interaction, however, still persists. While we address the heterogeneity problem in chapter 3, this chapter will focus on the interaction problem.

3D virtual worlds, as they are created by computer graphics applications, display non existing scenes and objects in a way that mimics our real three-dimensional world. While the look of such a virtual world can be very convincing, the "feel" of such a world usually is not. In order to interact with a virtual scene the user has to use I/O devices as interfaces between the user and the virtual domain. The I/O devices most often used for such a task are mouse and keyboard. These devices are best suited for the 2D desktop metaphor which is widely used for 2D applications, but not appropriate for 3D applications. There are several I/O devices which improve the navigation in a 3D world, like joysticks or the space mouse, but these devices are limited in their usage when it comes to moving, touching or manipulating virtual objects.

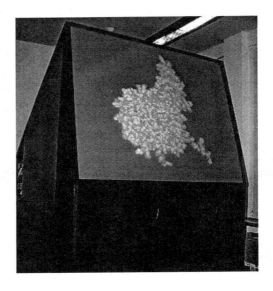

Figure 1: The Virtual Glove Box Device

Another drawback of those devices is that they lack the possibility to display a certain property that most of our real life objects possess intrinsically: Haptic behaviour and force-feedback. While it is commonplace in VR applications to walk or grasp right through solid objects, this is not so in real life. So called haptic or force-feedback devices try to overcome this limitation. But these devices still have several drawbacks, e.g. the PHANToM device [8][9] allows only an indirect manipulation of 3D content, while exo-skeleton devices like the CyberGrasp [1] are not sufficiently integrated in the visual representation of 3D content.

In order to overcome the problems and obstacles described above, we introduce a new I/O device called the Virtual Glove Box. A glove box is an apparatus used in chemistry or biology to work on targets in a closed atmosphere, without contaminating the substances or endangering the user. A glove box usually consists of a transparent hull in which the experiments can be performed. The user can reach into the box through gloves which are attached at the inside of the box. The gloves can be reached through corresponding holes in the box.

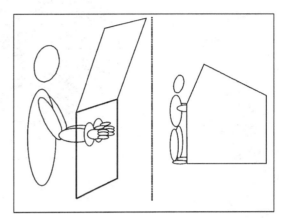

Figure 2: Using the Virtual Glove Box

We used the glove box metaphor to build our Virtual Glove Box. The Virtual Glove Box combines a stereoscopic display with haptic displays. The stereoscopic part consists of a two projector system with polarized lenses which continuously display images for each eye. The user views the scene through matching polarized glasses. Directly below the display frame are two openings through which the user can reach into the box. The user cannot see his real hands but rather a computer generated representation which mirrors the users motions (see Figures 1 and 2).

We achieve haptic feedback by using a Cyber Grasp exoskeleton for each hand [1] (see Figure 3). In this way the user can actively reach into the virtual domain to grab and move the virtual objects there. The whole Virtual Glove Box prototype consists of the glove box mock-up which holds two LCD projectors with polarization filters (one horizontal, the other vertical polarization) and the screen which does not interfere with the polarization of the beams. Inside the box is the emitter of a Polhemus Fastrak magnetical tracker [5] and the two exoskeletons for the hands each equipped with the corresponding receiver for the tracker. In this set-up the position of the hands can be consistently tracked and processed by the application. Both the visual

and the haptic display is driven by an IBM PC workstation running
Microsoft Windows 2000. The computer features an Intense3D Wildcat
4210 graphics card which is able to drive both projectors. One of the
technical challenges in this set-up was to avoid interference between the
users motions and the optical display.

Figure 3: Haptic Feedback Device [1] built in the Glove Box and hidden from the user

3.   SYSTEM INTEGRATION OF THE VIRTUAL GLOVE BOX

Although the new I/O device offers many benefits, it is unlikely that the
Virtual Glove Box supersedes the traditional infrastructure in an enterprise,
because of the space requirements of the device and the higher costs
compared to conventional hardware. Therefore care must be taken to
integrate the new device into existing platforms and structures through
dedicated software. It is important that results and models which were built
using the Virtual Glove Box are also available to applications which run on
traditional and possibly heterogeneous systems.

We address these problems by using an approach derived from
methodologies from software engineering: Applications for the Virtual
Glove Box are built as frameworks wherein so called components can used
to build complex virtual scenes and simulations [2][7]. Frameworks are
skeleton applications which provide the adaptor to different hardware
configurations or application backgrounds (e.g. Virtual Glove Box or
traditional equipment and a development application versus a trade fair
presentation). A framework comes to live through components. Components
are reusable software modules which encapsulate certain attributes and
features of the objects they describe. In the Virtual Glove Box context there

might be visual and haptic components describing for example pipes for a chemical plant or other non-visual components which encapsulate conventional simulators for example. Since the components are platform and application independent information losses can be minimized.

4. APPLICATION EXAMPLE

The Virtual Glove Box enhances the immersion since the haptic exoskeletons which are worn by the user are hidden from view by the apparatus and presents thus a holistic approach for immersive VR devices with haptic feedback.

The Virtual Glove Box offers a multitude of different fields of application in a digital enterprise. The foremost area of application as described above is the planning, modelling and simulation of three dimensional objects from production plants, machinery, pipeline systems etc. to structures on a microscopic level and smaller like molecular structures. The usage is not limited to real objects, abstract systems like work flow diagrams can be experienced likewise.

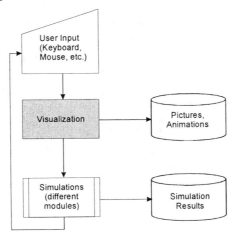

Figure 4: Traditional Approach

A very important aspect of the device is the communication presentation and visualization of new ideas or solutions, which opens the usage of the Virtual Glove Box to a many different sets of users. The expert benefits from the intuitive interaction because he can concentrate on his domain competences (e.g. design or mechanics) rather than learn to interact with complicated authoring systems. The same benefit is shared by users from the management, marketing or by customers: things are easier to communicate and to understand if you have a "hands-on "experience.

The usage of a real life interaction metaphor like a glove box drastically increases the user acceptance for such a new device. This is especially true for applications in a digital chemistry or pharmaceutical enterprise, where the glove box metaphor stems from. We demonstrate the usage and the benefits of the Virtual Glove Box device and concept in the field of molecular modelling applications. Molecular modelling describes the technique of designing chemical compounds with the help of the computer. These chemical compound, i.e. molecules etc. are modelled in order to achieve a predefined behaviour. Examples might be the design and research of protein folding structures or the design of specific molecules that interact with such proteins in a desired way (key-lock problem). The traditional notation of chemical structures uses characters and lines to describe a molecule. Such a description is essentially two dimensional. However the molecules and drugs encountered in the biochemistry or pharmaceutical industry cannot be fully described in such a notation since their functions derive from their three dimensional structure. Therefore the primary interest in molecular modelling is the three-dimensional nature of various molecules. In order to visualize such structures for molecular modelling the visualization has to be a 3D visualization which is expressive for the field of interest.

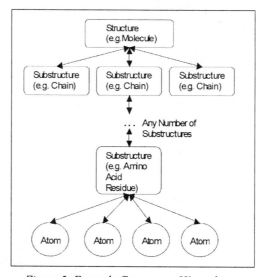

Figure 5: Example Component Hierarchy

In order to improve the process of molecular modelling and to find a solution to stated problems it is necessary to do an analysis of the role of the user in a molecular modelling application. The user, mostly a domain expert in the field of chemistry, biology or pharmacy, combines known chemical reagents, parts of molecules or even single atoms to form new structures. For this task exchanged forces or constraints like bond angles etc. have to be taken into account. Hence the composition of the 3D scene is twofold: it is determined partly by physically based behaviour of the entities, partly it depends on the user and his input (see Figure 4). Since the user is working with molecules and atoms as building blocks of the scene, a component based approach is very well suited to the task. In this concept atoms form basic components from which higher level components like molecules can be built (see Figure 5).

The process of molecular modelling is essentially an interactive simulation. The components used in this approach are therefore very similar to those in different simulation centred applications like e.g.. logistic simulations. When the user interacts with the components using the Virtual Glove Box device, he can make use of all aspects of the component, the visual representation, the simulation and the haptic behaviour (see Figure 6). The same components and the results from the work with the Virtual Glove Box can be transferred to another application running on conventional hardware. While the haptic behaviour is still available on the other platform the application framework might not access this feature, because there is no suitable I/O device. In another framework there might be no need for the simulation capabilities but only for the visual representation of the components. Thus we have shown that the Virtual Glove Box component concept adapts very well for different environments and is therefore well suited to be integrated into an established digital enterprise.

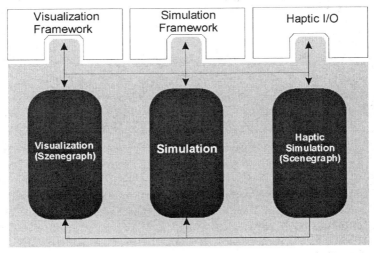

Figure 6: Inner Structure of a Component with Haptic Simulation

5.   CONCLUSIONS

In this paper we introduced a new interaction device for manipulating 3D content that responds to the necessities of a digital enterprise. We have shown that it is crucial to design these kind of high-level, holistic devices in order to take a competitive advantage in using 3D and virtual reality. We identified three main requirements for such an interaction device:

–   It has to be designed based on high-level interaction metaphors people are already used to and it has to actively fight acceptance problems

–   It has to be holistic and multi-modal

–   Not only the hardware-side of the interaction device needs to be taken into account during its design but also the software side (e.g. integration in existing software, application of component and reuse concepts).

If these requirements are met chances are high that productivity can be raised while time can be saved since there is a direct, intuitive interaction and a lower cognitive load associated with the usage of 3D. In addition, a digital enterprise can be achieved as acceptance in dealing with digital data is higher. And the benefits can be used for different groups of users (ranging from technical experts to end customers) and different purposes (ranging from product design to product presentation).

6.   REFERENCES

[1]   CyberGrasp by Virtual Technologies, Inc:
       http://www.virtex.com/products/hw_products/cybergrasp.html

[2]   R. Doerner, P. Grimm: Three-dimensional Beans - Creating Web content using 3D
       components in a 3D authoring environment, Proceedings of the Web3D-VRML
       2000 fifth symposium on Virtual reality modeling language, Pages 69 – 74, 2000

[3]   R. W. Lindeman, J. L. Sibert and J. K. Hahn: Towards usable VR: an empirical
       study of user interfaces for immersive virtual environments; Proceeding of the CHI
       99 conference on Human factors in computing systems: the CHI is the limit, Pages
       64 –71, 1999, Pittsburgh, PA USA

[4] V. Luckas, R. Doerner: Experiences from the Future: Using Object-oriented Concepts for 3D Visualization and Validation of Industrial Scenarios, ACM Computing Surveys, Vol. 32, Issue 2, p. 38-44, ACM Press, 2000

[5] Polhemus Fastrak, http://www.polhemus.com/ftrakds.htm

[6] K. Saliburym D. Brook, T. Massie, N. Swarup, C. Zilles: Haptic rendering: programming touch interaction with virtual objects; Proceeding of the 1995 symposium on Interactive 3D graphics, Pages 123 –130, 1995, Monterey, CA USA

[7] J. Sametinger: Software Engineering with Reusable Components. Springer Verlag, 1997

[8] M. A. Srinivasan, C. Basdogan: "Haptics in Virtual Environments: Taxonomy, Research Status, and Challenges", Computers and Graphics, Special Issue on Haptic Displays in Virtual Environments, Vol. 21, No. 4, IEEE Press, 1997

[9] Y. Yokokohji, R. Hollis, T. Kanade: Vision-based Visual/Haptic Registratration for WYSIWYF Display, International conference on Intelligent Robots and Systems, IROS ′96, Pages 1386-1393, 1996

Web-based Information Models to Support Product Development in Virtual Enterprises

Arturo Molina, José L. Acosta, Ahmed Al-Ashaab, Karina Rodríguez

Concurrent Engineering Research Group, CSIM/DIA, ITESM Campus Monterrey, E. Garza Sada 250, Sur. C.P. 64849, Monterrey, N.L. Mexico. armolina@campus.mty.itesm.mz

Abstract: Two information models have been defined to be key elements to support Product Development, namely: Product and Manufacturing Models. The Product Model contains all data related to a product's life cycle. The Manufacturing Model describes the capabilities and capacities of manufacturing facilities. The objective of these two information models is to assist simultaneous design teams by providing access to consistent sources of product and manufacturing information, and therefore allow computer applications to be developed to support concurrent design.

In a Virtual Enterprise, product development has to be carried out by distributed teams of different companies, therefore there is a need for these two models to be available in an environment which can be shared regardless where the teams are located. The use of Web technology has been defined to be useful to share product information and manufacturing technical capabilities among global companies. A Product and Manufacturing Models have been developed in prototype systems based on an object oriented database using Web technology to support the following application system: Design for Injection Moulding and Selection of suppliers for Product Development in a Virtual Enterprise.

Key words: Information Models, Knowledge Model, Virtual Enterprise, Web technology, Object Oriented Database.

1. INTRODUCTION

Increased competition in the global market place has forced companies to look for methods and technologies, which will improve the quality of their products, reduce the product development cycle and reduce production costs. Competitive advantage is driven by three major strategies: achieve mass customisation, become low cost suppliers and create leading edge products. These strategies in a global manufacturing environment must be implemented through the use of best practices (e.g. Concurrent Engineering) and computer technologies (e.g. Web-based engineering applications). However it requires effective collaboration and efficient management of product life cycle information in order to support engineering decision-making happening in geographically distributed enterprises [3]. Internet technologies addresses this need by providing services to share and transfer the knowledge and information across networked enterprises. Internet-based applications are designed to allow product information to be shared between extended enterprise partners. This comprises OEM (Original Equipment Manufacturers), supply chains, subsidiaries, consultants, and partners affiliated with the product's life cycle. Key issue in this context is Product and Manufacturing information and knowledge provision. The challenge is to allow accessibility to all parties involved in the process of a product development. This paper describes how Product and Manufacturing information/knowledge can be structured and organised in information models to support global product development. Web-applications based have been developed in order to demonstrate how these information models can be used in a global collaborative environment. Case studies are presented of how these models can support Product Development in a Virtual Enterprise, specifically: Design for Injection Moulding and Selection of suppliers for Product Development.

2. MODELLING INFORMATION AND KNOWLEDGE FOR INTEGRATED PRODUCT DEVELOPMENT

Modelling information and knowledge for product development is key for successful engineering collaboration in a global environment. Efforts in modelling information to support product development has concentrated in [9]:
1. Modelling product information to represent the important aspects of products throughout their life cycle i.e. requirements, design, manufacturing, production, packaging, distribution and recollection.

2. Modelling manufacturing information in order to capture the capabilities and capacities of manufacturing processes and resources of a particular facility.

The research reported here describes methodologies for modelling product and manufacturing information and knowledge.

2.1 Modelling Product Information

Research related to model product information has evolved in computer-based environments, which support the integration of automated applications. Details can be found in Krause et al. [6]. The Product Model should represent all the information needed throughout the life cycle of the product i.e. requirements, design, manufacture, production, assembly, packaging, distribution and recollection. The methodology developed and applied in this research has the following phases:

1. Design process analysis and documentation using IDEF0.
2. Analysis of product data and information
3. Product Model Configuration: function, solution and physical models.
4. Product Model Construction and documentation using EXPRESS
5. Product Model implementation using Object Oriented Database

The use of IDEFO model in the analysis and documentation is key to capture and formalise the company's design process. The graphical representation of the process established a common ground for discussion between the members of the engineering team in charge of the Product Model construction. Experiences with the application of IDEF0 in this task has shown that this model offers an overall picture of the process by representing the product information flow (e.g. reports, drawings, technical data), control points (e.g. checklist, formats, standards), responsibilities of engineers, and technology support (e.g. CAD, CAE, Databases). In addition, it is also useful for the identification of areas of opportunity for the improvement of the design process. Once the IDEF0 model has been constructed and approved by the SE team, the product model configuration takes place based on the product information analysed and recollected during the previous stage. Three important models have to be structured and constructed: function, solution and physical (figure 1). The description based on a function must satisfy user requirements and is expressed in those terms, in figure 1, for example in glass bottle design (e.g. volume, weight). The solution model has two descriptions: possible solutions to satisfy the function of a product and analysis required for satisfying different aspects of the life cycle. For example, the bottle design requires thermal and blow analysis Finally, the physical model describes how the product is structured in assemblies, sub-assemblies, components, parts and materials. The bottle is one product but have different regions for design: crown, neck, elbows, body and bottom.   Relationships between all these elements must have a parametric definition e.g. how geometric variables related to each other angles and radius. In order to comply with industry standards all information have to be documented using EXPRESS, in addition wherever possible STEP definitions have been used to represent geometric, material, tolerance and topology data. An EXPRESS model can easily be mapped into the Object Oriented Database language due to the flexibility of the language to represent all the information types defined by EXPRESS i.e. entities, lists, bags, arrays, etc.

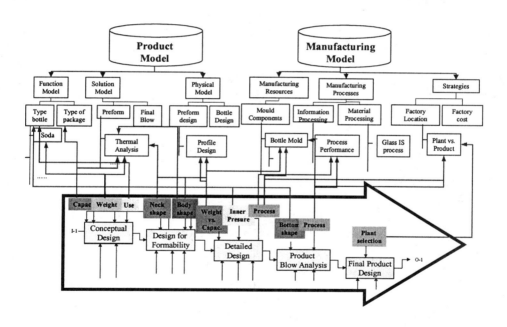

Figure 1. Structure and relationships of Product and Manufacturing Models.

2.2   Modelling Manufacturing Information

The rapid development of computer-aided information technologies has triggered the development of a second type of models, which complements the information required to achieve concurrent design. These information models represent factories i.e. machine tools, tools, manufacturing processes, material flow, orders, etc. The basis for the research into this type of data models came from original work done by the ESPRIT Project IMPPACT on a Factory Model [5]. Later work on Facility Model [7], Manufacturing Model [1, 10, and 11] has underpinned the work presented in this paper.

The Manufacturing Model must represent the manufacturing capability and capacity of a facility. The manufacturing information entities: resources, processes and strategies. The manufacturing resources and processes describe generic capability information. Different companies can use similar types of manufacturing resources and processes. In fact, two companies can have the same type of technology, i.e. manufacturing resources and processes. Nevertheless, the manufacturing facility of each of these companies could perform in a different way because of the company's decisions on how to organise and use those resources and processes. This aspect must be captured to truly represent the manufacturing capability of a company. The manufacturing strategies represent this company specific information which allows a company to specify how the resources and processes are organised and used in order to support the company's manufacturing function (figure 1). The methodology has the following phases:

1. Identification of manufacturing resources and processes.
2. Manufacturing Model configuration: resources and processes.
3. Identification of manufacturing strategies.
4. Establishing relationships between: resources, processes and strategies
5. Manufacturing Model construction and documentation using EXPRESS
6. Manufacturing Model implementation using Object Oriented Database

The IDEF0 process model has already defined resources and processes as mechanism used in product development activities. However there is the need to include relevant information regarding resources (e.g. physical dimensions and restrictions), processes (e.g. capacity, capability, and behaviour) and strategies (e.g. factory location, cost and performance). For example, continuing with the example of bottle design, it requires a mould to produce the bottle and the glass processing machine (e.g. IS machine). The mould has to be designed and manufactured according to bottle specification but historical information of moulds allows designers to design better mould using best practices guidelines. It is important to have information regarding the performance of a mould, machine and process. Then it is possible to predict the behaviour of a mould in a specific machine and process. This information is included in the Manufacturing Model. Strategies are related to the selection of the best facility to run a specific mould to product a bottle. This decision has to be based on different factors: localisation (i.e. close to customer), facility performance (i.e. historical performance of family of moulds), and facility cost.

The modelling methodology defined above allowed the authors to have a systematic approach to capture and represent information readily to be documented in EXPRESS and then be programmed in the object-oriented database. Details can be found in Molina and Bell [11].

2.3   Modelling Knowledge: Integrating Product and Manufacturing Information

The product and manufacturing information modelled can be used during the product development at different stages where a diversity of data and information can help engineering decision-making. However there is a need to create a higher level of support, i.e. knowledge based engineering. Functions that relate different variables of both models to construct the knowledge model. These functions are different types of relation that can describe: how the thickness of a bottle will affect the glass filling, or thermal behaviour of the mould, or how a radius or angle will created an area of concentration of glass. All these knowledge can be described in three forms:

1. Based on rules. There are two types of rules: Boundary rules and Conditional rules. Boundary Rules, which a specific pre-set range is determined for a particular variable (i.e. the pressure inside a pipe system has to be in between 80 to 120 psi). Conditional Rules, which are does who modify variables based on input parameters (i.e. the diameter of a drilling tool has to be the diameter of the hole minus 0.005 inch).
2. Based on Cases: Cases are the description of solutions of specific problem based on set of well-bounded variables. For example, best set-up of a machine based on historical data, forecast performance of a mould and solving production problems of low efficiency.
3. Based on Engineering Models: Knowledge based on mathematical relations that construct formal engineering models is considered within this group. This type of knowledge could be seen as a "black-box" that receives input variables and returns a specific scenario or analysis of a certain group of variables. Commonly knowledge based in models is related with knowledge based on rules, because the results of the first will need further analysis to make a decision. Some examples of this type of knowledge are fluid flow analysis inside a mould, calculating the maximum flow, the critical cross sectional area and the critical pressure location. The corrosion vs. time analysis at the inside walls of a glass

furnace; by using a mathematical model it is possible to obtain a curve with important critical values.

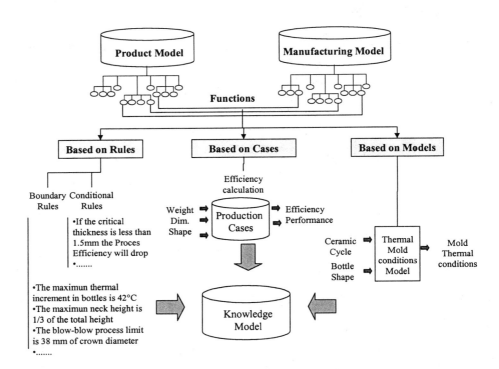

Figure 2. Building a Knowledge Model based on rules, cases and engineering models.

An example in relation to the glass bottle development is presented in figure 2. The rules mentioned are divided in boundary and conditional rules. The first one represents knowledge that could be included in further fuzzy logic analysis. The cases and models shown are simplifications of real analysis tools implemented in a glass company. This Knowledge model is embedded in the Product and Manufacturing Model implementation.

3. WEB-BASED INFORMATION TECHNOLOGY FOR VIRTUAL ENTERPRISES

The Product and Manufacturing Models can be used in product development in Virtual Enterprises. VIRPLAS, which stands for VIRtual industry cluster for PLAStics, was created to explore how the concept of Virtual Enterprises could enhanced the regional development of the plastic industrial sector of Monterrey, Mexico [4]. VIRPLAS has six members with different competencies within the plastic industry. One company is focused on product and mould design, another has capabilities to design and manufacture injection moulds components, two companies are on the business of injection moulding, and one company is specialised in commercialisation of plastic products and machinery. All these companies are Small and Medium Enterprises (SMEs). (www.mexican-industry.com.mx).

In order to create plastic products, one of the six members (Partner A) will be in charge of the part design; Partner B will be in charge of mould design. Partner A and B will use the Web-based Design for Injection Moulding System (SPEED) to design the product and mould. In order to select the companies which will produce the product, Partner A will use the Web-based Manufacturing Model to select the members of VIRPLAS which have the capabilities to inject the product. Partner C and Partner D will receive the mould from Partner B and will produce the part. As soon as the parts are manufactured they will be sent to the end-customer.

Both Web-based systems are based on the use of Internet and object oriented database technologies. The system was developed with Java and use Object Store OODBMS in order to provide efficient retrieval of the data and management of the complex data model.

The databases reside on a Sun Ultra 5 Server which works as a Web Server. The interfaces to access the SPEED and Manufacturing Model prototype systems were developed with Java applets and can be accessed through any common browser, i.e. Netscape or Explorer.

3.1 INJECTION MOULDING PRODUCT DEVELOPMENT

SPEED (Supporting Plastic enginEEring Development) is an information system based on INTERNET that captures the capabilities and characteristics of the injection moulding process to ensure the manufacturability of the plastic product and to define its production resources [2]. Three engineering applications within the SPEED prototype system have been developed. They are:

1. Design for Mouldability for the plastic product.
2. Support for the mould design and selection of components.
3. Selection of injection moulding machine.

The use of SPEED, is very important for the collaboration of VIRPLAS companies, this application can be customised according the requirement of each one of the companies and has the ability of extend as far as the user needs, see figure 3 for an example of product development (http://speed.mty.itesm.mx).

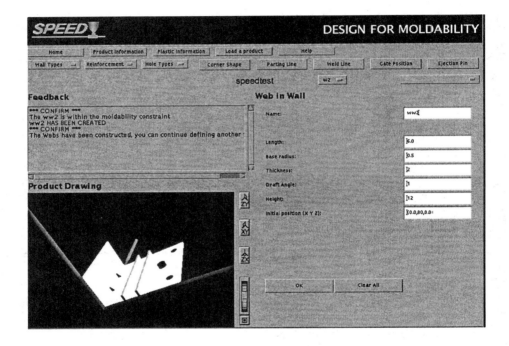

Figure 3. SPEED product development process

3.2 MANUFACTURING MODEL FOR SUPPLIERS SELECTION IN A VIRTUAL ENTERPRISE

The Web-based Manufacturing Model was implemented with the idea of supporting selection of suppliers in Virtual Enterprise. The prototype system allows capturing manufacturing information related to resources, processes and strategies of a company. The information about a factory can be capture at different levels within the organisation, for example Factory, Shop floor, Cell and Station. This is important because the model also supports the organisation of this information for the companies in the database. Figure 4 shows the interface for data entry in the Manufacturing Model.

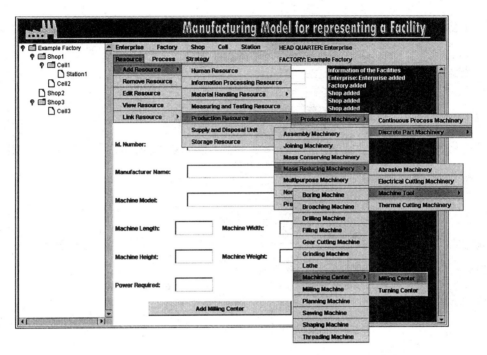

Figure 4. Manufacturing Model data entry interface

Queries can be performed to find machines and processes in the database, also companies that are capable of performing a process o have a specific resource. Sometimes the search for suppliers in a Virtual Enterprise requires to query in specific clusters, for example all companies that can process plastics or are die and mould makers, and so on. Figure 5 shows the query interface that can be accessed in the Virtual Industry Cluster homepage (www.mexican-industry.com.mx).

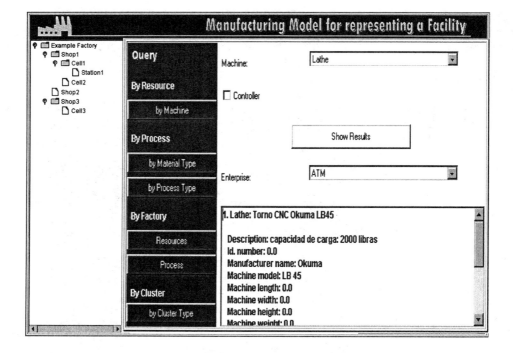

Figure 5. Manufacturing Model query interface

4. CONCLUSIONS

In summary this paper describes two methodologies developed to define and create two information models required to support integrated product development i.e. Product and Manufacturing Models. Also the definition of a Knowledge Model which relates both models has been described. These methodologies have been proved to be effective and feasible for the development Web-based prototype systems to support product design in Virtual Enterprises: SPEED and Web-based Manufacturing Model. These prototype systems have been developed using an object-oriented database (Object Store) and Java applets. These applications have demonstrated that Web-based applications are a powerful tool to support global collaboration in product development.

5. REFERENCES

[1] Al-Ashaab A., and Young R.I.M, (1992), "Information Models: An Aid to Concurrency in Injection Moulded Products Design", Presented in the Winter Annual Meeting '92 of ASME. Anaheim, California, Nov. 8-13.

[2] Al-Ashaab A., Rodriguez K., Cárdenas M., Gonzalez J., Molina A., Saeed M. and Abdalla H., (2000), "INTERNET Based Information System to support Global Collaborative Manufacturing", 6th International Conference on Concurrent Enterprising, ICE 2000, France.

[3] Bullinger H.-J., Warshat J., Fischer D., (2000), "Rapid Product Development - an overview", Computers in Industry, 42 (2000), 99-108.

[4] Flores M., Molina A. (2000), "Virtual Industry Clusters: Foundation to create Virtual Enterprises", in Advanced in Networked Enterprises - Virtual Organizations, Balanced Automation and Systems Integration, L.M. Camarinha-Matos, H. Afsarmanesh, Heinz-H. Erbe (Eds.), Kluwer Academic Publishers, 2000, pp. 111- 120.

[5] IMPPACT, (1991), Esprit No. 2165, Proceedings of the Workshop, Berlin, 26-27 February 1991.

[6] Krause F-L., Kimura F., Kjellberg T., and Lu S.C.-Y., (1993), "Product Modelling", Keynote papers, Annals of the CIRP, Keynote paper, Vol. 42/2/1993, pp. 695-706.

[7] Molina A., Mezgar I., and Kovacs G., (1992), "Object Knowledge Representation Models for Concurrent Design of FMS", in Human Aspects in Computer Integrated Manufacturing, G.J. Olling, F. Kimura (Editors), Elsevier Science Publishers B.V.(North Holland),IFIP, pp.779-788.

[8] Molina, A., Ellis, T.IA., Young, R.I.M. and Bell, R., (1994), "Methods and Tools for Modelling Manufacturing Information to Support Simultaneous Engineering", Preprints, 2nd IFAC/IFIP/IFORS Workshop Intelligent Manufacturing Systems - IMS'94, (Ed. P. Kopacek), Vienna - Austria - June 13-15, 1994, pp. 87-93.

[9] Molina, A., Al-Ashaab, A.H., Ellis, T.I.A, Young, R.I.M, Bell R., (1995), "A Review of Computer Aided Simultaneous Engineering Systems", Research in Engineering Design, 7: 38-63.

[10] Molina A., Ellis T.I.A, Young R.I.M, Bell R., (1995), "Modelling Manufacturing Capability to Support Concurrent Engineering", Concurrent Engineering: Research and Applications, Volume 3, Number 1, March, pp. 29-42.

[11] Molina A., and Bell R., (1999), "A manufacturing model representation of a flexible manufacturing facility", Proc Instn Mech Engrs Vol 213 Part B, 1999, pp. 225-246.

Fractal Management Systems for Extended, Holonic Enterprises

Pierluigi Assogna
Theorematica SpA
Via Simone Martini 143 – 00142 Roma
p.assogna@theorematica.it

Abstract: The complexity and the extension of the production environment are increasing at a fastening pace. This ever-increasing complexity requires new and more flexible Management Systems and decision supports that go above and beyond ERP's. Any information system designed and developed to support a manufacturing enterprise has as a (sometimes implicit) first step: the modeling of the enterprise. Four metaphors are briefly explored, in relation to the increasing complexity of the environment to be modeled, and to the parallel increasing flexibility of the resulting models:

- The econometric model
- The hydraulic model
- The cybernetic model
- The chain value (Porter)

In these last years a new enterprise organization called holonic or cellular has been defined. It is embodied by a multilevel network of holons, a network that has a high degree of flexibility, because it can easily restructure itself depending on the circumstances: in fact each holon has the capability of surviving even if detached by any network, and the ability to search for a new network to integrate into. The holonic structure has a fractal nature, so that each holon is in turn made of holons, and the single persons cooperating in the enterprise represent the lower level. This structure is the best one for enduring the chaotic evolution of the manufacturing environment. The paper concludes with the proposal of a modeling technique to be used when designing the Management System for a manufacturing organization.

Key words: Manufacturing, Holon, Fractal, Modeling

1. Foreword

Manufacturing Organizations are complex systems in complex environments.

Complex systems have the tendency (not completely explained) to drift towards places and situations where life is dangerous. They seem to be attracted by precipices.

This specific zone of the possible states of the system is called "the edge of chaos".

A Manufacturing Organization lives in turbulent environments, where turbulence is ever increasing. In order to survive and prosper it has to be able to anticipate troubles (cataclysms in complexity jargon) and take advantage as much as possible of relaxed periods to prepare for cataclysms.

Globalization is a typical tidal force that very often triggers a cataclysm in traditional manufacturing organizations, mainly because of the requirement of flexibility.

Flexibility has been the name of the game for some time now, but it is important to define what and where has to be flexible within the organization.

2. The harsh requirements of complex environments

Complex systems have a characteristic structure, which is that of fractals.

Fractals are mathematical formulas that, fed to an engine that builds geometric loci on a computer screen, come up with often spectacular images that look like vortices, complex spirals, clouds, embroideries.

Fractals (or to be more precise their computer renderings) have the characteristic that their main visual aspect (be it a vortex, a pattern of spikes, or any) repeats itself at different scale all the way down to infinity.

Real life examples of fractals are clouds, coastal shapes, water whirls, tree branches, river deltas, and others.

All these are complex systems, so that we can say that a complex system is always a fractal, that means that it is composed of sub-systems and sub-sub-systems and so on, all isomorphic.

Many different complex systems inter-related make a complex environment: at different levels different components (systems) inter-react in a very complex network of relations and feedbacks.

Because of this multi-level lattice of relations, everything is in constant chance of change.

This change, when it happens, has an interesting characteristic: it is an evolution, in the sense that the systems tend to become (at each major shift of their shape) more complex, with a richer structure. Systems that do not have this capacity of reorganizing at a higher structure complexity do not survive for long in such an environment.

From this viewpoint a complex environment is very demanding: either you are prepared, every now and then, to re-organize, or you perish.

Of course there are events that can completely destroy this sort of environments, but if we take as an example life on earth, you need a cosmic cataclysm to do the job.

This picture is very different from the mechanistic systems of reductionism ages (La Place's universe), where every sub-system had a precise configuration, and where its only possible change was from functioning to non-functioning.

No matter how accurately a system builds to itself a cozy niche, sooner or later the niche disappears, leaving the system with the old dilemma: evolve or perish.

3. How to keep evolving

Fortunately, this changing is not continuous, but discrete.

Complex systems do not evolve continuously, but in a discrete way: this is what S. Jay Gould calls "punctuated evolution". This happens when multiple systems co-evolve in a common environment, that is the typical situation with living organisms or social systems: each system remains in a quasi-static situation for some time, than is caught in some sort of cataclysm (usually short) and than again can rest for some more time. The lengths of calm periods are unequal, and the size of each cataclysm is variable.

There are two main reasons for this complex behavior.

The first comes from the natural tendency of every system to resist to changes: it is called homeostasis. A system tends to keep its configuration despite any external influence: it exercises all its mechanisms in order to absorb the disturbance.

Of course there is a limit to the capability of staying in shape under any amount of pushes and pulls. Once the threshold is reached, the system all of a sudden finds itself at a bifurcation: either it perishes, or it finds (fast) a new configuration that lets it survive under new conditions.

Every time a system pertaining to a complex environment passes through such a cataclysm, it sends shock waves all around, and this maybe triggers the turmoil of another system. Because of this interdependency, each system goes through the punctuated evolution.

The second reason of the discreteness of the size of cataclysms comes from the fractal-like multi-level structure of complex systems.

The major shifts (or evolution steps) that happen in a system involve one of the levels of its structure, if not the entire system. Depending on the level that participates in this cataclysm, its magnitude follows a quantum-like pattern of discreteness.

A given level enters a cataclysm when its direct components become unstable (generally because of exogenous disturbances), and as seen, before giving up it resists for some time.

Now, coming back to our Manufacturing Organization, it needs a mechanism that helps effectively in taking the opportunity, every time a blow strikes, to re-organize on a better stand.

4. Knowledge can help

The only basic mean that can help survival is knowledge. Only by knowing as much as possible about itself and its environment a system can survive and prosper.

In the case of a manufacturing organization the main tool for knowledge management is of course its Information System, that in recent years means its Enterprise Resource Planning (ERP) system.

Now, if we take a close look at the best sellers in this arena, we can see that all of them have as common and noble father the Material Requirements Planning (MRP I). This technique was devised in the 60's, when a manufacturing organization was supposed to stay stable within a stable environment. You can tell this by analyzing the way that these systems treat changes (be these engineering or material source, or whatever): changes are a nuisance, something that force procedures to amend situations, cancel transactions, revert flows, etc.

In a word, traditional ERP's are geared to a quasi-static world

If we take an integrated system such as an ERP, we can see it as a model of the organization and of its context (suppliers, customers, and partners). What is getting more and more inadequate is exactly this embedded model.

Modeling an organization means formalizing the relevant knowledge, and as seen, knowledge is the recipe for survival in complex situations.

Does this mean that we have to throw away ERP's? Definitely no, but we need probably to chop them (a situation similar to the broom of the "Apprentice Sorcerer") into interconnected mini-ERP's, as described later on.

5. *Modeling metaphors*

Let's start from the beginning, that is modeling.

When you start this exercise, you have in mind (explicitly or implicitly) a metaphor, a meta-model that provides the syntax of the modeling.

The resulting models essentially have to describe the interactions of Actors, Objects, Processes, and the granularity of this representation is very important for our talk.

In order to guide the modeling of the Objects, Actors, Processes of a manufacturing enterprise, different meta-models have been historically utilized, depending on the inclination of the domain experts, on the strengths and weaknesses of the meta-models, and on the dominating business paradigm of the day.

A meta-model is a metaphor, and as such is used to provide analogues for all the different aspects of the production environment that needs to be modeled. In order to model some specific aspects, the analogy needs to be stretched to a limit where the power of evocation, that is the main advantage of using metaphors, thins out; for this reason none of the typical metaphors is perfect in each and every respect.

We can briefly analyze four different metaphors, keeping in mind this granularity, and see for each what are the pro's and con's, in terms of:

- Degree of coverage of activities
- Formalism
- Evocation power
- Modularity
- Degree of complexity that can be represented

The econometric metaphor

It is based on the Cournot duopoly model. The adopted formalism is mathematically concise.

It is geared essentially to the general input and output of the organization, in terms of money.

For this metaphor the enterprise is basically seen as a monolith.

The hydraulic metaphor

It is a simplified application of Forrester's System Dynamics: the organization is seen as a whole system, where each part is "mechanically" connected to some other parts.

In this metaphor water is used to represent the connections because of its non-deformability, and because it does not change with time.

Mathematical formulas regulate the exchanges between these parts.

The cybernetic metaphor

Introduced by Stafford Beer [1], utilizes the organism and its mechanisms as the metaphoric analogy.

A very important step of this methodology is the analysis of the number of different states that a sub-system can reach. An axiom is derived from this analysis, i. e. that you cannot control effectively a system with n possible states with another system with m possible states where m < n.

Said in other words, the "knowledge" of the controller must be greater than that of the controlled.

Feedback and feed-forward circuits are very much components of the models created with this methodology.

The chain value (Porter) metaphor

Introduced by M. E. Porter [2], gives an economic value to every process (relation) between a system and its context.

There are 5 main set of activities that participate directly in the process, and 4 that support the process.

The scheme of this representation is the following:

Support activities		Infrastructural activities				Margin
		Human Resources management				
		Tecnology development				
		Purchasing				
	Inbound logistics	Production	Outbound logistics	Marketing & sales	Services	Margin

Direct activities

The Value Chain

This scheme can be applied both to the enterprise and to its components, provided that you can identify for each the described activity sets.

The resulting table of strengths and weaknesses is more or less the following:

	Metaphors			
	Econometric	hydraulic	Value Chain	Cybernetic
Degree of coverage	Good	Good for marketing and administration	Good for logistics and administration	Good
Formalism	Good for synthesis, not for analysis	Good	Good	Average
Evocation power	Scarce	Average	Good	Good
Modularity	Scarce	Average	Average	Good
Complexity	Scarce	Scarce	Average	Good

The table indicates that from the viewpoint of the complexity that can be described, and modularity of the models that you generate, the value chain and the cybernetic meta-models are the best ones.

There is also a trend, from the econometric to the value chain, from a monolithic to a composite view of the organization.

The system, as seen with the hydraulic meta-model, is mechanical, in the sense that the relations between the components are rigid and linear.

6. The Holonic Fractal Enterprise

The term holon was introduced by Arthur Koestler [3] and indicates a (sub)system capable of functioning alone, and at the same time prone to (and capable of working in) cooperation with other holons.

A holon can live by itself, but is better off when integrated into a system, a holonic structure.

It can well represent an organization component, at any level of a complex structure, that even if perfectly integrated in the business activity, is able of detach itself from the structure and attach to another one, if and when the circumstances should require this move.

And if the structure is all made of holons, it is not disrupted by migrations of this sort.

A holonic structure is normally made of holons of different kinds, and the more diverse the holons, the higher the repertoire of behaviors of the structure, and thus its survival capability.

Each holon has a core competence that justifies its presence in the structure, plus "a little bit" of all other competences that are used to negotiate its participation.

An holonic enterprise is made of holons, in turn made of holons, and can participate in higher level structures. As each holon has to have a common set of negotiating (or interface) capabilities, we can see this enterprise as a fractal enterprise, with isomorphic components.

This sort of enterprise is better equipped to afford the vagaries and blows of the global market, because of its inherent flexibility, and when a new technology changes some aspects of the stage, this network can re-organize itself, and the holons no more fit for the new picture migrate to other structures, and are replaced by new ones.

This flexibility and robustness involve the knowledge that each single holon has of its internal workings (that is its model) and of its context, that cannot be represented by the components it is related to.

It is exactly this the main difference between a holon and a traditional component: the former must be constantly aware of as much world as possible, while the latter just needs to know its surrounding. The same knowledge requirement applies to the supporting software of the holon

From the viewpoint of the modeling, this structure can be easily modeled both by the cybernetic and the value-chain, as each holon can be seen as a "mini-enterprise".

It could even be possible to integrate, in one final model, holons modeled with different methodologies, in order to exploit the different representation strengths of the metaphors for different sectors of an organization.

7. The fractal ERP

Going back to the fact that the ERP system is the working model of a manufacturing organization, a holonic enterprise needs an ERP structured as a fractal.

What is needed is an ERP made of mini-ERP's, each one made of micro-ERP's, on many levels.

It may sound bizarre, but in a sense we can see this structure in the material logistics planning and control of traditional ERP's: in fact these functions are normally embodied in long, medium, short term cycles that have recursive logic.

Even in the level-by-level planning of MRP this fractal nature is present.

But traditional ERP's come short of this granularity target: their implicit model is too much centralized. They need to be re-architected in fractal-like structures.

If we remember the knowledge requirement, it is not enough to take a traditional ERP module as the management mechanism of each holon.

Now the image of the broom of the Sorcerer, that chopped into mini-brooms performs its task more efficiently, becomes clear.

As a conclusion we suggest the following procedure, in order to develop a holonic Management System for a manufacturing organization:

1. Analyze the organization as made of a multi level structure of holons, finding for each component all the activities of an entire enterprise, even if only in germ.

2. Choose for each of these holons the most effective modeling metaphor, taking into account the "evocative" strength of the analogy that in this exercise is a very important asset. The resulting model has anyway to have Actors, Objects and Processes: the Object Oriented paradigm is too effective not to be adopted.

3. Assemble the holon at the enterprise level by putting in relation all the components.

4. At this point it is possible to find modules of ERP's that fit the picture at different levels, and supplement these with sort of "wrappings", that provide the whole sets of functionality that make the holon capable of staying autonomous.

8. *Bibliography*

[1]   *The managerial cybernetics of organization* (Allen Lane The Penguin Press Ed., 1972).
[2]   *Competitive Advantage: Creating and Sustaining Superior Performance*(Free Press, M. E. Porter , 1985)
[3]   *Beyond atomism and holism: the concept of the holon.* (A. Koestler and J.R. Smythies eds. Beyond reductionism. Hutchinson, London).

Service Federation in Virtual Organizations

L. M. Camarinha-Matos [1], H. Afsarmanesh [2], E. C. Kaletas [2], T. Cardoso [1]
[1] *New University of Lisbon, Quinta da Torre – 2825 Caparica, Portugal, cam@uninova.pt*
[2] *University of Amsterdam, Kruislaan 403, 1098 SJ Amsterdam, The Netherlands*

Abstract: The practical implantation of the concept of dynamic virtual enterprise is still far from expectations due to a number of factors such as the lack of appropriate interoperable infrastructures and tools, lack of common ontology, and the socio-organizational difficulties. However, the creation of industry clusters supported by advanced information and communication tools can meanwhile provide a basis for the rapid creation of dynamic virtual enterprises in response to the market opportunities. A federated service management approach is introduced in this context and its application to the tourism industry is discussed. Finally, the support for the aggregation of simpler services into value-added services, implemented by distributed business processes within different organizations, is presented.

Key words: virtual organization, service federation, virtual enterprise, industry cluster.

1.   INTRODUCTION

Virtual organization benefits and support technologies. There is already abundant literature about the potential advantages brought in by virtual organizations and virtual enterprises, and a lot of recent proposals address more advanced dynamic cooperative networked organizations. The idea of highly dynamic organizations, that form themselves according to the needs and opportunities of the market and remain operational as long as these opportunities persist, suggests a number of benefits such as:

- Agility: the ability to recognize, rapidly react and cope with the unpredictable changes in the environment in order to achieve better responses to opportunities, shorter time-to-market, and higher quality with less investment. The composition of a VE is determined by the need to associate the most suitable set of skills and resources contributed by a number of distinct individual organizations. When and if necessary, the VE can reorganize itself by adding / expelling some members or by dynamically re-assigning tasks or roles to its members.
- Complementarity: enterprises seek for complementarities (creation of synergies) that allow them to participate in competitive business

opportunities and new markets.
- Achieving dimension: especially in the case of SMEs, being in partnerships with others allows them to achieve critical mass and appear in the market with a larger "visible" size.
- Resource optimization: smaller organizations sharing infrastructures, knowledge, and risks.
- Innovation: being in a network opens the opportunities for the exchange and confrontation of ideas, the basis for innovation.

A large number of R&D projects try to establish some technological foundations for the support of Virtual Enterprises /Virtual Organization (VE/VO) [6]. Relevant examples can be found in the NIIIP program in US, the ESPRIT and IST programmes in Europe (e.g. projects such as PRODNET II, VEGA, X-CITTIC, VIVE, etc.), or inter-regional cooperation programmes such as the IMS (e.g. projects such as GLOBEMAN, GNOSIS, GLOBEMEN, etc.) and INCO (e.g. MASSYVE). Many of these development efforts were concentrated on the design and development of infrastructures and basic VE/VO support functionalities. But only a few of these initiatives correspond to *horizontal developments*, aimed at establishing the base technology, tools and mechanisms, while most others correspond to vertical developments, addressing certain needs of specific sectors such as the manufacturing, agribusiness, tourism, etc. Although it is natural that in the early phases of a new area, considerable effort is devoted to the basic infrastructures, the lack of a common and widely accepted reference model and infrastructure has forced the vertical development projects to design and implement their own infrastructures, deviating resources from their main focus.

On the other hand, the emergence of a number of standards and technologies represent potential enabling factors, such as for instance:
- Open interoperable underlying network protocols (TCP/IP, CORBA-IIOP, HTTP, RMI, SOAP),
- Open distributed object oriented middleware services (J2EE Framework, CORBA Framework, ActiveX Framework),
- Standardised modelling of business components, processes and objects (EJBs, OAG and OMGs Business Objects and Components),
- Business Process Modelling Tools and Languages (UML, UEML, WfMC XML-based Business Language)
- Open and standard business process automation and Workflow Management Systems (WfMC, OMG-JointFlow, XML-WfMC standards, many commercial products),
- Standard interfacing to federated multi-databases (ODBC, JDBC),
- Intelligent Mobile Agents (FIPA, OMG-MASIF, Mobile Objects),
- Open and standard distributed messaging middleware systems (JMS, MS-Message Server, MQSeries, FIPA-ACC),
- XML-Based E-Commerce Protocols (BizTalk, CBL, OASIS, ICE,

RosettaNET, OBI, WIDL),
- Web Integration Technologies (Servlets, JSP, MS-ASP, XSL).

However, most of these technologies are in their infancy and under development, requiring considerable effort to implement and configure comprehensive VE/VO support infrastructures.

Therefore, although the advantages of the Virtual Enterprise are well known at the conceptual level [9], [6], the practical implantation is still far from the expectations, except for the more stable, long-term networks or supply chains. Nevertheless, the potential agility of a VE in terms of fast reaction to business opportunities is certainly a desirable feature in a scenario of fast changing market conditions; but the early phase of VE planning and creation is still a difficult one that needs to be adapted even by advanced and competitive enterprises. Some of the obstacles include the lack of appropriate support tools, namely for partners search and selection, VE contract bidding and negotiation, competencies and resources management, well-established distributed business process management practices, task allocation, performance assessment, inter-operation and information integration protocols, etc. Further problems include the lack of common ontologies, and the proper support for socio-organizational aspects e.g. lack of a culture of cooperation, the time required for trust building processes, need for BP reengineering and training of people, etc. Furthermore, the fast evolution of the information technologies often represents a disturbing factor for non-IT companies.

Industry / service clusters as a basis for VE creation. The creation of long term clusters of industry or service enterprises represent an approach to overcome these obstacles and can support the rapid formation of VE / VO according to the business opportunities. The concept of **cluster of enterprises**, which should not be confused with a VE, represents a group or pool of enterprises and related and supporting institutions that have the potential and the will to cooperate with each other through the establishment of a long-term cooperation agreement. Buyer-supplier relationships, common technologies, common markets or distribution channels, common resources, or even common labor pools are elements that typically bind the cluster together. This is not a new concept as a large number of related initiatives have emerged during the last decades, namely in Europe and USA [4]. But the advances in information and communication technologies now bring new opportunities to leverage the potential of this concept, namely by providing the adequate environment for the rapid formation of agile virtual enterprises. For each business opportunity found by one of the cluster members, a subset of the cluster enterprises may be chosen to form a VE for that specific business opportunity. In this perspective, the expected competitive advantage of cooperative development of products and services becomes a more relevant tie among the cluster members. The more frequent situation is the case in which the cluster is formed by organizations located in a common region, although geography is not a major facet when cooperation is supported by computer networks.

The cluster enterprises are normally "registered" in a directory, where their core competencies are "declared". Based on this information, the VE initiator / creator, which is usually a member of the cluster enterprises, can select partners when a new business opportunity is detected. Clearly, several VEs can co-exist at the same time within a cluster, even with some members in common. A cluster represents a long-term organization and therefore presents an adequate environment for the establishment of cooperation agreements, common infrastructures and ontologies, and mutual trust, which are the facilitating elements when building a new VE. The concept of cluster is evolving in parallel with the emergence of other forms of relationships, such as "communities of practice" or "virtual communities", representing a general trend to the emergence of a kind of "<u>society of relationships</u>". The cluster does not need to be a closed organization; new members can adhere but they have to comply with the general operating principles of the cluster. Similarly, for the formation of a VE, preference will be given to cluster members but it might be necessary to find an external partner in case some skills or capacities are not available in the cluster. The external partner will naturally have to adhere to the common infrastructure and cooperation principles. In addition to enterprises, a cluster might include other organizations (such as research organizations, sector associations, etc.) and even free-lancer workers. The establishment and management of clusters through adequate infrastructures represent therefore an important support for the creation of agile virtual enterprises (Fig. 1).

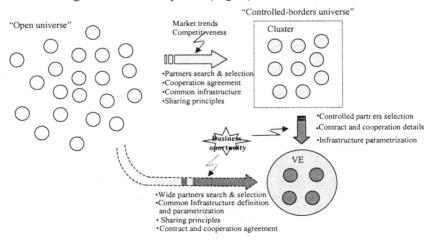

Figure 1. Two approaches for VE formation

This idea of using a cluster as the basis for the formation of virtual enterprises has been identified in other research works such as the COSME/VIRPLAS [10] or VIRTEC [5] projects. These projects have identified some of the major characteristics and needs of cluster management, but did not introduce the necessary IT infrastructure and support tools.

From a regional perspective, a well-managed cluster may offer the opportunity to combine the necessities of both "old" and "new" economies, and form a sustainable environment. The ICM (Industry Cluster Management) concept can support the exploitation of local competencies and resources by an agile (and fast) configuration of the most adequate set of partners for each business opportunity and therefore extending the scope of intervention of manufacturing companies into the services area. Furthermore, the local clusters can gather and empower a unique set of competencies tailored to regional culture and local customers' preferences, allowing a concerted offer of cooperation to global companies. In this way, members of the local industry cluster can play an important role in the customization and final assembly of products to local markets even though the basic components may be produced elsewhere. Therefore, in a time of tough competition provoked by the globalization, the organization and effective management of local industry or service enterprises clusters, focused on the characteristics of SMEs, provides a promising approach for regional sustainability. In addition to the mentioned benefits of cooperation on dynamic VE/VO, there is also the opportunity to share experiences and costs in the learning process of introducing new IT within an industry cluster, and to reduce the risk of failure.

Federated virtual markets. The concept of **federation** can support advanced management strategies for the services offered by a cluster. The federation mechanisms which have been developed in different areas such as information management (federated information management [1] [8]), multi-agent systems (federation of agents [11] [3]), or services management (services federation [12][2]) represent approaches to support the interoperation among heterogeneous, autonomous and geographically distributed entities through the definition of a set of common principles and mechanisms to allow a harmonized access to services and information, independently of the diversity of their sources.

Common features among the various materializations of the federation concept are the support for cooperation among distributed, possibly heterogeneous, and autonomous entities (for example, information resources or service sources) without the need for the clients of information / services to care about the communication details. In a federated system different authorities are in charge of different parts of the system and preserve their autonomy. The local autonomy of entities, which establishes access rights to local resources, is subject only to the agreements reached by the entities with other authorities. These agreements, although they may be seen as a restriction of each entity's autonomy, are required for cooperation. Therefore, the involved entities have to negotiate terms and conditions of their interoperation with other entities.

When applied to the management of a cluster, this concept allows a vision of the various cluster members as service providers that, independently of the way their services are implemented or located, make them accessible in a kind of controlled "virtual market". A client such as a VE creator can "shop" in this market for the best set of services to satisfy the needs of a given business opportunity. The service federation infrastructure shall provide the basic mechanisms for service registration, maintenance and removal, mechanisms for transparent (remote) access to services according to some agreed access rules, mechanisms to define and maintain these access rules to the services. Moreover, because the services in a service federation can actually represent any available sources, different federations besides a service federation can be established. For instance, a natural means of providing information management support to service federations can be realized by coupling the service federation with an information federation, which is established through federating a set of existing information resources within the cluster, even building on top of the agreements reached for the service federation.

This paper, in this direction, addresses the development of a federated approach to service management with a particular focus on the support to the creation of valued-added services.

2.   GENERAL ASPECTS IN SERVICE FEDERATION

One of the obvious components of a cluster management system is a directory where the cluster members are registered and their competencies / skills and capacities are declared. This directory is therefore a set of member profile records. Considering that competencies of cluster members are represented (to some extent) by services (service functions), this directory shall include a catalog of services (Fig. 2).

Due to the members autonomy (and legacy) there might be a large heterogeneity / diversity in the way services are implemented. However, in order to facilitate service selection ("shopping") and utilization, a common service interface needs to be agreed among the cluster members and, in case of legacy implementations service adapters have to be developed. A service interface can be decomposed in two parts: service specification descriptor and service invocation wrapper (or proxy), i.e. the service API.

The service specification describes the characteristics of the service such as service name and identifier, version, functionality, I/O specifications, applicability conditions, access rights, etc. Even temporary announcements of service providers can be supported through this component; for instance, a "hotel reservation service" could announce here a special offer for the just married couples. When a client is looking for a specific service in the catalog, the search criteria are expressed against the properties (attributes) of the service specification component. The service invocation wrapper, on the other hand, describes the programming interface of the service, that is, for instance the methods available, their parameters and return values, etc. It acts as a proxy to the actual service, providing transparent access to the service by hiding the implementation details, such as remote invocation, security and communication mechanisms. A client (application) will get a copy of this wrapper that will be linked to the application memory space to be used for the remote (transparent) invocation of the service.

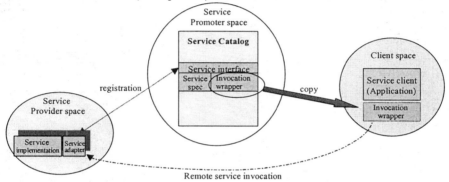

Figure 2. Services registration and invocation

From a computational point of view, the cluster can be seen as a network comprising one (or more) cluster management node(s) (or cluster promoter node(s) from a service market perspective) responsible for the catalog, and a number of member nodes representing the cluster members where the actual implementations of services may be installed (and running). Cluster members advertise their services by registering them in the cluster service catalog.

It is important to notice that the same service type can have a number of different implementations (given by different members of the federation). However a common interface has to be guaranteed (by the service providers) in order to facilitate the identification and use of these independently developed services. For this purpose a common ontology and common functional interface rules have to be agreed among the cluster members.

It shall be mentioned that several VE projects have tried to adopt / combine different standards such as EDIFACT, STEP or UDDI in order to eliminate the ontological mismatch among members of a VE. But in spite of these efforts it is difficult to have general-purpose solutions because:

- for different classes of information handled in a cooperation process there are no standards (e.g. quality-related information);

- EDIFACT or STEP cover wide application areas and therefore for any practical application subsets have to be selected and agreed among partners.

Although difficult to be solved in general terms, the problem is more tractable in a smaller (closed) universe as the one represented by a cluster. Common rules and principles can be agreed by the initial members of the cluster. Future members, in order to join the cluster, are required to accept those principles (part of the adhesion contract). As the cluster envisions a long-term relationship (in opposition to a single cooperation opportunity), the required investment / adaptation may be affordable and give the cluster true agility whenever opportunities for cooperation arise. The implementation of a harmonized representation of services in the service catalog does not necessarily mean that all cluster members have access to all services all the time. Service providers shall keep their autonomy and the right to specify whom and under which conditions, has access rights to their registered services. Therefore there is a need for an access rights manager component (Fig. 3) allowing the definition and validation of access rights. For example, a service provider can specify that a given service description is available for lookup only to a specific set of service requesters.

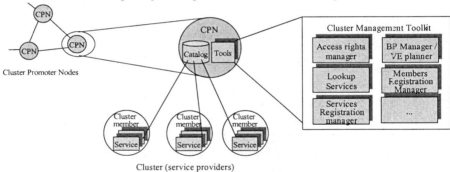

Figure 3. Cluster management functionalities

A service can be represented as an object class including a number of attributes and methods. For instance, a "hotel reservation service" could include:

Attributes:
- Service provider id
- Hotel category
- Location
- Other attributes
- Access rights list

Methods:
- Check_availability
- Check_prices
- Make_reservation
- Cancel_reservation.

When a business opportunity is found by a member of the cluster, this member assumes the role of broker (or VE initiator). This broker will typically elaborate a plan of activities (business process - BP) that are required to satisfy the business goal. In order to perform these activities, the broker, using a <u>BP manager component</u>, may search and select a number of service methods from the catalog. When service methods are assigned to the various steps of the BP, this BP may in fact become a distributed BP, as different parts of the process are performed by different service providers. The selection of a suitable set of service methods for a particular BP (in other words the creation of the VE) is a very important functionality that, in addition to the type of service required, has to take into account a number of other factors such as: performance history of the service provider (e.g. how reliable it is), compatibility (joint performance) between different service providers, visibility rules / access rights, and other specific requirements.

Therefore, the service catalog component shall offer intelligent **search / selection filtering lookup** functionalities, based on the service attributes, to assist the activity of the broker.

Higher level or value added services (VAS) might be created by a composition of different low-level services available in the catalog. Therefore, a service provider might be a service client as well as he can look for and select services from the catalog to compose his VAS. A new VAS can be registered in the catalog in the same way as the basic service. In section 4 a more detailed description of this process is presented.

In addition to the basic elements described there is a need for additional functionalities such as a members certification function (to perform an assessment of partners and how they behaved in past cooperation relationships), service assessment or certification function (handling issues such as reliability of the service, compliance with the common interface specifications, performance, etc., i.e. Quality of Service functionality), etc.

3. AN EXAMPLE – TOURISM INDUSTRY

Similar to the manufacturing industries, the need to remain competitive in the open market also forces the service providing companies to seek "world class" status. In the case of Tourism industry, companies remain focused on their core competencies, while realize the need to look for alliances, when additional skills / resources are needed to fulfill business opportunities. Cooperation among the actors / entities in Tourism industry is not a new phenomena. For instance, travel agencies typically offer aggregated or *value-added-services* (VAS) composed of components supplied by a number of different organizations. But to provide an *on-line value-added-service* for "booking a complete journey plan" that may include several means of traveling, several hotel bookings, car rentals, leisure tour bookings, etc., a networked cooperation must exist among many different organizations [2].

Fig. 4 shows an example of a possible business process for a traveling package where the various activities are supported by a number of services provided by different organizations.

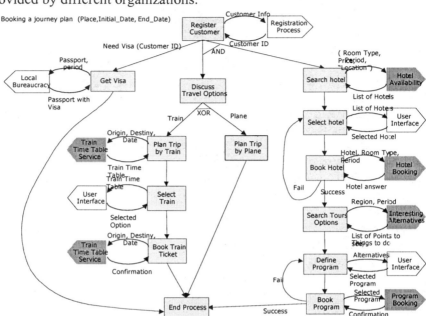

Booking a journey plan (Place,Initial_Date, End_Date)

Figure 4. Example of value added service in tourism

Clustering is a typical phenomenon in this sector, especially at regional level. Regional tourism promotion organizations try to offer an integrated view of the local resources (<u>local cluster promoter node</u>). In order to reduce the fragmentation and dispersion of information some trends to form <u>networks of clusters</u> are emerging. One example is the *enjoyeurope.com* initiative.

Following the general principles described above, the IST FETISH project is developing a computational infrastructure to support the tourism industry in Europe. The main goals pursued in this project are: (**i**) to increase the availability of tourism-related services via a mechanism of services federation; and (**ii**) to improve the interoperation among providers and clients, taking advantage of the new facilities provided by Internet. The intention is to promote the development (and sharing) of new tourism-related services, and the affiliation of organizations to a network of service federations. For example, if a travel agency wants to offer a very specific holiday package, the proposed infrastructure which is being developed in Java/JINI, is intended to provide a set of application tools to support it in a flexible way, combining different tourism-related services offered by different organizations.

In order to fulfill this goal, the adopted approach considers a three-layered architecture as presented in Fig. 5. It is assumed that the members of this federation (clients and service providers) will work cooperatively through the Internet, based on the interoperation mechanisms supported by the Jini technology. The layers defined in this architecture are the following: the Service layer, the Directory layer, and the VAS Support layer.

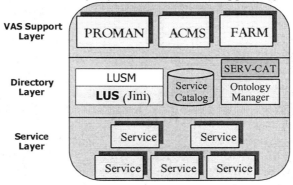

Figure 5. Three-layer architecture components

The Service layer contains the tourism services developed by the members of the federation. These services are publicized to the entire federation (e.g., a hotel reservation service, a rent-a-car service, etc.) or only to the members of a subgroup defined within a cluster. The Directory layer supports the procurement of the tourism services (advanced lookup service) implemented in the federation. This layer expands the Lookup Services (LUS) hierarchy, which is the core component of the Jini environment, by adding a specific component (the LUS Manager – LUSM, developed by SUN Spain) to handle the issues related to the service management needs. The Ontology Manager, developed by IASIS Italy, provides a common tourism ontology and mechanisms and rules for registering new service interfaces. The interfaces are stored in the SERV-CAT module developed by the University of Amsterdam. The VAS Support layer is intended to offer high-level applications that can be executed on top of the LUSM hierarchy offering advanced functionalities to the federation. Some of the main applications at this level are:
(i) The Process Management system (PROMAN) developed by Uninova, that aims at supporting the distributed business process (DBP) creation and management, as it will be described in next section.
(ii) The ACcess Manager System (ACMS), which serves as the entry gateway for the end users and service providers to FETISH, where some kind of human intervention is required. In this case, ACMS shall present an integrated user interface for certain *important* offered capabilities of

FETISH, such as the *"New service interface definition request"* and the *"Quality control and service operation maintenance"*.

(iii) Federated Access Rights Manager (FARM), developed by the University of Amsterdam. This module allows the definition and validation of access rights defined at the level of service proxy in the general context of FETISH as well as within the Virtual Enterprises dynamically created. For instance, a service provider can specify that a given service proxy can be available for lookup *only to a specific set* of service requesters or only to specific VE members. Therefore, this module must support the proper configuration of the access rights. Furthermore, the proper filtering and validation of access rights have to be accomplished at the moment in which the lookup process is being performed. This can be realized in two ways:

- There is a FARM module next to every LUSM node in the FETISH system, and each FARM node manages the access rights definitions for the proxies stored only in that LUSM node. In this case, after a lookup to the LUSM, the resulted proxies (or their IDs) are passed to the FARM for the validation of access rights against the requester, and only the validated proxies are returned to the requester. In this case, FARM must work in close cooperation with the LUSM.
- There is a single FARM module within the FETISH system to manage the access rights for all FETISH service proxies. In this case, the ACMS coordinates the activities within the FETISH system, and after a lookup passes the proxies returned by the LUSM network to the FARM for the evaluation of access rights. Then the validated proxies are returned back to the user.

The tourism services can be provided in different levels. The lower level contains the basic tourism services, which represent *atomic* services created by a tourism operator. On top of this level, it is possible to create another level in which the basic services are used as components of a *Value Added Service*. Let us consider that we can create (sub)VASs to be used as part of another VAS. In this way, we can foresee that it is possible to create an unlimited number of different levels of services. It is important to emphasize that all services are handled inside the federation in the same way, irrespective of the level they belong to.

In order to illustrate how this system works, let us consider that "Hotel Palma" is a member of one tourism cluster. Such a node wants to develop a new service called "HotelReservation". Therefore, the first step is to check if there is a specification for the service type in the Service Catalog. If the specification is found, the new service has to comply with that description. In case it is a new service type, it is necessary to create a new service specification and submit it to the ACMS module for possible registration in the Service Catalog (Figure 6a). Please note that a service specification contains the information that characterizes the service, such as: service identifier, input and output parameters, additional attributes, etc. In other words it is an abstract representation of the service.

After that, the service is developed by Hotel Palma. This development is an internal task that has to be performed by the node itself. Now, in order to put the "HotelReservation" service available to the federation, the Hotel Palma has to develop a "wrapper" or API that contains all methods and mechanisms that support the activation of the "HotelReservation", and register such a wrapper in the Service Catalog. In Figure 6b, for illustrative purposes only, this wrapper is called *Wrapper_HRP* ("HotelReservation Palma"). Please note that there might be several implementations of a service type (by different service providers). For instance, the service "HotelReservation" can be implemented by any of the hotels that belong to a given cluster. In this case, the only restriction applied to the different implementations is the compatibility with the interface registered for that service in the Service Catalog. For each implementation a wrapper (API) has to be registered in the Catalog.

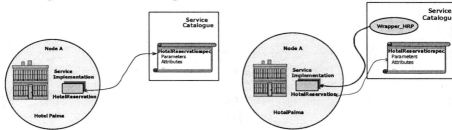

Figure 6a. Registering a Service Interface   *Figure 6b.* Registering a Service Wrapper

In addition, let us suppose that a travel agency "Paradise Tours" needs to reserve a room in Hotel Palma. If Paradise already knows that the "HotelReservation" service is available from hotel Palma, Paradise gets a copy of the "Wrapper_HRP" and uses it to make the reservation (Fig. 7). If Paradise Tours does not know about the availability of this service, it has to browse through the service catalog to eventually find that service interface.

The common interfaces defined for services are stored in the Service Interface Definitions Catalog (SERV-CAT). As mentioned earlier, common interfaces facilitate the service selection and utilization in a cluster of autonomous members. When a service provider is willing to develop and register a new service, it will first look in the SERV-CAT for the best suitable service interface among the available interfaces. Once an appropriate interface is found, the service provider will implement this interface to develop its service.

The SERV-CAT is implemented using the Oracle DBMS. The relational database model structures of SERV-CAT to store the service interfaces definitions correspond to the Java language structures. For instance, the types (classes and interfaces) defined within the SERV-CAT are grouped under packages. A type has a number of members, namely fields and methods. Each member has a type and a default value, while methods have parameters and return types. The parameter and return types of methods, however, are stored as XML files in the SERV-CAT.

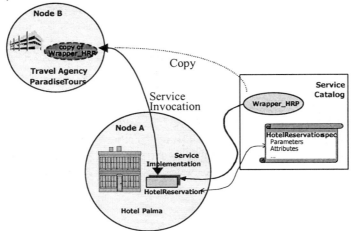

Figure 7. Using a tourism service

In order to separate the application logic from the presentation logic, an application server is developed, together with a Java FetishTypes package representing the data structures of SERV-CAT as Java classes to be used by the application server and client. The application server, SERV-CAT-AS, provides the functionality necessary for the client applications to browse and manipulate the catalog. The SERV-CAT-AS is an RMI server, which is continuously running and waiting for client requests. When a request arrives, it accesses the Oracle database using JDBC, and converts the result set into the corresponding FetishTypes objects, and returns it back to the client. The architecture for the SERV-CAT and its related tools is given in Fig. 8.

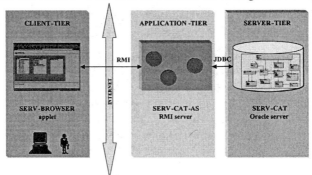

Figure 8. The SERV-CAT architecture

Other examples of capabilities to be supported by ACMS are:
1.*New service specification definition request.* Clearly, "new service type

definitions" require human intervention for intelligent decision making e.g. by a committee, that translates the request into "new concept definitions" to be further modelled and stored in the common ontology. The ontology definitions will be then translated into "new catalog definitions" to be submitted for storage within the Service Catalog.

2. *Quality control and service operation maintenance.* A procedure needs to be followed to handle for instance the users reports and/or complains on the service quality or availability. This procedure clearly requires human attention and intervention for intelligent decision-making. This will partially involve the directory administrator to monitor and /or control the proper availability and operation of services through the LUSM.

Besides the dynamic consortia established to execute each VAS, *long-term* cooperation agreements are also sometimes established among subsets of enterprises within a cluster or network of clusters to jointly support certain services. When a customer contacts a travel agency, he/she frequently notices that the agency shows clear preferences for certain specific providers, probably those that are involved in long-term cooperation agreement (formal or tacit), or those with whom the agency had good past experiences in other previous cooperation processes. Some other types of consortia may have a more *permanent* "institutional" organization, with or without a common ownership, for example the Best Western hotel chain. The three types of <u>consortia</u> (temporary, long-term, and permanent) require flexible common mechanisms namely in terms of access rights definition and control that are provided by FARM.

4.   VALUE ADDED SERVICES

As illustrated by the tourism sector, the aggregation of services into higher level or <u>valued-added services</u> is an important process in a cluster-based service market. Travel agencies or tourism intermediaries typically offer aggregated or value-added services composed of components supplied by a number of different organizations. The "materialization" of such a value-added service involves a process – business process (BP) – requiring the invocation and coordination of other services that may be provided by different members of the cluster. In such a case of distributed execution the BP becomes a <u>distributed business process</u> (DBP). The contributing members can therefore form a temporary organization or virtual enterprise. The DBP has to be properly managed in order to guarantee that it is successfully performed. This process management activity involves many issues including:

(i)   The definition of the DBP and its structural decomposition in terms of BPs, sub-BPs, and activities;

(ii) The distribution of such BPs/ activities among the VE members;

(iii) The execution of each BP/ activity by the VE members; and

(iv) A supervision process capable of handling the problems detected during the execution of BPs/ activities that may affect the performance of the whole VE.

Proper coordination policies supported by flexible coordination mechanisms are necessary to ensure the cooperation among partner enterprises. A typical approach, usually found in the workflow management systems, and the work on the coordination languages, focuses on the separation between the *coordination* and *computing* (or *service methods processing)*. At the same time, the sharing and exchange of information in a cooperative virtual organization is always inter-related to both the business processes running at each enterprise, as well as the distributed business processes that run at different sites. Therefore information management aspects cannot be independent of the business process coordination issues.

In order to support these requirements a Process Management System (PROMAN) based on the concept of <u>federated market of services</u> is developed by the Robotics and CIM group of the New University of Lisbon / Uninova. The Fig. 9 presents the architecture of PROMAN, which is composed of three modules, namely: the Business Processes Administrator (**BPA**), the Business Process Editor (**BPED**), and the Business Process Executor (**BPEX**). The BPA is the PROMAN interface to the external entities (other components of the cluster management system) and the manager of BPED and BPEX. Both BPED and BPEX provide functionalities to the BPA, respectively "edition" and "execution/monitoring" functions. The BPED is a graphical editor tool that supports the BP edition-related needs in PROMAN. The BPEX is a multi-thread execution system that supports the execution and monitoring-related needs in PROMAN.

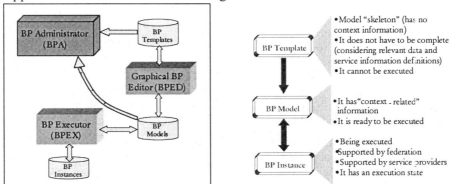

Figure 9. General structure of PROMAN   *Figure 10.* BP templates, models, and instances

PROMAN handles three main types of entities: BP templates, BP models, and BP instances (Fig. 10). A <u>BP template</u> represents a "skeleton" that can be used to support the creation of a <u>BP Model</u>, which represents a VAS. The main difference between a template and a model is that the latter contains "context-related" information that allows its execution. Besides that, some construction rules can be relaxed in the templates, since they will be further extended and analyzed before being executed. Each execution of a BP model is represented by a <u>BP instance</u>.

BP templates can be shared with the entire cluster or a subset of partners. This means, a cluster node can create a BP template and store it in a kind of "central repository of templates", from where it is downloaded and used by any other allowed cluster node. Similarly, a BP model (as a VAS implementation) may be put available to the entire cluster (or network of clusters), a subset of this cluster, or even to outside clients. A BP model is created and registered following the procedure already described for services registration. This means that the BP interface is registered in the Service Catalog, the "BP proxy" is registered in the LUSM network, and such a BP must be ready to be executed when remotely activated by a client. The main difference between the publication of *BP template* and *BP model* is that the template itself is available to be downloaded from the "central repository of templates", whilst the model is available to be only executed remotely. In other words, while a template itself is offered to the community becoming a "shared property" among the cluster members, a model is developed and remains as a "property" of its developer.

Both BP templates and models are created using a graphical language (Fig. 11) provided by BPED (Fig. 12). The main entities used in BP Templates and Models are the following: Begin/End, Activity, Transition, Service (remote, local, or sub-BP), Relevant data, Split, and Join. Begin/End are the symbols used only to mark both the starting and the ending points of a template/model. An activity represents a task to be performed, which can be a single (atomic) task or a complex one (implemented as a sub-BP). A single task is represented by the basic service methods, which are invoked directly from the service providers, via the proxies registered in the LUSM network. A complex activity invokes, in fact, another BP. This mechanism supports a hierarchical utilization of the BPs. A transition is the connector that links two activities. Local functions correspond to proprietary functionalities provided by the VAS implementer. The entire set of transitions defines a path to be followed when a BP model is executed. A relevant data set is used to handle the flow of information among the activities that belong to a BP model. Splits/Joins are used to define an additional logic mechanism to work together with the transitions, in order to define the path to be followed during the execution of a BP model.

Some characteristics of a BP model are the following:

- o It is composed of a set of activities linked by transitions following a parallel and/or sequential flow;
- o Each activity invokes a service method (basic, sub-BP or a local function);

- The transitions can be conditioned and temporized;
- Loops and cycles are allowed in a BP model;
- It can be partially defined, which means, it is not previously completed in terms of activities, services and data involved in. These can be dynamically provided during the instantiation of this model;
- Supports the definition of exception handling procedures.

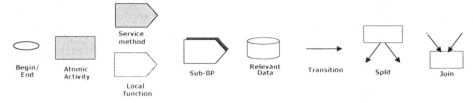

Figure 11. Graphical BP definition language

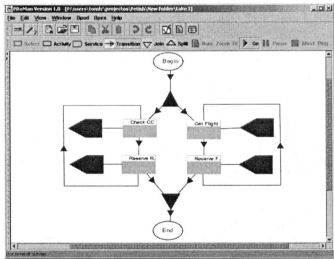

Figure 12. Example of BPED use

The main functionalities of the BPA component include:
- Support the management of BP templates, models, and instances;
- Save/Load of BP templates and models;
- Upload/Download the BP templates to/from the Service Catalog;
- Find service interfaces;
- Find service proxies (in interaction with the LUSM module);
- Register VASs interfaces in the Service Catalog;
- Register VASs proxies in the Service Catalog;
- Provide a persistent storage for each BP instance;
- Support the recovery process of BP instances;
- Support dynamic changes in BP models (temporary and permanent).

BPED functionalities include:
- o Edit BP templates/models: Insert, eliminate, select, copy, move entities, etc.;
- o Graphical facilities: Visualization resources (Grid, Zoom, etc.); Configuration of BP services icon;
- o Validate BP templates/models (graphics, syntax, and semantics);
- o Trigger the execution of BP models;
- o Support a "constrained edition" of BP models (i.e., partially "lock" the edition of a BP model, which has one BP instance being executed).

Finally, the list of BPEX functionalities includes:
- o Load & Parsing BP models;
- o Validate BP models (check syntax and semantics);
- o Creation of the BP instance;
- o Accept a remote activation of a VAS;
- o Execute services (remote or local invocations);
- o Manage exceptions;
- o Optional pre-validation of models;
- o Support Dynamic Changes: temporary & permanent; a "pause" mechanism during the execution of a BP instance; interaction with BPED in order to support the changes;
- o Save/Load of BP instance status information, events information;
- o Color-based representation of the BP instance status;
- o Support the intervention from a human operator.

It shall be noted that although a BP model assigns a service method to each activity, it might happen that this service is not available at the moment of its invocation. This might be due to a temporary network problem or because the service provider, as an autonomous entity, decided to change its policy. Therefore, before actually invoking services, BPEX has to check their availability (e.g. through the leasing mechanism of Jini) or the network connections. In order to reduce the number of exceptions requiring specific recovery, a BP model may associate a list of alternative service methods to each activity. PROMAN extends the coordination model first developed in the PRODNET project [7] by considering a service federation model.

5.   CONCLUSIONS

An industry cluster, when supported by adequate management functions, forms a proper environment for facilitating the rapid creation of dynamic virtual enterprises in response to business opportunities.

A federated services approach as a basis for a virtual market of services represents a flexible solution for managing the services offered by the various members of an industry cluster while preserving their privacy and autonomy.

On top of the service federation infrastructure, a complementary functionality is offered for value-added services creation and the corresponding distributed business process modeling.

Further work is being carried out on the integration of the value-added services definition and the federated access rights management mechanisms.

Acknowledgements. The authors thank the European Commission and IFIP for the partial support for this work and their partners of the FETISH-ETF, MASSYVE and COVE projects for their valuable contributions.

References

1. Afsarmanesh, H.; Tuijnman, F.; Wiedijk, M.; Hertzberger, O. – Distributed schema management in a cooperation network of autonomous agents, Proceedings of DEXA'93, Prague, Czech Republic, 1993.
2. Afsarmanesh, H.; Camarinha-Matos, L.M. - Future smart organizations: A virtual tourism enterprise, *Proceedings of WISE 2000 – 1st ACM/IEEE International Conference on Web Information Systems Engineering*, Vol. 1 (Main Program), pp 456-461, IEEE Computer Society Press, ISBN 0-7695-0577-5, Hong Kong, 19-20 June 2000.
3. Beasley, M.; Cameron, J.; Girling, G.; Hoffner, Y.; Linden, R.; Thomas, G. – Establishing Co-operation in Federated Systems, *Systems Journal*, Vol. 9, Issue 2, Nov. 1994.
4. Bergman, E. M.; Feser, E. J. – Industrial and regional clusters: Concepts and comparative applications, www.rri.wvu.edu/WebBook/Bergman-Feser, 2000.
5. Bremer, C. F.; Mundim, A.; Michilini, F.; Siqueira, J.; Ortega, L., 1999 - A Brazilian case of VE coordination, in Infrastructures for Virtual Enterprises - Networking Industrial Enterprises, L.M. Camarinha-Matos, H. Afsarmanesh (Ed.s), Kluwer Academic Publishers, ISBN 0-7923-8639-6, Oct 1999.
6. Camarinha-Matos, L. M., and Afsarmanesh, H., 1999 - Infrastructures for Virtual Enterprises - Networking Industrial Enterprises, Kluwer Academic Publishers, ISBN 0-7923-8639-6, Oct 1999.
7. Camarinha-Matos, L. M.; Lima, C., 1999 – PRODNET coordination module, in [6].
8. Garita, C.; Ugur, Y.; Frenkel, A.; Afsarmanesh, H.; Hertzberger, L.O, 2000 - DIMS: Implementation of a Federated Information Management System for PRODNET II, Proceedings of 11th International Conference and Workshop on Database and Expert Systems Applications - DEXA '2000, London, England.
9. Goranson, H.T., 1999 – The Agile Virtual Enterprise – Cases, metrics, tools. Quorum Books, ISBN 1-56720-264-0.
10. Molina, A.; Flores, M.; Caballero, D., 1998 – Virtual enterprises: A Mexican case study, in Intelligent Systems for Manufacturing, L.M. Camarinha-Matos et al. (Eds.), Kluwer Academic Publishers, ISBN 0-412-84670-5.
11. Osorio, A.; Camarinha-Matos, L.M. – A federated multi-agent infrastructure for concurrent engineering (CIM-FACE), Studies in Informatics and Control, Vol. 5, N. 2, June 1996, pp.143-156.
12. SUN, 1999 - JINI Technology Architectural Overview, Jan 1999, http://www.sun.com/jini/whitepapers/architecture.html.

FUNCTIONAL ANALYSIS AND SYNTHESIS OF MODULAR MANUFACTURING SYSTEMS USING THE HOLONIC THEORY: APPLICATION TO INTEGRATED ROBOTIC WORKCELLS

D.M. Emiris, D.E. Koulouriotis, N.G. Bilalis
Department of Production Engineering & Management
Technical University of Crete
73100 Chania, GREECE
e-mail: emiris@dpem.tuc.gr

Abstract - This paper investigates the application of the concept of holons in manufacturing and business organizations. Holons are defined as an identifiable term of a system that has a unique identity, yet is composed of subordinate parts and in turn, is part of a larger whole. The generality and simplicity of Koestler's pioneering ideas for holons and holarchies, triggered rather recently the interest of numerous researchers, who suggested their use for the development of a suitable framework for designing the architecture of next generation manufacturing systems. The holonic analysis approach as a first step to modular building of systems exhibits significant advantages, such as, faster application design, re-usability of components, optimized interface with the providers, and ability of the system to evolve. The holonic synthesis approach as the second step for synthesizing new systems also exhibits attractive advantages, such as, ability to generate applications by using elements from a continuously enriched design template, time and cost savings due to the determined consistencies and interfaces among the components, and ability to add new elements to the template, thus extending the design capabilities. This paper capitalizes on the theory of holons, thus taking full advantage of its generality, and investigates new fields where this theory can be applied effectively. The presented research investigates generic manufacturing and business aspects, such as the modular design of products, manufacturing cells and business process modeling, since it is estimated that the holonic organization can provide valuable innovative concepts and solutions for improving the efficiency of such systems. A real production system with embedded robot systems is examined to reveal the full potential of the proposed approach.

1. INTRODUCTION

Holons, were introduced by Koestler [7], in an attempt to develop a framework for analyzing the internal mechanisms of all living organisms in nature and to study the evolution and organization of life mainly in biological and in social systems; they were defined as an identifiable term of a system that has a unique identity, yet is composed of subordinate parts and in turn, is part of a larger whole. Holons exhibit both autonomy and the ability to cooperate; that is, they are able to produce and to execute decisions through mutual agreements and co-coordinated actions. Apart of being autonomous and self-contained units, holons may also come together in order to form hierarchically organized structures called *holarchies*. These result from the integration and co-operation of various holons to a system for performing a specific function.

The generality and simplicity of Koestler's pioneering ideas for holons and holarchies, triggered rather recently the interest of numerous researchers, who suggested their use for the development of a suitable framework for designing the architecture of next generation manufacturing systems. Several research projects, such as the HMS, investigated the synthesis of systems consisting of highly flexible, reusable, and modular units, able to reconfigure quickly and easily in order to produce various products. Similar research directions were adopted in the case of business systems, where products are substituted by services. The common denominator in all approaches is the intention, and thus ability, to provide customized products/services at a low cost, as dictated by the production trends of the 21st century.

The general approach of designing and constructing modular systems comprises of two main steps, namely, (a) the analysis of existing systems in order to reveal functionalities and to generate a "palette" of modular components, and (b) the development of a coherent methodology for interfacing and integrating such modules into new systems, which may execute different functions depending on the way modules are synthesized.

The holonic theory fits precisely in this philosophy: holons, being autonomous self-reliant units that exhibit co-operative capabilities, may be regarded as suitable blocks that can be synthesized in various ways to provide systems/holarchies with different degrees of complexity and size. The holonic analysis approach as a first step to modular building of systems exhibits significant advantages, such as, faster application design, re-usability of components, optimized interface with the providers, and ability of the system to evolve. The holonic synthesis approach as the second step for synthesizing new systems also exhibits attractive advantages, such as, ability to generate applications by using elements from a continuously enriched design template, time and cost savings due to the determined consistencies and interfaces among the components, and ability to add new elements to the template, thus extending the design capabilities.

The concepts of holons and modular systems and products, may also be well applied for designing generic products. The term is used to describe products designed in a generic Bill of Materials (BOM), which allows some flexibility over certain features of the generic product. Such an approach enables a corporation to deliver variable customized products, thus achieving product differentiation and higher customer satisfaction; nevertheless, the production of a very large variety of different products may cause significant problems in production planning and delivery times. These problems may be eliminated if: (a) a large number of common components is used, which facilitates production and materials requirement planning, and (b) if routing procedures are attached to each component of the end item. Both requirements are satisfied through the holonic approach; the routing procedures, in particular, may be expressed as interfacing rules.

This paper capitalizes on the theory of holons, thus taking full advantage of its generality, and investigates new fields where this theory can be applied effectively. Extensive research is carried out in the exploration of new potential application areas and a systematic approach is followed for the examination of the different facets of the holonic theory in a coherent manner. The presented research investigates generic manufacturing and business aspects, such as the modular design of products, manufacturing cells and business process modeling, since it is estimated that the holonic organization can provide valuable innovative concepts and solutions for improving the efficiency of such systems.

A real production system with embedded robot systems is further examined to reveal the full potential of the proposed approach. The methodology for obtaining a modular design template from which larger and more complicated, robot-integrated systems can be built is outlined herein, while sample rules for synthesizing such systems are exhibited. These systems are considered as the best paradigms for illustrating the feasibility and applicability of the holonic approach. The reason is that robotic work cells are typically composed of various components, which depend on the specific application to be handled. Such components are sensors, end-effectors, actuators, etc. Depending on the application, sensors may be used, for example, to measure proximity, force/torque, or to acquire an image; end-effectors may possess fingers or be simple grippers, while external actuators may be glue nozzles, welding torches, etc.

The use of the holonic approach permits the synthesis of various robotic applications by combining the distinct components, assuming of course, that their interface properties and technical specifications are known in advance. This is achieved if an appropriate "palette" is available a priori; such a palette may be created if certain basic robotic applications are analyzed and the holons that compose them are identified and described. This palette may then be enriched as new applications are examined and created.

2. HOLONIC MANUFACTURING SYSTEMS

The world of manufacturing has experienced radical changes during the twentieth century. The first thing to notice is *mass production and standardization of products*. After the Second World War, when world-wide capacity began to come into line with world-wide demand due to technology improvements, this situation changed. This caused a shift to manufacturing strategies, where manufacturers began to capture market share by differentiating their products in new way; thus, in addition to price, another competitive factor emerged, that of *product differentiation according to customer's needs*. The advance of technology led to the development of sophisticated manufacturing equipment, permitting production at significantly higher rates and lower cost. Additionally, manufacturing management techniques such as Just-in-Time and Total Quality, added to over-capacity by making manufacturing operations even more effective and efficient. At the end of the previous century, the manufacturing emphasis was thus to compete *on prices, choice and quality*. The above trends, justified the need to proceed through a review of the existing architectures for manufacturing systems so as to provide new solutions compatible to future expectations. In order to meet diverse customer requirements and maintain manufacturing competitiveness, next generation manufacturing systems must exhibit such features as *rapid development and deployment, flexibility with respect to product quantity and variety, and re-usability of manufacturing equipment and systems* [3,6]

In this environment, the concepts of open hierarchical systems (OHS), holons, holarchies, and holonic systems have been put forward as the key concepts for designing the architecture of the next generation manufacturing systems. Believers of the *Holonic Manufacturing Systems* (HMS) theory, believe that the best way to address the needs for next generation manufacturing systems is to *create open, distributed, intelligent, autonomous and co-operative manufacturing systems*. Such systems should be *highly flexible, reusable and modular manufacturing units* [4,5]. These units should also be able to reconfigure quickly and easily in order to produce various products, and to co-ordinate among themselves according to needs and to respond intelligently to unforeseen disturbances in the external environment, and to better maintain smooth and seamless production.

A **Holonic Manufacturing System** (HMS) is defined as *a highly decentralized manufacturing system consisting of co-operative, intelligent, autonomous modules, called "holons", that together yield an agile and self-organizing manufacturing system and support global optimization* An HMS is comprised of holons, people, communication network and methods for cooperation, including procedures for negotiation and resource sharing [8].

In a HMS, its key elements, such as machines, work centers, plants, parts, products, human operators, departments or divisions are regarded as holons. Yet, holons are elements which exhibit both autonomy and the ability to cooperate, that is, to make and carry out decisions through mutual agreements and coordinated actions. Therefore, each holon must have the data necessary for deciding its own actions, means of communicating with other holons, algorithms and procedures for negotiating and executing mutually agreed actions, and means of carrying out such algorithms, procedures and actions, whether by manual or automated means. For instance, a "virtual" (information only) product holon which needs to become an actual" (physical) product must be able to request other holons in the HMS to cooperate to carry out the manufacturing processes [9].

These necessary attributes for manufacturing holons lead to the conclusion that, in order to perform any actions, they need to exhibit intelligence and possess their own centralized command system [10]. Thus, *in a holonic manufacturing system, intelligence, information and control must be distributed among all holons*. This attribute differentiates HMS from other relevant manufacturing concepts, such as CIM, where the main control and intelligence resides mainly in the central computer, which controls the complete network; furthermore, this attribute is very useful to increase the capabilities of the entire system, because:

- Multiple, even conflicting goals can be achieved, since individual holons can be working on individual goals concurrently; and
- The system is readily extensible; a new holon can be added to an already working system, increasing the level of competence of the latter.

The issue of system's extensibility, establishes perhaps the greatest challenge for the HMS project. As a matter of fact, the vision of the HMS is the *design of highly decentralized architectures, built from a modular core of highly standardized, autonomous, co-operative and intelligent building blocks (holons)*. These modules exhibit all the attributes of holons; therefore, they are exchangeable, extendible and reusable and will exhibit skills such as self-diagnosis, self-repair, self-learning and self-control and organization [1,2].

Summarizing, the main distinguishing features of a HMS are: (a) the system is built from a modular core of highly standardized, autonomous, co-operative and intelligent building blocks (holons); (b) the elements are reusable; (c) the elements are self-configuring; (d) the system is developed from bottom to top; (e) the system is easily extendible; (f) a distributed database exists, containing goals for and state of, all holons updated in real-time; and (g) a communication network exists to allow information exchange with neighboring and remote holons.

3. APPLICATION

In this section, an industrial application which will be used as an example for the *analysis of a production system*, is described. The aim of this analysis is to develop a pool of hardware and software re-usable components to enable fast and efficient generation of new applications from the analysis of existing ones. A complete description of components should contain rules of composition/ interfacing, properties, technical characteristics, limitations, etc. The application examined herein has resulted from an effort to create a basis for systematic generation of robotic applications in automotive industry; these applications involve sensory feedback, as it is expected to provide significant profits such as, accuracy improvement, uniform quality, production increase, reduction of errors and of lost material, etc.

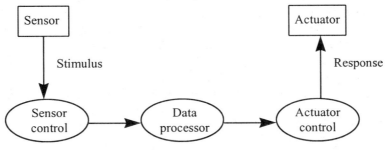

Figure 1: *Sensor-System-Actuator Model.*

The basis for applications generation is expected to have the form of a *blueprint*. The term blueprint is used to describe an aggregate of entities (hardware and software components) that are combined to form a robotic application. The description of the entities may contain properties, rules of composition/interfacing, technical characteristics, limitations, etc. As the analysis and study of the blueprint proceeds, the exact contents of the entities description will become clearer and more precise. In this context, an attempt will be made to generate a pilot blueprint in order to assess the feasibility of the concept and the evolution directions. The form of a blueprint for a robot system is based on the "sensor-system-actuator" model, depicted in Figure 1.

The implementation of a pilot blueprint proceeds through the following steps: (i) selection of a small number of robotic applications with visual feedback; (ii) first analysis of the applications into decomposable elements; (iii) further decomposition into primitive elements to the finest possible degree; (iv) identification of relationships of interaction and interfacing between the components; (v) identification of the common elements between applications and hierarchical composition of the common elements to groups; (vi) specification of the communication and interfacing rules between components; (vii) generation of a blueprint.

The blueprint generation in the final form will include the components, rules of composition, interfacing, etc. for a large number of applications. Among the profits of this implementation are the following:

(1) One will be able to generate new applications by using elements of the blueprint, so as to check for consistencies during generation, thus saving a large amount of time.

(2) New elements will be added to the blueprint if necessary. The blueprint will thus be enriched as new applications evolve.

(3) The need for new design and customized solutions from the providers will be minimized. Customization requests will be more specific. This will result to shorter implementation time and significant economy in development funds.

(4) The weak points of components, methods, etc. will be easier to identify and thus easier to correct, while the strong points will be more obvious and more easily estimated, and thus easier to profit.

(5) The time for application design and test will be minimized. This, in combination with the economic benefits of customized implementation, are expected to provide significant economic profits.

(6) The time for the implementation of a production line will be shortened. New products will enter the market sooner than expected, thus providing competitive advantages to the company.

(7) The applications will be more formally specified, described, and recorded. This will facilitate the development planning and will clarify the new research directions.

The application examined herein is the *sunroof placement application*. In this application, the workcell contains two industrial robot arms (IR1 and IR2), two conveyor belts (C1 and C2), and a glue nozzle (GN). Conveyor C1 brings the sunroof parts to the workcell, while conveyor (C2) brings the car bodies (Figure 2). The car bodies are assumed to be ready for the sunroof placement, that is, an appropriate opening has been created. The sunroof is placed in the inner part of the car body, that is, on the lower side of the opening. When the sunroof is positioned on the car body in order to be fixed, considerable pressure is exerted on the metal of the car body. This pressure may cause irreversible deflections to the car body. A rigid "negative" is positioned on the upper side of the sunroof opening, that is, in the outer part of the car body, n order to account for this deflecting pressure. Both the sunroof *and* the negative are pressed simultaneously, and for the same time period, against the car body in opposite directions and pressure magnitudes, so that fixation is performed without metal distortions.

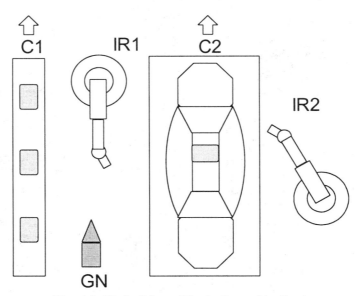

Figure 2: *Workcell Layout for the Sunroof Application.*

The work phases for this operation are separated to those performed by IR1 and by IR2. The work phases for IR1 are the following:

P1.1 *Pickup of the sunroof from conveyor C1*: The robot moves to the conveyor and picks up the sunroof. Proper grasp of the sunroof is ensured by the function the end-effector (EE1) of IR1. When the EE1 senses that the object is properly grasped, the robot picks up the object.

P1.2 *Glue application on the perimeter of the sunroof*: The robot moves the object close to the glue nozzle to begin gluing. Uniform and precise glue application is ensured by the use of a contour following sensor. When gluing finishes, IR1 removes the sunroof from the glue nozzle.

P1.3 *Placement of the sunroof on the car body*: The sunroof is moved close to the car body and placed to the opening. Precise placement is ensured by the use of visual sensors that monitor the sunroof position relatively to the opening and guide robot motion. After placement, the sunroof is fixed to the car body using clamps. After fixation, EE1 releases the sunroof and IR1 goes back to its initial position.

The manipulator IR2 has attached on its end the "negative" and clamps. There exists only one work phase for IR2:

P2.1 *Placement of the "negative" on the car body*: IR2 moves the "negative" close to the sunroof opening,and places it precisely using visual sensors to monitor the "negative's" position relative to the opening and to guide the robot's motion. After sunroof fixation on the car body, IR2 goes back to its initial position.

The analysis method that is used is the "Structured Analysis". This method is used for determining the requirements for a system, and to construct logical system models. These models are simplified depictions of the function of an object or the sequence of a process. They are easier, quicker and cheaper to create than the final system, and modifications can be effected here at a fraction of the cost required to alter a finished system. The tool that is used for creating Structured Analysis descriptions is ProMod, a CASE environment that contains a Structured Analysis tool.

Hierarchical structures are an integral part of the Structured Analysis method, enabling information to be represented on various levels of abstraction. All data can therefore be included without losing the diagram clarity, by successively "refining" diagram elements into further diagrams. In Structured Analysis Flow Diagrams are used. Flow diagrams model processes which transform data, and the interfaces between those processes. Diagrams are built up of *Nodes* (processes), *Stores*, *Terminators* and *Data Flows*.

- Data flows indicate the flow of data in a Flow Diagram to and from Nodes, Terminators and Stores.

- Nodes symbolise the processing of data. They transform incoming data into outgoing data. A Node representing a complex process can be additionally depicted in a separate Flow Diagram, its "child diagram". This process is called "refining", and is the basis for the system hierarchy.

- Stores hold data for use in more than one process (Node). Information is written to and read from Stores by Nodes. This is represented by incoming and outgoing Data Flows respectively.

- Terminators are data sources or sinks which show the exchange of information between the SA model and the "outside" world.

Structured Analysis in its simplest form represents the flow and transformation of data. A Flow Diagram does not take account of the chronological order in which events occur. Real Time elements can then be used in a Flow Diagram to model the system's dynamic behavior. These elements reveal the conditions under which the processes are triggered.

As regards the sunroof placement application, firstly, it is decomposed into two *robot systems*. Each robot system contains a robot (and eventually other actuators), sensors and the corresponding data processors, as shown in Figure 3. Each robot system is treated as an operational unit which performs a well defined task (production step). For example, a robot which grasps an object, applies glue on it, and then places it on another object is such a robot system.

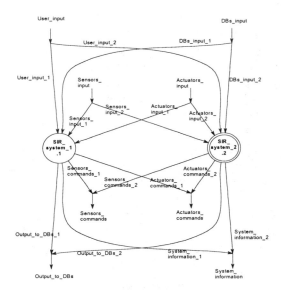

Figure 3: *Sunroof Placement Application.*

In the second decomposition step each robot system is described with a data processor, which communicates and exchanges data with the *control units* of sensors and actuators (Figure 4). The control units are the interfaces of sensors and actuators. The hardware capabilities are accessed through them. For each class of sensor, a corresponding sensor control unit class is introduced (Figure 5). For each sensor control unit class, there is a corresponding sensor processing entity (Figure 6). Such a flow diagram description does not imply that, in an implementation, for all sensor control units of a sensor class, only one sensor processing unit is used. It just describes the kind of data processing for a sensor class. In an implementation, more processing units of the same kind may be used in parallel.

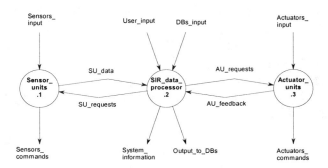

Figure 4: *Robot System Analysis.*

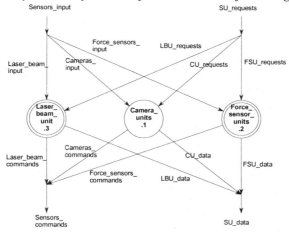

Figure 5: *Sensor Units.*

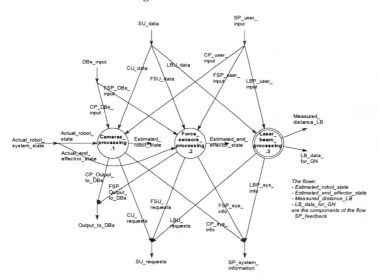

Figure 6: *Sensors Processing.*

4. CONCLUSIONS

From the previous analysis and the study in modular manufacturing systems using the holonic theory, the following conclusions result:

- The holonic architecture is a highly promising solution for the future, since it can combine the best features of hierarchical and heterarchical architectures is a flexible manner.
- The holonic architecture appears consistent to the needs for the design and organization of production systems and enterprises in the future.

- There exists a large number of possible applications of the holonic theory; however, the most promising areas seem to exist in designing modular production systems and products, in business organization of companies in holonic networks, in analyzing an enterprise to its business processes, and in biological systems.
- The holonic theory can be considered as a very effective framework for studying, analyzing and designing complex systems. The concepts of holonic organization can be exploited effectively for building a modular design template (*autonomy*) and the rules for synthesizing, later on, these modules in larger systems (*co-operation, interface rules*).

The contributions of this study include the following:

- development of a systematic basis for analyzing, structuring, and decomposing robotic applications with sensory feedback and formulation of a complete set of decomposition rules for these;
- establishment of the a CASE tool to analyze and structure robotic applications along with presentation of the holonic software concept and suggestion of parallelisation implementation;
- identification of the correspondences between software entities and system components.

BIBLIOGRAPHY

[1] Bongaerts L., Wyns J., Detand J., Brussel H.V., Valckenaers, "Identification of Manufacturing Holons", in *Proceedings of the European Workshop for Agent-Oriented Systems in Manufacturing*, pp. 55-73, Berlin, 1996.

[2] Bongaerts L., Valckenaers P., Brussel H.V., Peeters P., "Schedule Execution in Holonic Manufacturing Systems", in *Proceedings of the 29th CIRP International Seminar on Manufacturing Systems*, May 11-13, 1997, Osaka University, Japan.

[3] Carr D.K., Johansson H.J. *Best Practices in Reengineering*. McGraw-Hill, 1995.

[4] Fukuda T., Ueyama T., "Autonomous Behavior and Control", in *Proceedings of the IEEE International Conference on Robotics & Automation*, 1993.

[5] Huang C.C., Kusiak A., "Modularity in Design of Products and Systems", *IEEE Transactions on Systems, Man and Cybernetics*, **28**(1), January, 1998.

[6] Kidd P.T. *Agile Manufacturing: Forging New Frontiers*. Addison-Wesley, 1994.

[7] Koestler A. *The Ghost in the Machine*. Hutchinson & Co, London, 1967.

[8] McHugh P., Merli G., Wheeler W. *Beyond Business Process Reengineering: Towards the Holonic Enterprise*. John Wiley & Sons, 1995.

[9] Van Brussel H., "Holonic Manufacturing Systems, the Vision Matching the Problem", in *Proceedings of the First European Conference on Holonic Manufacturing Systems*, December 1994, Hanover, IFW-Hanover.

[10] Wyns J., Brussel H.V., Valckenaers P., Bongaerts L., "Workstation Architecture in Holonic Manufacturing Systems", in *Proceedings of the 28th CIRP International Seminar on Manufacturing Systems*, May 15-17, 1996, Johannesburg, South Africa.

A social actors network approach for the design of networked and virtual enterprises

António Lucas Soares[1,2], César Augusto Toscano[1], Jorge Pinho de Sousa[1,2]
[1]INESC Porto and [2]Faculty of Engineering of University of Porto
Rua José Falcão, 110, 4050-315 Porto, Portugal. e-mail: {asoares@inescporto.pt,
ctoscano@inescporto.pt, jsousa@inescporto.pt}

Abstract   This paper describes an approach to the technical and organisational design of networked (and virtual) enterprises using a social actors network perspective. The goal is to outline a set of methods, tools and models that support a more effective analysis, design, implementation and (re)configuration of of the organisational structure and information systems supportting networks of enterprises. This paper introduces the social actors network concept, as well as its principles and scope. The application of social actors network in the domain of networked/virtual enterprises is explained. The paper concludes with an example derived from the EU supported Co-Operate IST project.

Key words: networked and virtual enterprises, social actor networks, agent systems

1.   INTRODUCTION

Over the past decade much attention has been paid to the relations between enterprises in order to cope with an increasingly global environment. "Supply Chain Management", "Networks of Enterprises", "Virtual Enterprise", are terms that entered into the vocabulary of every manager, engineer and scientist interested in the production of goods or services. These terms also stand for a broad and active area of research. On a technical side, the research work has been split over topics such as distributed information technology infra-structures and distributed information systems architectures, novel forms of organisational structures facilitated by those technological advances, new business processes based on inter-organisational relations, etc.

On the social and organisational side, much less research work has been done. Particularly, when it comes to consider inter-organisational cooperation, the focus is at the business process level, with the actors (individuals and groups) being diluted in very aggregate roles. Nevertheless, relationships between enterprises will continue to involve people in key managing activities, taking operational, tactical and strategic decisions supported by conveniently distributed information systems. In this context, we support the argument that there is a need of methods that consider the design of inter-organisational co-operation including explicitly the social-actors involved in such co-operation processes. Therefore, we propose an approach to the analysis and design of networked/virtual enterprises (N/VE) centred around the concept of social actor's networks (SAN) [3]. The main goal is to provide a conceptual and methodological tool that is able to:

1. Support the modelling of N/VE focusing on the structural aspects i.e., to know which organisational elements (social actors) are involved, how these elements are related to each other, what kind of relationships exist between elements;
2. Support the modelling of N/VE focusing on the relational aspects i.e., to characterise social actors, to characterise the relationships between social actors;

This approach is intended to be a companion to a conventional business process analysis (including modelling). While a business process modelling method (for example UML with Erikson-Penker extensions) provide a tool to analyse the distributed business processes structure in a N/VE, it does not so for issues that are most important in implementing and operating these processes: social actors and their relationships. SAN is intended to be used as a conceptual tool to question and plan a desired network of social actors concerning the structural, relational and individual aspects.

We argument that the SAN approach can be an effective approach to the analysis and design of N/VE, unifying the organisational and information technology aspects.

2.   PRINCIPLES AND SCOPE OF SAN

SAN belong to a category of sociological methods (structural methods) where the goal is to highlight the structural relationships between social actors, enabling the contextualization of their actions in a systematic way. It is a simple representation of a system of complex relationships and exchanges between social actors [2].

More specifically, SAN are networks that connect individual and/or collective social actors. This concept is more specific than "social network" that names a very active social sciences area of research (a social network is not necessarily a SAN: take the example of a public transportation network).

Social actors can be in contact through physical infrastructures (phone networks, data networks) and this contact is mediated by several kinds of artefacts, from phone sets to computer programs. Here our main interest is obviously computer mediated communication, but SAN can encompass, in principle, every possible form of social contact.

Within the N/VE domain, there are several types of structures (networks) that are interesting to model, going from more formal types such as client/supplier or co-operative planning, to less formal ones such as technical co-operation, informal information exchange, interest groups, etc.

2.1 Definitions

A set of social actors constitute a network if there is some sort of relationship between them. Social actors are represented as the nodes of the network, while the relationships are represented as arcs. Social actors are individuals or groups capable of performing activities and of interacting with other actors in order to pursue goals. Below, we will consider software agents as a class of actors eligible to be included in a SAN.

It can be useful to distinguish three general classes of relationships in a SAN : transactions, controls and links. *Transactions* include every kind of exchange between two social actors within a network. This can go from well defined transactions such as exchange of data to ill defined transactions such as the exchange of influence. Transactions between two actors can be *symmetric* if both actors can establish similar types of transactions or *asymmetric* if it is not the case. *Control* can be defined as the capacity of a social actor to decide on performing or not a given transaction with other social actor. Control can be *unilateral* or *shared* (this can be further characterised by a level of control over a given transaction). *Links* are the more general class and include every possible kind of relationships not falling into the previous categories. Examples of links can be attitudes such as "awareness", "hostility", "trust", being possible to distinguish between *positive, negative* or neutral links (or using another terminology *identification, differentiation* or *indifference* links).

2.2 Basic levels of analysis

The analysis of a SAN can be done at several levels. The choice of the adequate level depends obviously upon the purpose of the analysis. One possible division (inspired in [3]) is to consider the individual, relational and structural levels. This provides several abstraction or detail levels.

The *structural level* tries to describe and analyse globally a given network, requiring complete data about the different actors and types of relations. For example, for a complete supply chain to be studied at the structural level all the participating companies must be known, as well as the whole of their client-supplier relations. Such analysis would provide for example an identification of companies with the same supplying characteristics (types of products, delivery performance, supply chain tier, etc.).

The *relational level* concerns the characterisation of the relationships between the social actors. Some types of measures taken from graph theory (e.g., distance, accessibility) can be used to characterise these relationships. Analysis at this level can be either qualitative and quantitative. Examples of this are the evaluation of logistic issues between two supply chain commercial partners (such as delivery lead times, quality performance), the degree of trust between partners, the degree information exchange concerning several subjects (such as production plans, informal information concerning production status).

Finally, at the *individual level* the social actors are analysed and characterised based on the relationships in which they participate. This should be complemented with other types of characterisation of the social actors, resulting from the application of other analysis methods. Examples of analysis results are the evaluation of the importance of an actor (e.g., measured by the centrality of his position concerning information exchange in the network), the potential for an actor to establish relationships in the network, etc.

2.3   Basic types of analysis

At the basic levels described above, SAN approach can be used to analyse different aspects of a network of social actors. In social sciences research, SAN approach has been used for several types of analysis: communication networks, support networks, enterprise networks, policy networks, commerce networks, etc. (see e.g., [3]).

The basic types of analysis of N/VE reflect the perspective under which the analysis is done. The possible analysis types fall in two categories (1) organisational and/or technical structure analysis and (2) social interaction analysis. These two categories promote a socio-technical approach, as they can be identified with a technical and social sub-systems analysis.

3.   ANALYSIS OF SAN IN THE N/VE CONTEXT

A network of enterprises can be made up of several different entities and types of relationships according to their relative position within e.g., the supply chain, and consequently may have several interaction forms working simultaneously. Interaction among participants in the network, their co-ordination and their access to information are becoming increasingly important. Setting up a network and managing it in an optimised way, whilst balancing customer needs with increased performance along the whole chain, may be a key factor for the competitiveness of a specific company and in overall the entire network. For a clear understanding of what are the enabling and restraining factors of effectiveness in an established network or one setting up, SAN modelling will help to analyse the complementary perspectives.

This approach is intended to be sufficiently powerful to be used in both organisational and technical design. In fact, the broad goal of our research work is to investigate tools to support the joint and optimised development of information systems and organisational processes in N/VE. Organisational processes encompass business processes as well as the individual and social processes.

We will describe now how the SAN approach can be used in the main phases of virtual/network enterprises life-cycle: creation and operation.

3.1   Creation

According to [1], the major steps in creating a VE or network of enterprises are: (1) setup infra-structure resources, (2) manifestation, (3) creation/evolution, (4) contract negotiation/re-negotiation, and (5) enterprise configuration/re-configuration. Phases 1 and 4 will not be considered as the SAN approach would have little to do about them. Phase 1 concerns to the definition and installation of generic computational resources and phase 4 has to do mainly with the establishment of contracts between enterprises.

Manifestation regards basically the individual profile definition and the manifestation of interest in participating in a VE by disseminating the profile. *Creation/evolution* involves in the first place the identification of a business opportunity. After this, a viability study that includes a rough plan for the VE should be made. Partner search and selection as well as the VE definition are the remaining steps. *Configuration* is the step were the VE processes are detailed and defined. VE processes have to do basically with inter-enterprise co-operative activities, which include the communicative elements definition such as formats and ontologies, as well as access rules. VE processes are configured in the creation phase of the VE, but reconfiguration is likely to be needed at some points of the VE life.

3.2 Operation

In N/VE operation we are concerned with the execution of the distributed business processes. The main issue in this phase is to guarantee that the appropriate infrastructure (or execution system) is deployed to fully satisfy the design and configuration resulted from the creation phase.

Nevertheless, during the operation of a N/VE it might be necessary to do some adjustments (due to either operational or tactical reasons) such as change partners, include a new partner, change a partner role, etc. This originates a reconfiguration of the N/VE. This is a (re)design activity similar to the one in the creation phase.

3.3 SAN approach to configuration/reconfiguration

SAN role in the N/VE creation phase is mainly to give support to the network formation. This covers mostly the configuration step but also touches manifestation and evolution. Furthermore, some support can be given to reconfiguration activities in the operation phase. In the research work done so far we have found particularly useful the structural and relational levels of analysis.

3.3.1 Structural analysis

This analysis involves three steps: identification of the social actors, definition of the type(s) of SAN to be analysed and the modelling of the SAN.

The identification of relevant social actors cannot of course be dissociated from the type of analysis to be carried out. Furthermore, during the modelling and analysis, additional actors can be identified. This can be a complex topic and it is outside the scope of this paper. For now, it is satisfactory to consider as social actors the individuals or groups that are relevant in operational terms, i.e., the personnel, departments, etc., involved in the execution of the distributed business processes.

There are many types of analysis that can be done to support the configuration/reconfiguration of a N/VE. Again, the distributed business processes preliminarily defined set a first set of possible SAN. For example, if the N/VE is particularised in a supply chain, a logistic analysis (in a broad sense) is an obvious type of SAN to be considered (see figure 1). Less obvious would be a SAN depicting a "co-operative planning" process, involving several actors co-operating in responding to customer demands in the supply.

3.3.2   Relational analysis

SAN Relational analysis focus on the relationship between two social actors in a SAN. The main goal is, within a given SAN, to characterise how two particular social actors interact. By classifying the links between nodes (either qualitatively or quantitatively) in any of the types above referred, it is possible to evaluate relationships against some pre-defined criteria. Examples of the later are "degree-of-interaction", "degree-of-control", "degree-of-dependency", "connection potential", "distance", etc.

In the example of figure 1, the relationship between the logistic dep. of Ent-A and logistic dep. of Ent-C is quite obvious in terms of co-operative planning for the supply chain. In the design of a co-operative planning business process for the N/VE several relational issues between the two social actors should be analysed for an optimised configuration of these processes. For example, operational issues (stated in the desired activity models for the business process) imply the exchange of production plans by the two departments. Consequently, either side should be able to access to this information timely and easily. Thus, in terms of configuring the N/VE, either of these actors should be characterised with an high level of accessibility regarding plans information.

On the other side, "co-operative planning" requires an attitude of trust between social actors (regarding particularly to information sharing). If the links in a network model this attitude, e.g. by assigning to each link a "degree-of-trust", this SAN could give the relative degree of trust between any two (connected) actors.

4.   EXAMPLE: SUPPLY CHAIN CO-OPERATION FOR ORDER ACCEPTANCE AND CHANGE

4.1   The CO-OPERATE project

Co-OPERATE is an IST project supported by the European Commission, that aims at developing solutions to enhance the entire supply network, providing an advanced information and communication infrastructure to support general co-operation, as well as particular methodologies for co-operative planning and for network set-up and support. The project has the general goal of improving the performance of networked enterprises by achieving: co-operation and supply chain integration, higher responsiveness, lead time, reduction throughout the supply chain and reduced inventory levels, improved flexibility in the supply chain, increased capability to solve unexpected problems in the supply chain. In order to achieve these goals a number of "business solutions" (or general functionalities) have been defined, that will focus on the network aspects of the business processes and will try to accommodate and support the current internal tools and processes of companies. These business solutions are the following: feasibility studies for new order or change requests across the network, long term business planning for the network, standard operational order and planning processes, exception handling process, multi sourcing co-ordination, process visibility, performance management information.

4.2   General requirements for a business solution

In this work we are mainly concerned with the first of the above business solutions - ReFS - "feasibility studies for new order or change requests across the network". It should provide a set of functionalities that support the request for new orders or large order changes across the network and co-ordinate the feasibility checks at the individual companies within the shortest possible time. This includes checking of capacity and materials from suppliers. For the general functionality of ReFS, it will be considered a scenario that assumes that a core company with their own customers - downstream in the chain, is customer of a number of suppliers. This company holds the order data and ensures the availability of them to their suppliers. The suppliers can request actualisation of orders either by a pull or push process. Information about orders should be sent automatically to all suppliers using the best communication channel. All orders from the core company to each supplier should be presented together with their actual status and the supplier should have the choice to view them sorted according to a customisable filter. All new and changed orders (which not yet have been acknowledged by the supplier) should be highlighted. The planner in the supplier companies should be able to acknowledge each new or changed order and register the acknowledgement data such as date/time, user identification, etc. All changes are stored in a history table. Feasibility checks at the individual companies should be made within the shortest possible time including checking for own capacities and requesting needed materials from suppliers, which need to do the same process themselves. This deals with the feasibility of acceptance of new orders, with the production of a quotation and with a real-time check of capacity and materials availability. Results should be co-ordinated and consolidated to come to a complete solution. This process achieves a co-operative production plan for the network. The end customer or a supplier, should be able to track, along the production chain, the positioning and the status of completion of each related order.

Figure 1 – A "logistic SAN" including IS agents

4.3   SAN analysis for the business process configuration

The first network developed to analyse this business solution was a "logistic" SAN (see figure 1). The structural analysis enabled, in the first place, a common understanding of how the actors involved would interact with each other in what concerns to the operational aspects of order acceptance and change.

4.3.1   Information system specification

We consider now the specification of the information system supporting

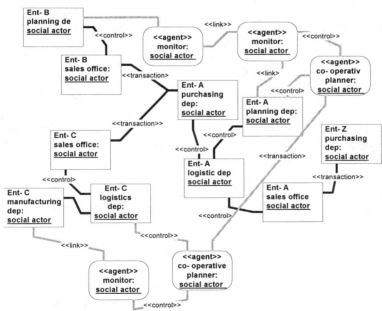

the "logistic" SAN of figure 5. A first specification of a community of agents would imply to define how the agents are conceptually integrated in the network. Two options can be considered:

1. Agents are social actors, i.e., are considered as behaving minimally as the human counterparts. This implies that agents establish relationships with human social actors and between themselves.

2. Agents are extensions of the human social actors, modifying the characteristics and behaviour of the later. This leads to the modification of the types of relationships between social actors.

In a first approach we decided by the first option because it is easier to carry out the conceptual integration. Nevertheless, further research is needed in order to support with rigour this decision.

4.3.2   Defining the infrastructure

To support ReFS business process, an information system relying on a multi-agent infrastructure was specified. We considered this particular agent based system structured around the following basic architectural principles:

- each business unit in the network of companies is served by a set of agents;
- each of these sets forms a "node" in a community of agents distributed by the several "units" in the network;
- in each node, the agents co-operate to achieve local goals;
- in each node, each agent performs one or more functions, and co-ordinates its decisions with the other agents in the node;
- different nodes in the network co-operate to achieve global or local network goals; co-operation between nodes is carried out through the interactions of the individual agents in the different nodes;
- the functionality across the network or supply-chain is achieved through the interaction of the different nodes;
- the number and types of agents present in each node is variable, i.e. we can have all kind of agents or only some of them.

SAN models contemplating several aspects were built for this organisational reality (involving three companies in a supply-chain). The analysis of these models, together with technical UML based business models, led to a first specification of community of agents, identifying mainly the agent models and agent co-operation models.

4.3.3   Basic types of agents

From the above analysis, the required functionalities where assigned to a set of basic agent types (the new social actors). Basically, the design of the system was viewed as a set of roles that (1) interact with each other according to some protocol, (2) have responsibilities, which in a direct or indirect way determine the functionality inherent to that role, (3) have a set of permissions which identify the "rights" associated with the role, and subsequently identify the resources that are available to that role when its responsibilities are placed into practice, (4) perform local computations or activities (local in the sense that no interactions are carried out with any other role in the system).

Table 1 – Two basic agent types

Agent	Role
Co-operative Planner	• Co-operates with other planners in other nodes, in orde to develop plans that achieve the network goals • Interacts with the local Capacity Agent;
Monitor	• Acquires and aggregates the network or supply-chain state variables; compares current state variables wit target values; • Identifies deviations or disruptions, e.g. resource failur (loss of production capacity), logistics problems (transport delays), quality issues; • Propagates warnings to other agent types.

The resulting SAN is depicted in figure 1. This SAN shows clearly (in qualitative terms) the new relationships between the human social actors. For example, it can be concluded that Ent-A planning dep. will be able to better know what is happening in the suppliers, although potentially Ent-B would be "more accessible". Ent-C doesn't want (or can't) to make available information about the production plans and status.

5.   CONCLUSION AND FURTHER WORK

This is a preliminary report of an on-going work. Although the conceptual tools provided by the approach revealed useful in complementing a distributed business process analysis and supporting information systems specification, there are more interesting possibilities open by a further sophistication of the network models. In particular, the use of graph theory can provide a set of quantitative results that can support the analysis even knowing that it is basically qualitative. Measures such as the "connection capital", "centrality", "groupability", can be calculated from the graph, if appropriate values are assigned to the connections. These measures will help to characterise in more detail the social actors and their relationships.

The example focused exclusively in the structural analysis. Current research work will result in a future paper describing how to fine tune the relationship between two social actors, using a relational analysis. In the same way, social interaction analysis is object of current research.

References

[1] Camarinha-Matos, L., Afsarmanesh, H., Rabelo, R., 2000, Supporting Agility in Virtual Enterprises, E-Business and Virtual Enterprises, Camarinha-Matos et al. (eds). Kluwer Academic Publishers.
[2] Lazega, E., 1997, Network Analysis and Qualitative Research: a Method of Contextualization, in G. Miller and R. Dingwall (eds.), Context and Method in Qualitative Research, London, Sage.
[3] Lemieux, V., 1999, Les réseaux d'acteurs sociaux, Presses Universitaires de France.

Biographies

António Lucas Soares is a senior researcher at INESC Porto and Assistant Professor at the Faculty of Engineering of University of Porto.
César Augusto Toscano is project leader at INESC Porto.
Jorge Pinho de Sousa is a senior researcher at INESC Porto and Assistant Professor at the Faculty of Engineering of University of Porto.

Virtual Enterprises and Security

István Mezgár[a] and Zoltán Kincses[b]

[a] *Computer and Automation Research Institute, Hungarian Academy of Sciences,*
   Budapest 1111 Kende u. 13-17, Hungary, E-mail: mezgar@sztaki.hu
[b] *SEARCH Laboratory, Budapest University of Technology and Economics*
   Budapest 1117 Pázmány Péter st. 1, E-mail: kincses@ludens.elte.hu

Abstract: In the industry the communication security has got an increasing role due to the increasing number of virtual enterprise (VE) realizations. As the basic characteristics of VEs is the communication on the Internet, and during these transactions huge amount of extremely valuable technical data, and information (product, process data besides the business information) are moving between the members of the networks, the security has became a vital problem of this field.

The paper introduces shortly the main characteristics of VEs and some VE reference architectures and the communication architectures and protocols they can use. In the second part the hardware and software security methods, approaches and tools are shortly described that can be applied during the transfer of technical related data, information on the network, and on the computer while storing or handling the information. The main goal of the paper is to call the attention of the industrial community with some examples to the importance of security in (computer) network communication.

Keywords: communication security, reference architecture, virtual enterprise

1.   INTRODUCTION

Secure identification, authentication of the users and communication security are main problems in network communication. In the industry the communication security has got an increasing role since the growing number of virtual enterprise (VE) realizations. The spread of virtual enterprises (VE) and VE-like manufacturing architectures is on the doorstep of the industry. There are many applications, but because of the lack of proper co-operative, effective and secure software tools in the filed the massive spread of real VEs has been pushed into the close future.

As the basic characteristics of VEs are the communication on the Internet, and during these transactions big quantity of important technical data and information (development-, product-, process data- besides the business information) are exchanged through the networks, so today security is in the focus in these environments.

The paper aims to introduce the existing networked manufacturing architectures, their communication protocols and standards, the connections among the different fields of the networking technology. The description of the connected applications is focusing on the virtual enterprise, on the importance of security in "manufacturing communication", and outlines the possible secure forms of communication for VEs.

The paper doesn't intend to give a full overview, or a detailed description on networking, or on security, rather wants to flesh the dangers of sending valuable information through networks and how to avoid these traps, and push the users in the field of manufacturing into the direction of secure communication.

2. VE CHARACTERISTICS AND RAs

2.1 Characteristics of VE

There are different forms of the distributed enterprise, but the virtual enterprises are one of the most up-to-date forms of production. Based on the different definitions of VE it can be stated that the intensive use of computer networks and the high-level organization flexibility are main parameters of virtual enterprises. Enterprises forming a holonic manufacturing system are potentially enabled to co-operate with each other to achieve a common goal. This happens in reaction to an external stimulus, taking the form of a new business opportunity which can be better exploited by more joined enterprises than by an individual firm. In these circumstances a *virtual enterprise* is created [3].

In its current definition, the virtual enterprise is formed by a proper combination of specialized nodes, including financial and engineering firms, manufacturers, assemblers and distributors. This structure can be seen as a holarchy, in that it is a temporary, goal-oriented aggregation of several individual enterprises. Each virtual enterprise is created to pursue a specific business objective, and remains in life for as long as this objective can be pursued. This temporary aggregation is supposed to involve enterprises from different sectors and categories.

After that, the individual nodes resume their independence from each other. Node resources that were previously allocated to the expired business are re-directed toward the node individual goals, or toward other virtual enterprises it may have joined.

To allow this kind of dynamic re-configuration of the whole system in response to market changes significant requirements must be met. On one side, individual enterprises must improve their flexibility and extend their connections with the other members of the system. On the other side, the manufacturing infrastructure must support fast interaction as well as information sharing between the nodes. Some of the most significant benefits expected for enterprises joining a virtual organization are:

- New business opportunities become available, by combining the productive capacity and marketing strength of all nodes in the virtual organization.
- Design and development capacity is increased by knowledge sharing between nodes with complementary skills.
- Cost and risk factors for the development of new products are shared among the nodes.
- Due to the specialization of roles within the network, each individual enterprise is enabled to focus on its core processes, thus optimizing and improving them.

2.2 VE Reference Architectures

There are different projects to aim some type of general (reference) VE model, or more complete reference architecture. A general model offers limited possibilities but it can have some special characteristic at the same time. The network architecture can be defined as the design principles, physical configuration, functional organization, operational procedures, and data formats used as the bases for the design, construction, modification, and operation of a communications network. Reference architecture is a combination and arrangement of functional groups and reference points that reflect all logically possible network architectures in a structured, flexible, and modular way. Reference architecture gives the general representation of the architecture from which all individual architectures can be derived.

The reasons, or goals to develop reference architectures are to give a complete, and well-defined frame (applying the standards, the ready-to-use of new technologies) for the practical applications. By using the reference architecture the actual individual architectures can be configured easily, and through applied technologies and communication protocols they are platform independent.

The National Industrial Information Infrastructure Protocols (NIIIP) project initialized by the National Institute for Standards and Technology (NIST) is one of the most complete realizations of a VE architectures [5]. It intends to bring together the product realization process integration efforts, by developing general global protocols for the technical standards of product data definition, communication, and object technology and workflow management. The NIIIP doesn't intend to develop a new system, rather applying existing standards to consolidate, harmonize, and integrate the many sets of existing protocols. The main goals of the NIIIP reference architecture is to help establishing and operating of VEs in the industry, by applying standardized solutions for VE connectivity, for industrial information modeling and exchange and for Management of VE projects and tasks.

Another architecture has been developed in the PRODNET-II project. The main goal of the PRODNET project was the development of infrastructures to support industrial VEs through the design and development of an open platform and the adequate IT protocols and mechanisms [1]. PRODNET was focused mainly on SMEs in order to support them with tools to inter-operate with other networks. The architecture employs the new emerging standards and advanced technologies in communication, co-operative information management, and distributed decision-making. PRODNET deal with a number of VE environment requirements.

3. COMMUNICATION REQUIREMENTS AND PROTOCOLS IN VE

3.1 Communication requirements in VE

The basic characteristic of VE is the flexibility both in information- and in material flow. All main events of its life cycle are connected to communication on the network. The communication requirements for a VE can be summarized in the followings:

1. Integration of different communication forms and resources
   Communication through connected telephone-, computer- and cable networks, and application possibilities of different protocols, connecting wired- and wireless equipment.
2. Reliable and high quality communication services
   Reliability covers the high on-service time (technical reliability), the high availability (well designed/balanced network – resource reliability), the HW and SW security both for equipment and communication lines (access reliability), well controlled/organized networks (organization reliability), all these with reasonable cost.
3. Global time co-ordination
   It is essential the exact co-ordination of the different actions in time during the life cycle of the VE, so a general time has to be declared for communication.

4. Traceable communication

Traceable means to document and audit the communication in a way that fulfils the requirements of bookkeeping (e.g. delivery report and receipt notification) and legal aspects (e.g. digital signature).

As the goal of the present paper is the description of the virtual enterprises from security aspects, in the following the HW and SW security of equipment and communication lines (access reliability) and the control/organization of networks (organization reliability) will be discussed. The security requirements for a VE can be listed as follows:

1. Protection of all types of enterprise data (for all company forming the VE)
2. Privacy and integrity of all types of documents during all phases of storage and communication (Data and communication security – Certification, Encryption),
3. To enable for companies confidential access control,
4. Authorization and authentication of services (digital signature).

These services need to be flexible and customized to meet a wide array of security needs, including specific high-level requirements. In order to fulfill the communication and security demands some basic aspects have to be taken in consideration while selecting security and communication technologies:

1. Platform independent SW tools have to be applied,
2. Standards have to be applied (accepted and "de facto" standards as well),
3. Appropriate architectures with ability to integrate different resources.

Fulfilling all types of the introduced requirements for individual enterprises would be very hard if not possible, so different general network- and organizational structures have been developed, that have been carefully designed and tested. These structures can be defined as reference architectures, and they are available both for the organization and for the information infrastructure of virtual enterprises.

3.2 Communication Protocols

There are two basic sets of general communication protocols that are organized in multi-layer architectures.

The Open Systems Interconnection (OSI) - Architecture is a communication system architecture that adheres to the set of ISO standards relating to open systems architecture [4]. The Open Systems Interconnection-Reference

Model (OSI-RM) is an abstract description of the digital communications
between application processes running in differing systems. The model employs a hierarchical structure of seven layers. Each layer performs value-
added service at the request of the neighboring higher layer and, in turn, requests more basic services from the next lower layer. The TCP/IP is two interrelated protocols that are part of the Internet protocol suite. TCP operates
on the OSI Transport Layer and breaks data into packets, controls host-to-
host transmissions over packet-switched communication networks

In manufacturing communication, there are several protocol architectures
that are based on the network reference models and on their protocols. In the
followings only a short list is given the most familiar protocols applied in
manufacturing environment. The MAP (Manufacturing Automation Protocol), TOP (Technical and Office Protocols), CNMA (Communications Network for Manufacturing Applications), OSACA (Open System Architecture for Controls within Automation Systems), PRO-
FIBUS and Fieldbus are worth to mention protocols. These architectures, or
protocols use, or are based on the ISO/OSI reference model, and partly also
use its protocols. Recently the techniques and tools supported by OMG are
also involved in manufacturing automation.

4. SECURE COMMUNICATION IN VE

4.1 Computer systems and security

Security can be defined as the state of certainty that computerized data
and program files cannot be accessed, obtained, or modified by unauthorized
personnel or the computer or its programs. Security is implemented by restricting the physical area around the computer system to authorized personnel, using special software and the security built into the operating procedure
of the computer. When applied to computer systems and networks denote
the authorized, correct, timely performance of computing tasks. It encom-
passes the areas of confidentiality, integrity, and availability.

Security is a conscious risk-taking, therefore in every phase of a computer system's life cycle must be applied that security level which costs less
than the expense of a successful attack. With other words security must be
so strong, that it would not be worth to attack the system, because the investment of an attack would be higher than the expected benefits. At different levels different security solutions have to be applied, and these separate
parts have to cover the entire system consistently.

In Table 1 the main practical fields of security are summarized in order to
better understand the content of the following sub-chapters. The abbrevia-

tions of SW and HW have a broader purport in the table not only referring to the computer science.

In the field of security standards and quasi standards have an important role. In the followings some of the most relevant ones are introduced shortly, only to show the directions and status of these significant works.

In order to classify the reliability and security level of computer systems an evaluation system has been developed and the criteria have been summarized in the so-called "Orange book" [6]. Its purpose is to provide technical hardware/firmware/software security criteria and associated technical evaluation methodologies in support of the overall ADP system security policy, evaluation and approval/accreditation responsibilities promulgated by DoD Directive 5200.28.

Table 1. Main fields of computer security

	Organization security	Personal security	Network (channel) security	Computer (end point) security
SW secu-rity	Definition of security pol-icy (e.g. ac-cess rights)	Employ-ment of trained and reliable staff	Using tested net-work SW tools, and continuously checked communi-cation channels and well configured network elements	Using tested applica-tion SW tools, and continuously checked operation system, and properly configured HW systems
HW secu-rity	Placing the computers in secure loca-tion of the building and offices	Physical identifica-tion tech-nologies (finger-prints, etc.)	Prevent direct, or close access to net-work cables, or application of spe-cial technologies	Prevent direct physi-cal access to comput-ers by unauthorized persons, or a close access in electromag-netic way

The ISO/IEC 10181- [2] multi-part (1-8) "International Standard on Se-curity Frameworks for Open Systems" addresses the application of security services in an "Open Systems" environment, where the term "Open System" is taken to include areas such as database, distributed applications, open dis-tributed processing and OSI. The Security Frameworks are concerned with defining the means of providing protection for systems and objects within systems, and with the interactions between systems. The Security Frame-works are not concerned with the methodology for constructing systems or mechanisms. The Security Frameworks address both data elements and se-quences of operations (but not protocol elements) that may be used to obtain specific security services. These security services may apply to the commu-nication entities of systems as well as to data exchanged between systems, and to data managed by systems.

4.2 Secure Architectures

The security architectures represent a structured set of security functions (and the needed hardware and software methods, technologies, tools, etc.) that can serve the security goals of the distributed system. In addition to the security and distributed enterprise functionality, the issue of security is as much (or more) a deployment and user-ergonomics issue as technology issue. That is, the problem is as much trying to find out how to integrate good security into the industrial environment so that it will be used, trusted to provide the protection that it offers, easily administered, and really useful.

The critical security points for virtual enterprises are the access points, the improperly configured systems, the software bugs, the insider threats and the physical security. The following two examples shows how part of these problems are solved in the VE reference architectures.

In NIIIP, secure communication can be implemented at three levels:

- IP level — discussions of protocol-level security are underway within industry organizations.
- OMG level — a Request For Proposal (RFP) for the Security Object Service has been issued.
- NIIIP level — solutions for VE security. NIIIP is proposing an object interface with an extra security tag to activate the enforcement of secure communications over networks. These security requirements can be implemented using well-known data encryption methods available on the Internet. These include Public Key Cryptography from RSA Data Security Inc. and Data Encryption Standard (DES).

The goal of the PRODNET Communication Infrastructure (PCI) is to fulfill the security and the legal requirements besides the functional ones. By using the PCI it is guaranteed that no one, other than the owner, can access to a document (privacy), the content of a document can't be changed without detection (integrity), each received document is unambiguously connected to an identifiable sender (authentication) and the logging information is maintained for auditing and communication management.

4.3 Security in the Network

At the beginning of networking there was a need mainly for the reliable operation, but the secure and authentic communication has became a key factor for today. According to Internet users, security and privacy are the most important functions to be ensured and by increasing the security the number of Internet users could be double or triple according to different surveys. The main reason of the increased demand is the spread of electronic commerce through the Internet, where money transactions are made in a size of millions of dollars a day. It is not just the question of our letters content or our user account, but it is the question of money. False transactions in the real world are not so easy than make them in the insecure virtual world, where the speed and the effect of these false transactions are not only dangerous for the individuals but it is highly dangerous also for big organizations.

Every part of a network could be attacked just the level of expected success and effort determines the targets. From server break-ins (change a Web-page, create false user account, steal information) through DNS spoofing (the attacker falsify the IP address, and the user thinks is surfing a trusted site, but it is not the case, and if the credit card number was given on such a site there is no guarantee that others will not use the card numbers later for their aims) till password or other information stealing every element of network could be dangerous without attention to security.

Experts are arguing with each other and with outsiders too about the strength of each crypto algorithm, but in the first step every algorithm is better than nothing. Instead of FTP there is SFTP (secure FTP) or SCP (secure copy), instead of HTTP there is SHTTP, which is HTTP over SSL (Secure Socket Layer). Instead of simply e-mail there is the PGP (Pretty Good Privacy) signed e-mail. With these techniques it can be guaranteed that the information in e-mail, file or on the Web page will be reached only by authorized parties. These solutions are SHOES (security help-tools over existing solutions), which helps not to walk barefoot on the information superhighway.

In cryptology-based algorithms a key-role has the key-management from issuing state through storage and use till revocation of keys. At issuance it is possible to sign a key by the issuer, and then the issuer must be a well-known party for others who want to check the signature on this key. If a key is not signed, it is up to us the trustee decision. On each way there is the problem of revoke, because these keys are not forever. Sometimes there are time-stamps on the keys or signatures, sometimes they are used for just one time (for example the One Time Password system), and exists algorithms with merging these two ideas. In this case the One Time Password is a ticket, which has an expiration time.

4.4 Security of the Computer

The first line of defense is the physical access of a computer. There is a phrase in security experts' community that every computer is possible to break in, if its console is accessible. Electromagnetic emission is another key factor in securing physically a computer from being monitored from distance with specific antennas. Here the tempest room is applied, based on the well-known Faraday-cage idea.

The password protection is the first step in computer access management. As the next step the private files have to be protected from other users and even from the supervisor, who can access private files on a server even if the owner has not logged in. Therefore in high security systems and today for PC users too there are different CFS solutions (Crypto File System) available, where the content of the hard disk (HD) or home directory is encrypted. After the logon procedure the user has to enter another password than the logon password, which will deactivate the lock algorithm of HD or home directory.

In case of logging into a computer system it is recommended to use an interactive memory-resident virus scanner (under some operating systems, like Windows it is base requirement against viruses) to protect our resources from viruses coming from Internet or intranet or from a colleague's floppy.

The best is when the prevention techniques are applied on different levels. Routers help us in selecting the information by its source of IP address. The firewall filters different services and protocols. The server filters the break-in trials, and user operations from outside and inside too. Local programs are scanning for viruses. There are solutions to filter viruses on the firewall too, but these elements of security services are occasional solutions in specific systems. The required security technologies have to be applied on each level to result the whole system evenly consistent (also called robust).

The training of users has a significant role in that domain. Users will never apply a security technology if there is any less complicated procedure (which is insecure!) than the secure solution. It must be explained for the users, what they can lost without security and what threats they can produce against their user colleagues.

In academic sphere it is not a big problem to trust in system administrators, because there is a non-financial work management. For a bank or for a company it is not allowed to put any other parameter before security, because the profit and the long-term trust is based on the availability and non stop service.

5. CONCLUSIONS

Security techniques and technologies have become high priority as different kinds of networking are approaching to each other, sometimes are integrated. This gradual integration is called convergence, and the technology convergence is based on the common application of digital technologies to systems and networks associated with the delivery of services.

The global nature of communication platforms, particularly the Internet, is providing a key, which opens the door to the further integration of the World economy. The distributed production systems with different sizes will play a definite role, but originating from their openness and flexibility their information systems will be a security risk. They will need complex, flexible security systems that are user friendly and platform independent at the same time. The developments of hardware and software elements of such systems are going on and the potential users have to get acquainted with them. The main goal of this paper was to help this process.

ACKNOWLEDGEMENTS

Part of the work included in this paper has been done with the support of the OTKA (Hungarian Scientific Research Found) project with the title "The Theoretical Elaboration and Prototype Implementation of a General Reference Architecture for Smart Cards (GRASC)" (Grant No.: T 030 277).

REFERENCES

[1] Camarinha-Matos, L.M., Afsarmanesh, H., Garita, C., Lima, C., Towards an architecture for virtual enterprises, Keynote paper, Proc. 2nd World, Congress on Intelligent Manufacturing Processes and Systems, Springer, Budapest, Hungary, June 1997, pp. 531-541. - The PRODNET II project, http://www.uninova.pt/~prodnet/

[2] ISO/IEC 10181-1:1996 Information technology -- Open Systems Interconnection -- Security frameworks for open systems: Overview.

[3] Mezgár, I. Communication Infrastructures for Virtual Enterprises, position paper at the panel session on "Virtual Enterprising - the way to Global Manufacturing", in the Proc. of the the IFIP World Congress, Telecooperation, 31 Aug.- 4 Sept. Vienna/Austria and Budapest/Hungary, Eds. R. Traunmuller and E. Csuhaj-Varju, pp 432-434.

[4] Tanenbaum, A.S., Computer Networks, Third Edition, Prentice-Hall, 1996.

[5] The NIIIP reference architecture, final document, http://www.niiip.org/public-forum/index-ref-arch.html

[6] Trusted computer system evaluation criteria, Orange book, DoD 5200.28-STD, Department of Defense, December 26, l985, Revision: 1.1 Date: 95/07/14.

Intelligent Open CNC System Based on the Knowledge Server Concept

János Nacsa
(nacsa@sztaki.hu)
Computer and Automation Research Institute (www.sztaki.hu)
Budapest, Hungary

Abstract: In an ideal scenario of intelligent machine tools [18] the human mechanist was almost replaced by the controller. During the last decade many efforts have been made to get closer to this ideal scenario, but the way of information processing within the CNC did not change too much. The paper summarises the requirements of an intelligent CNC evaluating the different research efforts done in this field using different artificial intelligence (AI) methods. In the second part of the paper a low cost concept for intelligent systems named Knowledge Server for Controllers (KSC) is introduced. It allows more devices to solve their intelligent processing needs using the same server that is capable to process intelligent data. In the final part this concept is used in an open CNC environment to build up an intelligent CNC. The preliminary results of the implementation are also introduced.

Key words: intelligent CNC, knowledge server, open systems

1. INTRODUCTION

There are many definitions of the intelligent machine tool in the literature. In a well known book [18] Wright and Bourne said that „*We must therefore acknowledge that the degree of intelligence can be gauged by the complexity of the input and/or the difficulty of ad hoc in-process problems that get solved during a successful operation. Our unattached, fully matured intelligent machine tool will be able to manufacture accurate aerospace components and get a good part right the first time*". They told that an intelligent machine tool had the CAD data, the materials and the set-up plans as inputs and could produce correctly machined parts with quality control data as outputs.

It is clear that AI techniques are necessary to apply if one wants intelligent NC machine, but - of course - the usage of them is not adequate in intelligent behaviour.

Table 1 summaries the features of an intelligent CNC (Wright and Bourne collected them more than ten years ago) and shows two further things: the positive changes done in the recent years and the still existing gaps where - according to the scientific community - AI offers solutions with its information processing methods.

Table 1. Commercial needs for the intelligent machine tools

Features (forecasted in 1988)	Big advance by 2001	AI methods still needed
Reduce the number of scrap parts following initial setup.		✓
Increase the accuracy with which parts are made.	✓✓	✓
Increase the predictability of machine tool operations.	✓	✓
Reduce the manned operations in the machine tool environment.		✓✓
Reduce the skill level required for machine setup and operations.		✓✓
Reduce total costs for part fabrication.	✓	✓
Reduce machine downtime.	✓	
Increase machine throughput.	✓✓	✓
Increase the range of materials that can be both setup and machined.	✓	
Increase the range of possible geometries for the part		✓
Reduce tooling through better operation planning	✓	
Reduce number of operations required for setup	✓	
Reduce setup time by designing parts for ease of setup	✓✓	
Reduce the time between part design and fabrication	✓✓	
Increase the quantity of information between the machine control and part design operations	✓✓	
Increase the quantity of information between the machine operations and the machine control		✓
Increase the quantity of information between the human and the machine control	✓✓	✓

Analysing the above list, it is clear that many features do not require direct AI methods. We can state that the main reasons of the advancement were:

(1) the development of the hardware elements (more sensitive sensors, more precise actuators, quicker and stronger computers etc.) even in higher requirements.

(2) The development of the software and the methodology mainly in the preparation phases of the manufacturing (in design, planning, scheduling, resource management etc.) and in the user interface issues (more comfortable and informative 'windows-like' screens and menus).

Even there is a big advance in the technology of the CNCs, the knowledge processing and other AI methods have not appeared within the controllers, so in some points there is no real development. Special heuristic rules, problem-solving strategies, learning capabilities and knowledge communication features are still missing from the recent controllers available on the market. It is also true for many new, open or PC-based CNCs, where DSP add-on-boards provide the necessary computation power and speed.

Further requirements of intelligent CNCs can be find out and defined (Table 2.): based on other papers [9,10,16] and different discussions. The second column indicates whether the different AI techniques (mainly rule base systems, neural nets and fuzzy logic) would provide methods and solutions. One can find positive answers to all these issues in the resent literature:

Table 2. Future requirements of an intelligent CNC

Further features	AI would help
Model based on-line path generation	(✓)
Automatic tool selection	✓
Technological based settings of the operational parameters	(✓)
Automatic compensation of machine limits	✓
Automatic back-step strategies	✓
Detection and compensation of geometrical deflection	✓✓
On-line selection of control algorithms	(✓)
Intelligent co-operation with other devices to solve problems together	✓
Detection and correction of tool wear and breakage	✓✓
Automatic handling of rejected workpieces	✓✓
Detection and managing of emerging situations of the machine tool	✓✓
Complex self-diagnostics	✓✓

The list may be continued with the learning capabilities and others. In the users' point of view these features rough in a controller, that "recognises the problem" and "efficiently and reasonable solves them" with minimal disturbance of the environment of the controller.

2. RESULTS IN INTELLIGENT CNCs

On the one hand (1) in the recent literature one can find many different topics related to intelligent CNCs. Unfortunately they often do not mean intelligent behaviours but the application of intelligent methods. Sometimes it is the case that authors call their devices "intelligent" if one module of the system contains AI based method. On the other hand (2) the key controller vendors leave everything to the users or machine tool builders offering PC/Windows based CNCs. With these systems any software modules (e.g. even AI based ones) can be coupled into the controller but they do not offer real solutions or methodologies, but only software possibilities. Both facts are far from the user wishes stated in the previous chapter.

The following list summarises the most important active research topics in this field. A real intelligent CNC would contain most of these issues.

- Fuzzy logic based concurrent control of some operating parameters (e.g. cutting speed, depth of cut, feed rate) independently from the given tool and the workpiece.
- Neural nets and fuzzy rules in the CNC's control algorithms.
- Optimal path planning, real-time correction of the trajectory.
- Compensation of temperature (and other) deformations.
- Life time management of the tools and other parts of the machine tool including self-diagnostics.
- Tool breakage detection (maybe forecasting) and tool wear monitoring (maybe compensation) with AI methods.
- The utilisation of CNC management (setup, orders, etc.) via intelligent agents.

Monostori [9] classifies the possible intelligent parts of a CNC into three groups, namely: (1) tool monitoring, (2) operation/machine tool modelling and (3) adaptive control.

A general problem in all the three groups is, that the AI based solutions are typically limited and valid only in a very narrow field. If one changes some parameters of the operation or the environment, the earlier successful methods become false.

A special type of adaptivity partly helps on this hard and well-known problem. If it is possible to replace the different modules of the controller time by time, than one can guarantee, that a given AI module can run within its limitation, and over it another module (e.g. a much simpler one) covers the same functionality. It can be realised (among others) if the controller is open to allow this replacement.

There are some research efforts where more than one intelligent modules are built into the controller. So Cheng at al. [3] developed a PC based controller where some DSP cards serve intelligent functions (e.g. adaptive control of cutting force using fuzzy logic; knowledge based self-diagnostic and error recovery/management; multisensor based neuro-fuzzy tool monitoring).

3. OPEN CONTROLLER INITIATIVES

Manufacturing has constantly been a technological domain, in which the industry was driven to apply the current high-tech in the computer and control area. It is no surprise, that the Open Systems concept has also diffused into the manufacturing area, and factory managers are often referring to the open manufacturing systems. The terms and definitions are far less exact than the terms applied in the operating systems environment, but by now, the change of the global manufacturing paradigms (e.g. see Kovács [7]) are directing our focus on the key user aspects of openness.

The need for a new and open CNC architecture was emerging at many places around the world. One of the most important work was done from 1992 within the frames of the European project named OSACA (Open System Architecture for Control Applications) [15]. The main results of the project are: an analysis of the state-of-the-art and future requirements of NC controllers, a reference architecture, a general and platform independent API for inner CNC and outside communication and a configuration system supporting the possible machine tool vendors. Similar efforts are going on in Japan under the IROFA Consortium [14] and in the U.S. within the OMAC projects [13].

In all these projects the module structures and hierarchies (reference architecture) of the NC controllers and the APIs of any defined modules were published. Unfortunately even if the aims of these efforts are nearly the same the resulted controllers are incompatible in many sense [12]. But because of the open modules and the precisely defined APIs, any of these efforts could be a very good starting point of building intelligent CNCs. It is possible to implement any modules of the reference architecture using AI methods or to add further advanced modules to provide extra features for the controller.

4. DIFFERENT APPROACHES OF KNOWLEDGE SERVERS

The features of World Wide Web led Eriksson [5] to introduce knowledge server to easier solve the installation and version control problems of expert systems and to provide a web based interface of the knowledge base for the different users.

Some advanced knowledge based systems are based on this concept. So the Cyc system, the most important research on the common sense, is organised as a knowledge server [8]. Also the Istar knowledge server [1] provides on-line advises in many different topics (e.g. stock exchange, Internet security). There are also some applications of knowledge servers in manufacturing (e.g. Váncza at al. [17] uses it in a robotic inspection planning system).

In the HPKB (High Performance Knowledge Environment) [4] some hundred thousand rules are performed in an intelligent knowledge environment. In this project the different intelligent components are called knowledge servers. The components are communicating with each other via the OKBC (Open Knowledge Base Connectivity) protocol [2] specified at Stanford.

5. KNOWLEDGE SERVER FOR CONTROLLERS

Knowledge Server for Controllers (KSC) is defined as a server providing capability of intelligent data processing for other systems. It allows the basic system to reach external intelligent processing resources, because it does not have any. The KSC contains a high performance reasoning tool, and different knowledge based modules. All the modules have their special rules and procedures. The client system calls these modules, passes them specific data if necessary, and the KSC module can collect data if the knowledge processing requires. All the data acquisition and user interaction is done by the client system. It is clear that in KSC the clients have much more tasks than a simple browser based user interface and in the applications listed in the previous chapter.

It should be stated that KSC does not deal with fuzzy and neural net based AI modules. The computing power and the necessary software costs and complexity of these methods are less than the rule or model based ones. (In the case of the neural nets it is true only if the net is not trained on-line.)

The KSC allows the different modules to run independently, to co-operate as agents or to control each other. The third case means that one module is started by another one because either the second one uses the results of the first one or the inference of the first one led to the need of the second module.

Generally the resources of the KSC can use more clients (controllers or SCADA systems) simultaneously. It leads to a cost effective AI solution, because one costly AI tool can solve all the intelligent problems in a distributed environment. The overhead of the KSC (network connection, one more computer, some delay etc.) is much less comparing to the advantages (AI tool licensing, less computing power in the clients/controllers, one server module may used by more clients etc.).

Using the KSC together with the component based software technology (e.g. CORBA) gives a very adaptive software frame to solve complex problems.

In the Fig. 1 a CNC with an embedded PLC controls a machine tool. The modules of both controllers are open and some of them are also clients of a knowledge server (KSC). It means that these modules can run special AI methods during their work that is an independent service is implemented in the KSC.

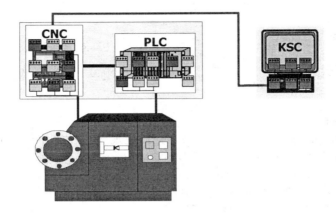

Figure 1. Intelligent CNC using KSC

6. PROTOTYPE INTELLIGENT CNC BASED ON KSC

In an early prototype of the intelligent open CNC that is using KSC, the axis module (Fig. 2) was implemented based on the OMAC module specification [13]. As a basic configuration the OMAC *Axis* module uses two *ControlLaw* modules to have position and velocity control over the real axis that is reached via the *IOPoints* module. The *Axis* is manipulated from the *AxisHMI*.

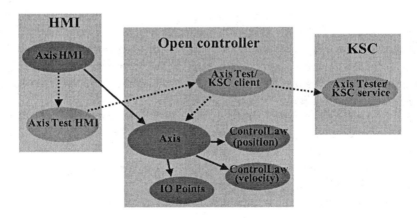

Figure 2. Intelligent tester of a machine tool axis

An advance axis tester is put on the top of this. *AxisTest* module handles all the tests but it gets the necessary position and velocity values from a knowledge based general tester running as an application on the KSC.

The KB tester determines some goal positions and motion speeds, that the *AxisTest* module executes with the axis using jog commands. The results (execution time, tuning in errors etc.) are sent to the KSC that analyses and qualifies the axis.

In the prototype the modules are built in CORBA, the controller and the HMI is programmed in Java, while the KSC is based on G2 environment [6].

7. CONCLUSIONS

The different research results and the open problems of intelligent CNCs were shortly introduced. It was stated that many wishes from the 80's are still unsolved and many of the existing open issues claim AI solutions.

The features of the open controllers were also summarised with the most important open system controller initiatives.

The knowledge server concept of Eriksson was introduced in the field of controllers with some important modifications comparing with the original idea. The features of KSC were discussed and an early prototype (axis tester) was introduced. Further works are going on to develop a complete intelligent CNC for a 3D milling machine using KSC.

It should be mentioned that a KSC was successfully implemented in an advisory system of an electrical substation of a nuclear power plant [11]. In this application the KSC was connected to a SCADA system and it supported 5 different intelligent functions.

8. REFERENCES

1   Basden A: The Istar Knowledge Server, (2000)
    http://www.basden.u-net.com/pgm/Istar/index.html
2   Chaudhri F. V., R. Fikes, P. Karp, J. Rice. (1998) OKBC: A Programmatic Foundation
    for Knowledge Base Interoperability. Proc. of AAAI-98,
3   Cheng, T., J. Zhang, C. Hu, B. Wu, S. Yang (2001) Intelligent Machine Tools in a
    Distributed Network Manufacturing Mode Environment, Int. J. Adv. Manuf. Technol.
    17:221-232
4   Cohen Paul R., Robert Schrag, Eric Jones, Adam Pease, Albert Lin, Barbara Starr, David
    Easter, David Gunning and Murray Burke. 1998. The DARPA High Performance
    Knowledge Bases Project. In Artificial Intelligence Magazine. Vol. 19, No. 4, pp.25-49
5   Eriksson H: Expert Systems as Knowledge Servers (1996) IEEE Expert, Vol. 11, No. 3
6   Gensym (2000) A Strategic Choice for Improving Business Operations: G2 Classic,
    http://www.gensym.com/products/g2.htm

7   Kovács, G. L. (1996) Changing Paradigms in Manufacturing Automation. Proc. of IEEE International Conference on Robotics and Automation, Minneapolis, USA, Vol. 4. pp. 3343-3348

8   Lenat, D: CYC: A Large-Scale Investment in Knowledge Infrastructure, Comm. of ACM, 38(11) pp. 33-38, 1995; + Cyc Knowledge Server: http://www.cyc.com/products.html

9   Monostori L. (2000) Intelligent Machines, Proc. of 2nd Conf. on Mechanical Engineering, Budapest, Hungary, May. 25-26, pp. 24-36

10   Nacsa J, G. Haidegger (1997) Built-in Intelligent Control Applications of Open CNCs, Proc. of the 2nd World Congress on Intelligent Manufacturing Processes and Systems, Springer, Budapest, Hungary, June 10-13., pp. 388-392

11   Nacsa J., Kovács G. L. (2001) Alert State Detection and Switching Order Generation at a 400/120 kV Substation using Knowledge Server, ISAP2001 Conference. (Intelligent Systems Application to Power Systems), Budapest, 18-21 June

12   Nacsa J. (2001) Comparison of three different open architecture controllers, IFAC MIM, Prague, 2-4 Aug.

13   OMAC API Work Group (1999) OMAC API SET, Version 0.23, Working Document, http://www.isd.mel.nist.gov/projects/omacapi/

14   OSE Consortium, (1996) OSEC: Open System Env. for Controller Architecture Draft, http://www.mli.co.jp/OSE/

15   Pritschow, Sperling (1994). Open System Controllers - A Challenge for the Future of the Machine Industry. CIRP Annals

16   Teti R, S. Kumara (1997) Intelligent Computing Methods for Manufacturing Systems, Annals of CIRP, Vol 46/2, pp. 629-652

17   Váncza J., Horváth M., Stankóczi Z.: Robotic Inspection Plan Optimization by Case-Based Reasoning, Proc. of the 2nd World Congress on Intelligent Manufacturing Systems and Processes, June 1997, Budapest, Hungary, pp. 509-515

18   Wright P.K., D.A. Bourne (1988) Manufacturing Intelligence, Addison-Wesley

9.   BIOGRAPHY

János Nacsa received his diploma degree in electrical engineering from Technical University of Budapest in 1987. Since that he is a research associate of the CIM Laboratory of the Computer and Automation Research Institute, Budapest, Hungary. He published more than 50 papers and led different research projects. His research topics are open systems and application of artificial intelligence in manufacturing.

An Ontology-Based Collaborative Framework for Decision Support in Engineering

Hauke Arndt, Rüdiger Klein
DaimlerChrysler Research Department, Knowledge Based Engineering Group

Abstract: In this paper the MOKA framework as an expressive and generic model of design problem solving is extended by a model of design rationale. The MOKA framework is based on a product and process model which allows an explicit description of requirements, design descriptions, problem solving steps, decisions, etc. In this paper it is shown how decisions and their rationale can be integrated into this explicit product and process description. This tight conceptual integration of design decisions and their rationale into the MOKA framework is achieved by specifying a design rationale ontology which fits into the MOKA framework.

Key words: Product-Model, Process-Model, Design Rationale, Decision Support, MOKA-Framework, Ontology of Design Rationale

1.   INTRODUCTION

A typical technical product today tends to be very complex, and so does the corresponding design and engineering process. Information systems of any kind are ubiquitous now in engineering in order to manage this complexity. Though in the meanwhile many of them (like CAD and PDM systems) reached a considerable maturity, there are challenging problems left. One of them is *interoperation*: though very important for collaboration in engineering, the various IT systems follow different approaches to information modeling and problem solving. This makes their communication and collaboration complicated and error prone. Another such challenging problem is adequate IT support for decision making (by *systems or humans*). Both issues are closely related to each other: interoperation is not primarily a data exchange problem - at most it is an issue of collaborative problem solving, i.e. of coordinated decision making. And decision making is one of the main elements in engineering problem solving - with many relations to practically all aspects of the collaborative engineering process. Both aspects - interoperation and decision support - gain special importance in the *Internet/intranet* context. Using these new technologies to their full extend and benefit requires - among others - a more sophisticated approach to interoperation and decision support.

IT support of decision making is by far not a new problem. There is a number of approaches describing design rationale (see for example [Blessing1994; M.Klein1992; MacLean1996; Shum1996]). Although these approaches are quite expressive they are of limited use, because in most cases the rationale is described *separated* from the information about the product and the design process. Currently, there is no satisfying practical approach [Lei et al. 1995; Brazeir et al. 1995; Gero 1998]. This is quite natural taking the complexity of product and process knowledge into account - as well as the *many different aspects* of design decisions and their rationale.

In this paper we describe an *integrated* approach to design rationale bringing together an explicit model of design and a sophisticated model of design rationale. We use the MOKA framework [Klein2000] as an expressive and well-founded model of design and an adaptation of the QOC design rationale model [MacLean1996]. Combining both will result in an expressive and usable model of design and design rationale.

In this paper we describe first the MOKA framework [Klein2000] (chap. 2) and important aspects of the design process (chap. 3). In chap. 4 an *ontology-based* model for representing the process- and product-model of the MOKA framework is presented. These chapters build the basis for chap. 5 where the MOKA framework is extended in order to integrate design rationale. In chapter 6 we present an example for the design rationale framework. Chap. 7 contains a summary and our conclusions.

2.   A FRAMEWORK FOR DESIGN DECISION MAKING: THE MOKA FRAMEWORK

In this paper we describe a model of decision making in engineering and the rationale behind it. It is based on and integrated into the MOKA framework [Klein2000]. As a pre-requisite for modeling decisions and their rationale, this framework allows us to capture the complexity of product and process knowledge and the interrelation between them (see *figure 1*).

The MOKA framework comprises an *expressive product model* which allows us to represent the whole complexity of product information in a clear, well structured and formalized way. Especially, not only the completed design can be modeled but also all related and intermediate product information like *generic* knowledge, design *requirements* of any kind, *function-structure* relations, behaviors, etc.

In the MOKA framework, this product model is closely related to a *process model*. This is essential because design problems can normally not be solved in a pre-defined manner - they need complex processes which are essentially determined by the product to be designed. Consequently, engineering problem solving is an incremental, interactive, and mostly hierarchical decision process with propagation of decision consequences.

3. THE DESIGN PROCESS

In order to get a better understanding of what design decisions are and what their rationale can look like we need a more detailed description of the design process:

Typically, the design process starts with a set of initial requirements[1] which may be incomplete and even contradictory. It ends with a complete design description which fulfills the (adjusted) requirements and is consistent with the general design and engineering knowledge. This process transforms the initial state into the final result. It is a kind of complex dynamic search based on a continuous interplay of synthetic and analytical steps.

Synthetic steps elaborate the design description, add more and more information to it in order to fulfill requirements and make it complete. Analysis steps are used to investigate the various relations between design and requirements, among conflicting requirements, etc. which may result in relaxed or new requirements. This is essential because a key property of design problem solving is its non-monotonic nature: adding new information to a design description may *change* other properties (for instance, behavior, geometric shape, or total weight).

Search is needed in order to escape from conflicts, dead ends, or inappropriate design states. Inappropriate are those states which fulfill all "hard" requirements and constraints but are non-optimal with respect to some "soft" requirements. Optimality is important in many design problems: frequently we are not only interested to find any solution but one which is optimal (or at least close to optimal) with respect to some criteria.

Due to the complexity of products and processes in design the corresponding search spaces tend to be huge. In practice, there is frequently no explicit description of all possible alternatives in a given state of design

---

[1] Leaving aside case adaptations or variational design which can be "reduced" somehow to the basic case.

problem solving. The human designer takes heuristic knowledge in order to focus on those problem solving steps which look promising - leaving other alternatives unattended.

Currently, it seems not to be realistic to have an automated design system: engineering knowledge is partially so complicated that it can not be modeled adequately with current means. So, essential parts of relevant knowledge have to be left with the human user. The system only has parts of the knowledge available. Typically, a human designer will use such a design support system as an intelligent assistant. This system should propagate consequences from decision through the whole product and process model, it should analyze conflicts, make proposals to the human about how to proceed, etc. The integration of design rationale into this process is an essential pre-requisite for the practicability of collaborative problem solving in design.

4. AN ONTOLOGY-BASED MODEL OF PRODUCT AND PROCESS KNOWLEDGE

This all together forms a complex set of requirements to information modeling. In order to fulfill them, the MOKA framework allows us to structure this complex information by *product and process ontologies* [Gruber1993]. An ontology is a set of interrelated formal definitions of all notions relevant in a certain domain. This allows us to describe explicitly the main information categories and their relations in a domain - an important issue for the integration of different information models.

In the MOKA framework we have the following information categories (see UML diagram in *figure 1*):
- a meta model (upper structure) as the core of the ontology [Klein2000]. This allows us to capture the *main knowledge categories* and their relations, and to structure the whole bunch of product and process knowledge accordingly;
- following this upper structure, the product model (*product description* in fig. 1) consists of *generic* as well as *concrete* information about functions, structures, and behaviors using all kinds of classes, attributes, relations, hierarchies, constraints, and different views. *Requirements* and *design descriptions* are related to each other in a logical way, as they are to the generic design knowledge;
- the process model (*process* object) is described by synthesis and analysis tasks to elaborate requirements and design descriptions [Klein2000];
- the process model is connected to the product level by dynamic knowledge categories. They reflect the *logical relations* on the product

level: for instance, fulfilled, unfulfilled or conflicting requirements, etc. (not all relations are shown in *figure 1*).

A strategic layer allows us to represent explicitly knowledge about the way decisions should be made: which criteria are more important then others, etc.

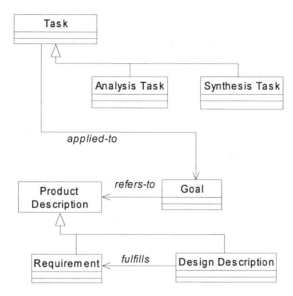

Figure 1. Simplified Product and Process Model of the MOKA framework (UML notation)

5. AN ONTOLOGY FOR DECISIONS AND THEIR RATIONALE

5.1 A General Model of Design Rationale

Currently, there are several models of decisions and their accompanying rationale (for example in [Blessing1994; M.Klein1992; MacLean1996; Shum1996; Chung1994]). Our approach is primarily based on QOC [MacLean1996] because it is self explaining as well as sufficiently expressive.

QOC uses the following information categories (ontology) for the description of design rationale: *problem*, *criteria*, *alternative*, *decision* and *argument* (see *figure 2*). Each category is specified in more detail by several attributes, e.g. a name, a textual description, references, remarks, etc. The *problem* category describes the 'what' of the decision to be made. *Criteria* are used to model the way in which decisions are made. Each *criterion* has an individual importance - a kind of weight - with respect to a given *problem*. These weights are essential for an appropriate decision making. *Alternatives* describe the several different solutions possible for the actual *problem*. The *alternatives* can either be technical solutions (product oriented alternatives) or distinct tasks of the problem solving process (process oriented alternatives). The *decision* describes the chosen alternative for a certain problem. An *argument* may (or may not) provide support to a certain alternative with respect to a relevant criterion.

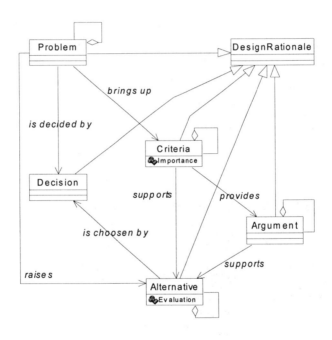

Figure 2. General Model of Design Rationale (UML-like notation)

In *figure 2* also the relations between the several information categories of our QOC ontology are shown. All categories are related to each other, and most of them have additionally a breakdown structure (*consists of, consists in* - not shown here).

As mentioned earlier, an essential aspect of design rationale are their *dependencies*. Decisions and their rationale can depend on or conflict with each other. These dependencies can be modeled by two relations (*affects* and *conflicts*) on the *problem* category (also omitted here by space limitations).

5.2   An Integrated Model of Products, Processes and their Design Rationale

In order to support the understanding of a complex technical product, it is important to describe design rationale together within their context. In engineering, this is product and process related information - as modeled in the MOKA framework.

The MOKA framework can straightforwardly be extended such that the various kinds of decisions and their rationale can be modeled in an coherent way. Decisions can be characterized along *four dimensions* which build the core *ontology of the decision and rationale* information model:

1. The role within the *design process* (types of decisions): synthesis, analysis, conflict resolution, default, optimality decision, strategy of decision making;
2. The relation to *product* knowledge: to the involved requirements, functions, structures, design descriptions, hierarchy, etc.;
3. *Dependencies*: decisions are not independent from each other - one may be the pre-requisite of another one, a decision may be in conflict with other decisions, hierarchical decomposition of decision making[2], etc.
4. *Scope*: in which context is a decision made, and how far should the consequences be considered.

So, we gained an ontology of decisions and their rationale integrated into the MOKA framework: a *well-formalized*, *explicit, expressive* and *integrated* approach. *Well-formalized* because it is based on a formal ontology of decisions and their rationale. *Explicit*, because all relevant relations to product, process and decision making are modeled in an explicit way. *Expressive*, because the whole expressiveness of the MOKA framework and the underlying ontology (explicit meta level, dynamic knowledge categories etc.) is available. And *integrated*, because decisions and their rationale are conceptually related to all other aspects of knowledge.

Figure 3 shows the ontology integrating our model of design rationale into the MOKA framework (describing both process and product model).

The MOKA framework describing the product and process model (*see figure 1*) on one side and the model of design rationale (*see figure 2*) on the other side are integrated by four relations. *Tasks* on the process level describe typical problems in the context of design rationale. This forms the

---

[2] Hierarchical decomposition of the problem and its problem solving process is a typical strategy in engineering.

basis of a design rationale. Therefore the *task* is associated with *alternatives* (relation: *raises*), *criteria* (relation: *brings up*), and *decision* (relation: *decides*). All possible *alternatives* are *product units* (resp. *requirement* or *design description*). For this reason *alternative* is a super-class of *product unit* (all relations are inherited).

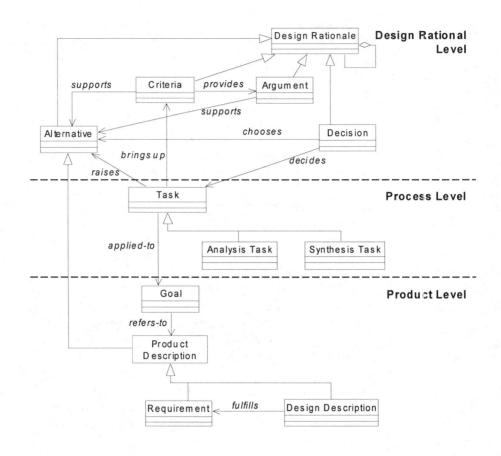

Figure 3. Model of Process, Product and Design Rationale (UML-like notation)

5.3   How to use the information about decisions and their rationale

The model presented here (see *figure 3*) allows us to describe the design product and its relation to the design process integrated with the underlying design rationale in their interdependencies. Also the scope of a design rationale is represented, as the knowledge about the process and the product is described in the same model.

This integrated well formalized approach allows us to use information about decisions and their rationale *in the same way* as other engineering information: it may be stored, retrieved, exchanged, etc. For instance, a *query* may be asked to a corresponding *decision management system* about all decisions currently influencing the length of a certain part X. The product relationship is obvious: length of part X; the process steps involved can be, for instance, default value assignment to the length of part Y and relation to length of X via a geometrical constraint; this also clarifies the dependency aspect; and the scope may be determined by the structural or functional hierarchy to which part X belongs. Alternatively, the decision management system could be queried for all decisions based on any geometrical default assignment.

6.   AN EXAMPLE: COOLANT PUMP

Designing a coolant pump for vehicles is a typical engineering problem. It is a suitable example to be presented here as an illustration of our approach, because it is small enough to be understood easily but complex enough to demonstrate the usability of our approach.

In principle, a coolant pump (see *figure 4*) consists of spiral, impeller, bearing and a seal. The whole pump is located in a casing.

During the routine design process of a coolant pump a decision about the kind of the drive has to made. The drive of the pump could either be an *electric*, a *belt* (using a belt and a flange to transform mechanical energy from the vehicle's motor) or a *direct drive*. The decision about which drive to take is based mainly on cost aspects and on space and shape considerations of the pump housing.

In our approach this would be modeled as follows (see *figure 5*[3]): During the design process an *analysis task* "*select*" with the associated *goal* "kind of drive for coolant pump" is generated. This *task* forms also the basis for the design rationale. Associated to this *goal* are all possible *design descriptions* "*electric drive*", "*belt drive*", and "*direct drive*", which correspond to the possible *alternatives* on the design rationale level. The *criteria* needed to solve the actual problem properly are modeled with the help of the relation between design descriptions and decision arguments. With this knowledge *arguments* are generated which evaluate the *alternatives* with respect to the relevant *criteria* - providing the foundation for the *decision* to be made.

---

[3] The whole information related to this application has been modeled in our ontology modeling tool OntoWorks (to be described in a forthcoming paper).

Figure 4. Picture of a Coolant Pump

In this way all decisions occurring during the design process can be described in an adequate way.

7.   SUMMARY AND CONCLUSIONS

Design information tends to be quite complex: information about products, their requirements, behaviors, functions, properties, constraints, compositions, etc. - as well as information about problem solving: synthetic steps like instantiation of components or value assignments to properties and analytical steps to identify new requirements, find conflicts, etc. The MOKA framework provides expressive and well formalized means for the description of product and process related design information *and* the relations between them.

In this paper we presented an approach to integrate decision information and the accompanying rationale into this framework. Starting with an QOC like approach to model design decisions and their rationale, we suggested a way how these two models can be brought together. An ontology describing the MOKA framework and an ontology-based model of design decisions and the rationale behind them were integrated by defining the relations between information categories in both models.

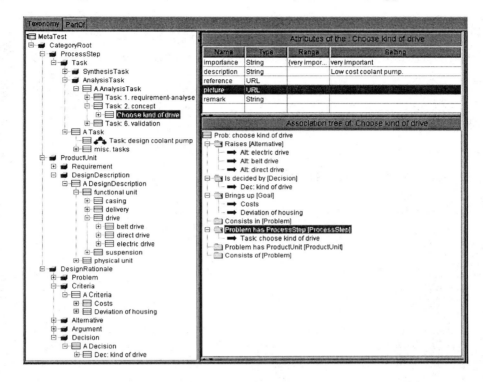

Figure 5. Modeling the Coolant Pump (Screenshot, not all details are actually shown)

This allows us to use information about decisions and their rationale in the same way as other engineering information: it may be stored, retrieved, exchanged, etc. With this explicit information about decisions and their rationale the collaboration in design problem solving can be supported much better then without it. Each partner (human or system) can use this information about the decisions made by others in the own decision process. This is essential for conflict resolution, optimality considerations, reuse of former cases, etc.

Of course, the approach presented here is only a first approximation. Its strength results from its tight conceptual integration into the MOKA framework as an explicit and well formalized model of design. Many aspects of design rationale - as used, for instance in teaching, explanation generation, or negotiations - are not fully covered here. A more sophisticated approach with more expressive models is under preparation. There we will focus especially on the integration and usage of *background knowledge* into more advanced applications of design rationale.

This gets special importance in an *interconnected world*. To use the Internet or intranet for product *data exchange* is one thing (important enough) - another, more important one is to use it for distributed *collaborative problem solving*. The exchange of complex information about the design process, about decisions and their rationale is *indispensable* for a distributed interconnected virtual enterprise. For this purpose, an explicit, expressive, and integrated model for decisions and their rationale is essential. Only this enables the partners really to exchange meaningful information and not just data. The *Semantic Web* will provide the necessary expressiveness and formalization [SemWeb].

ACKNOWLEDGEMENTS

We would like to thank Ute John, Michael Heinrich, Henrik Weimer, and the other members of the FT3/EW team for their useful comments on this paper.

BIOGRAPHIES

Hauke Arndt studied Mechanical Engineering and received his Diploma in 1999 from the Technical University of Berlin. Currently he's working on his Ph.D. thesis in the R&T department of DaimlerChrysler. His main research interests are design rationale, engineering knowledge management and knowledge based systems.

Rüdiger Klein received a Diploma in Physics in 1975 with honors from Humboldt University in Berlin and a Ph.D. from the Academy of Sciences in Berlin in 1979. Since 1991 he has been a senior researcher in the R&T department of the DaimlerChrysler AG, responsible for Knowledge Modeling research including ontologies, knowledge management, and knowledge based engineering.

From 1998 to 2000 Dr. Klein took part in the MOKA Esprit research project (No. 25418). Dr. Klein is a vice chair of the International AI in Design conference and a member of the program committee of the International ACM Symposium on Solid Modeling. He is the author and co-author of more than 30 scientific publications.

REFERENCES

Brazier, F. M. T., van Langen, P. H. G., and Treur, J.: 1995, *A logical theory of design*, in J. S. Gero, and F. Sudweeks (eds), Preprints Advances in Formal Design Methods for CAD, Key Centre of Design Computing, University of Sydney, pp.247-271.

Blessing, L. T. M.: 1994,. *A process-based apporach to computer-supoorted engineering design*, thesis university of twente, enschede, the netherlands, ISBN 0 952350408.

Chung, P. W. H., Goodwin, R.: 1994, *Representing Design History*, in J. S. Gero and F. Sudweeks (ed.), Artificial Intelligence in Design '94, Kluwer Academic Publishers, Dordrecht, pp.735-752.

Gero, J. S.: 1998, *Towards a model of designing which includes its situatedness*, in H. Grabowski, S. Rude, and G. Grein (eds), Universal Design Theory, Shaker Verlag Aachen, pp.47-56.

Gruber, T.R.: 1993, *A translation approach to portable ontology specifications*, Knowledge Acquisition 5:199-200.

Klein, M.: 1992, *DRCS: An integrated system for capture of designs and their rationale*, in J. S. Gero (ed.), Artificial Intelligence in Design '92, Kluwer Academic Publishers, Dordrecht, pp.393-412.

Klein, R.: 2000, *Knowledge modeling in design – the MOKA framework*, in J. S. Gero (ed.), Artificial Intelligence in Design '00, Kluwer Academic Publishers, Dordrecht, pp.77-102.

Klein, R.: 2000, *Towards an Integration of Engineering Knowledge Management and Knowledge Based Engineering.* 2000 International CIRP Design Seminar: Design with Manufacturing: Intelligent Design Concepts Methods and Algorithms, May 16-18, 2000, Haifa, Israel.

Klein, R.: 1998, *A Knowledge Level Theory of Design.* G. Jagucci (ed.): Proceedings of the 10th International IFIP WG 5.2/5.3 Conference PROLAMAT 98, Trento, Italy, September 1998.

Lei, B., Taura, T., and Numata, J.: 1995, *Representing the collaborative design process: a product model-oriented approach*, in J. S. Gero, and F. Sudweeks (eds), Preprints Advances in Formal Design Methods for CAD, Key Centre of Design Computing, University of Sydney, pp.274-291.

MacLean, A., et al: 1996, *Question, options, and criteria: elements of design space analysis*, in Moran, T.P., Carrol, J.M. (eds.), Design Rationale: Concepts, Techniques, and Use, Lawrence Erlbaum Associates, Publichers, Mahwah, New Jersey, 1996, ISBN 0-8058-1567-8

SemWeb: http://www.w3.org/2001/sw/

Shum, S. B. 1996: *Analyzing the usability of a design rationale notation*, in Moran, T.P., Carrol, J.M. (eds.), Design Rationale: Concepts, Techniques, and Use, Lawrence Erlbaum Associates, Publichers, Mahwah, New Jersey, 1996, ISBN 0-8058-1567-8

Takeda, H., Nishida, T.: 1994, *Integration of aspects in design processes*, in J. S. Gero and F. Sudweeks (ed.), Artificial Intelligence in Design '94, Kluwer Academic Publishers, Dordrecht, pp.309-326.

Protzen, J.-P., Harris, D., Cavallin, H.: 2000, *Limited computation, unlimited design*, in J. S. Gero (ed.), Artificial Intelligence in Design '00, Kluwer Academic Publishers, Dordrecht, pp.43-52.

Order Management in Non-Hierarchical Production Networks using Genetic Algorithms

Horst Meier, Tobias Teich, Harald Schallner

Dept. of Mechanical Engineering
Ruhr-Universität Bochum
D-44780 Bochum
Germany
meier@lps.ruhr-uni-bochum.de
schallner@lps.ruhr-uni-bochum.de

Dept. of Economics
Technische Universität Chemnitz
D-09107 Chemnitz
Germany
t.teich@wirtschaft.tu-chemnitz.de

Abstract: The global economy of the 21st century will be characterized by competing corporate networks rather than by competing enterprises. Competition outside of and cooperation within supply chains will be the important factors. The tendency of companies to concentrate on their core competences causes a narrowing of their strategic focus, while at the same time these companies widen their focus by seeking alliances. Specialization and global networking are intensifying each other. The results are sinking costs in larger markets. The special research area 457 of the German Research Community is dealing with these topics. Another research project called INTERKON proposes an inter-organizational framework for collaborative manufacturing control in production networks. This paper combines both approaches. A short introduction gives a first insight into the problems that arise with the management and control of production networks. The purpose of this paper is to put forward a specific functionality to schedule production considering constraint networks.

Key words: production management, production networks, genetic algorithms

1.   INTRODUCTION TO THE PROBLEM

A customer contacts a potential node of the network and releases a demand for the production of a specific good. The node has access to the front end of a model core and stimulates the genesis of a temporary network for the production of the good demanded. The information core is able to build directed graphs which stand for every adequate technological alternative for the production of the demanded good. For this purpose it uses information which is stored as fact- and rule-knowledge in an ontology-based database. Firstly, the different paths through the graph are built according to technological routings. Secondly, competence cells are assigned to the nodes of the network according to technological constraints. The objects of the graph are attributed. As a result, each node contains information on costs and times, while each directed edge contains information on logistics and quality.

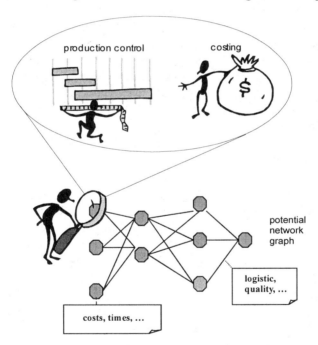

Figure 1. ASP-Functions

The task is to find a way through the graph which approximates to an unknown optimum for a time- and/or cost-based objective function. For the choice of efficient algorithms is necessary to know which class this problem belongs to. At first glance this may look like a Travelling Salesman Problem (TSP). A more detailed observation reveals that precedence relations between the nodes have to be modelled. Although sequencing problems can be modelled as generalized TSPs in this case, the directed edges have to be weighed with distances which influence the value of the objective function. The calculation of these distances is difficult since all predecessors in the network influence the value of a directed edge. At the same time the nodes represent alternatives which influence the succeeding nodes. The problem described consists of elements of the CTSP and the JSP. Additionally, there is the problem of the alternative routings mentioned above. Thus it is an NP-hard problem. The routing within the network is managed by Ant-Colony-Optimization. This aspect will not be discussed in this paper. Instead of this, emphasis is put on the ASP-functionalities manufacturing control and costs (figure 1).

The routing must not be understood as a static optimization. It aims at getting process information on times and costs within a competence cell instead. In the following manufacturing control is focussed. Once the network of competence cell alternatives is generated, the nodes have to be attributed with times. At this point the network management should not only be allowed to ask for offers because this might result in an informational delay which might lead to the well-known bullwhip effect. The solution proposed here is a direct contact with the competence cell by means of a special ASP-functionality. The network management stimulates the reaction of a front end via standardized interfaces (like e.g. XMI). The front end schedules the order flow. This can be done by a virtual loading of the new request into the real production system. Fast intelligent technologies have to be used to compute the earliest due date. Genetic Algorithms (GA) could be one such method.

It is the task of shop floor scheduling to find sequences in which *n* jobs are to be processed on a set of *m* resources (machines, staff, special tools etc.) so that several constraints (technological, organizational etc.) are satisfied and a given objective function (e.g. mean flow time, mean lateness etc.) is maximized. Each job can consist of one or more operations. The technological sequence of the operations within a job is not necessarily set. Thus real-world shop floor scheduling comprises many different combinatorial optimization problems like Job Shop, Flow Shop, Open Shop, Parallel Machines Problems and many other. While models for these problems and heuristics to solve them had been developed in operations research, it was still not possible to handle shop floor scheduling as a whole. Even today it is mostly done manually. With the help of evolutionary methods and other meta-heuristics like Simulated Annealing or Threshold Accepting, there seemed to be hope for a change of this situation for the first time.

During the last ten years a number of attempts has been made world-wide to solve hard combinatorial optimization problems in the field of production control with the help of Genetic Algorithms (GA) [2, 3, 8, 13, 14]. However, real-world applications in the field of shop floor scheduling are still rare, although the results shown in benchmark tests have been far better than those of the best priority rule heuristics. What is the reason for this?

There is the problem of a difference between theoretical research work and the development of a real-world application. However this is not the main point. More important is the fact that benchmark problems are rather small compared to real-world ones and that they do not contain those many, little extra problems and degrees of freedom which increase the complexity of real-world shop floor scheduling. This complexity is raised to the power of ten compared with e.g. the job shop benchmark problems that are often used for testing scheduling GAs. Regarding this high complexity the question of the time needed to calculate a satisfying solution is of highest interest for a practical use of EA-based scheduling tools. Often there are demands for calculation times no longer than five minutes, as shown in a study carried out by the authors [12]. There is no reason to believe that the solution an Evolutionary Algorithm delivers after five minutes could not be improved further. Therefore we decided to let the algorithm run continuously. There are two problems that have to be solved to do this.

Firstly, there has to be an organizational framework which allows companies to make use of a continuously running Evolutionary Algorithm for shop floor scheduling. Since in today's organizations scheduling is usually done just once at the beginning of a planning period, while afterwards only minor changes are allowed, companies are not prepared to handle new improved schedules delivered by an EA after the first schedule has been fixed and its sequencing information has been distributed to every worker. To change this, the authors developed an organizational framework in which these new improved schedules can easily be handled. However, since these matters do not lie within the field of Evolutionary Algorithms, they are not discussed further in this paper.

Secondly, there is the Evolutionary Algorithm itself. Since situations in production systems may change frequently, no set parameter constellation seems appropriate. Parameters should be the matter of a co-evolution process to make them adaptable. Last but not least, the computational resources companies can use for shop floor scheduling are often rather limited. Therefore a simple approach of parallelization using intelligent terminals of a plant data acquisition system (PDA) has been developed and tested successfully.

2.   EVOLUTIONARY SCHEDULING ALGORITHM

The shop floor scheduling EA introduced by the authors consists of two parts: a Genetic Algorithm for the schedule optimization and a Evolutionary

Algorithm for the parameter co-evolution. A GA, developed by the authors in 1998 for job shop problems, was used for schedule optimization [5]. It was built upon a genetic representation of permutations with repetition. This approach was developed by Bierwirth et al. in 1993 [1].

For each operation the corresponding job occurs within a chromosome. Thus it is possible to avoid invalid solutions without a complicated evaluation procedure. In order to build a semi-active schedule, the chromosome has to be read only from the beginning to the end. Each operation has to be placed on the necessary resources at the earliest possible starting date at which all resources have to be available. The predecessor of the operation within the same job must be finished. The latter condition could be weakened because sometimes there are operations whose sequence is not necessarily set. For crossover we used the Generalized Order Crossover (GOX) developed by Bierwirth et al. in 1993 and the Generalized Position Based Crossover (GPBX) developed by the author [12]. Therefore possible disadvantages of the use of just one crossover operator shall be overcome. Experiments carried out by the author [12] seem to justify this approach because the mean results reached by combined use of the two operators were better than the results reached by using single operators. Both GOX and GPBX are generalized versions of crossover operators which were originally developed by Syswerda [11] for the Travelling Salesman Problem. Syswerda's attempt to preserve order and absolute position information respectively is unfortunately partly undermined by Bierwirth's genetic representation. The limitation is that actual order/ position information of operations shall be saved, whereas the chromosome contains job numbers to avoid invalid solutions. This dilemma on the one hand and the high redundancy of the representation which increases the dimension of the search space by some additional powers of ten on the other hand make it worthwhile to use a more powerful scheduling GA in the future. Therefore the authors are currently comparing the concepts of van Bael [13] and Mattfeld [8] with regard to their suitability for real-world shop floor scheduling. As regards mutation the scheduling GA uses a reservoir of no less than six operators which differ in their power to destroy the genetic information. A shift operator which takes only one operation out of the chromosome and moves it to another position and a scramble operator which changes the whole chromosome at once are the two opposite ends of a scale these operators can be placed on. For mutation one of them is chosen according to the same principle as employed with the crossover operator. The same concept is applied with selection schemes. The idea behind the implementation of this variety of operators is to produce a robust algorithm which avoids a problem that GAs often have when using only a few operators. They do well with one problem but fail to deliver satisfying solutions with another one. With its variety of operators this algorithm proves to be robust, yet at the cost of an inferior quality of solutions for benchmark tests. A first version of this algorithm was implemented in 1998 as a generational replacement GA. Results concerning this algorithm can be found in [5]. For easier parallelization it was later changed to a steady state GA. New solutions are accepted if they are better than the worst individual or they are accepted with a certain probability if they are worse. This concept was originally used in Simulated Annealing [7]. The probability is lowered during the course of the optimization according to what is called a cooling schedule in Simulated Annealing. The parameters which determine the cooling schedule can themselves be changed by co-evolution. The idea of linking the Simulated Annealing Concept with a steady state GA was developed because of the excellent results the authors reached with a simple Simulated Annealing procedure on the Fisher-Thompson benchmark problems for the JSP [6, 10].

The schedule optimization GA uses a weighed combination of mean lateness and mean flow time as objective function. Job priorities can easily be taken into account. For the simulation tests both objectives were weighed equally. The results are shown at the end of this paper. Parameters for the schedule optimization are represented as integer values within the chromosome which also carries parameters for the co-evolution process. The schedule optimization GA is run until there are no improvements within the last 400 iterations. Then its parameters are altered by a normal distributed mutation operator as part of a simple (1+1)-Evolution Strategy (ES). The scheduling GA is run again with the new set of parameters until there are no improvements throughout the last 400 iterations. Then the former parent and offspring are compared and the better one is chosen for a new mutation.

3.   INTER-ORGANIZATIONAL FRAMEWORK

The non-hierarchical structure of the production networks examined here demands a collaborative decision-making. In this case coordination is not carried out through hierarchical instructions, but is established through bi- and multilateral decision processes respectively. A multitude of competence cells with different interests and objectives takes part in these processes. Consequently, inter-organizational manufacturing control is carried out via collaborative agreements on detailed order conditions and across several manufacturing levels.

The order conditions (e.g. due date, used material, reserved capacities, and reserved stock etc.) are formulated as constraints that influence the manufacturing control of more than one enterprise. This means that bilateral constraints have to be communicated and committed by all decision-makers involved. However, identifying and quantifying the implications of each decision that must be taken in conjunction with all but the smallest manufacturing level is not humanly possible. This makes an ASP-functionality necessary that performs "what-if" analyses for each competence cell before making local production decisions. The functionality also supports the interaction process with Internet as the medium among different companies.

A developed model editor offers intelligent support to help companies model their manufacturing process and the features they require. This approach is based on the "Reference Model for Shop Floor Production" according to the ISO-Standard (Workgroup TC184/SC5/WG1) [4]. This model comprises variables for each relevant manufacturing issue. Each process step is modelled on a very general level as an activity with the input or output of information, material, resource and control information. Thus there is the differentiation between the four generic processes of "transform", "transport", "store" and "verify". In addition, the relationships between manufacturing issues are formulated as constraints and the desired objective is defined. The ASP-functionality checks such factors like manufacturing capacities and parts orders to decide whether a new order can be inserted into the production schedule and still satisfies the resource constraints and the order conditions, e.g. the requested date. Scheduling is understood as a process of assigning activities to resources in time.

Besides, an interaction protocol has to be defined for each customer-supplier relation in order to coordinate the supply of materials with work-flow and resources to optimize production. The manufacturing control is specified individually by the coordination strategy and the degree of integration.

The different degrees of integration classify five inter-organizational coordination patterns according to the customer order decoupling point, see [9]. The customer order decoupling point determines the temporal and structural limit between the production with a concrete customer order assignment and the production for an anonymous customer on a forecast basis. The limit is often selected according to the delivery times required on the part of the market. These must not be exceeded by the throughput times of the customer order assigned production and delivery processes. There is no direct customer's reference to the production steps before the customer order decoupling point. Therefore the maximum degree of integration of the customer-supplier relation - which can be selected sensibly - derives from the customer order decoupling point that depends on the individual organization structure. Thus we differentiate between the following five coordination patterns:

1. Integration degree: Deliver-to-stock

   Current and future requirements of the customer are communicated to the supplier so that an early planning of the demand coverage can take place. Thus the responsibility for the material availability is assigned to the supplier. Consequently, the customer can reduce the complexity of his order coordination since the material availability concerning bought-in parts is contractually guarantied.

2. Integration degree: Make-to-stock

   It is the task of the classic customer-supplier relation of order-related delivery to agree upon order conditions concerning the product, the price, the delivery date, the quantity and quality as well as the place. The customer orders are mainly processed via stockpiled shelf inventories which have been produced beforehand, based on forecasting.

3. Integration degree: Assemble-to-order

   A closer involvement of the supplier is possible with the order-related final assembly since the final assembly of the components does not begin before the order placing. The necessary components are produced in pre-ceding manufacturing steps on the basis of sales forecasts. They can be stored temporarily. Apart from the order conditions of the classic customer-supplier relation the reservation of capacities of the assembly can be committed to.

4. Integration degree: Make-to-order

   The synchronization of the order coordination with the order-related manufacturing goes a step further. Here a customer-specific manufacturing is possible, a reservation of the manufacturing capacities needed for the customer order can be committed to. The lot size that is to be produced is thus determined via the order quantity. The necessary material is provided through ordering processes or through the storehouse.

5. Integration degree: Source-to-order

   In this case bought-in parts are individually ordered, according to the customer orders. Alternatively, a reservation of bought-in parts can be arranged with the customer who also has the possibility to consign the material necessary for the production. Thus the supplier can be completely integrated into the manufacturing control. Therefore the customer arranges the complete order-specific manufacturing process as regards delivery dates, capacities and material.

With every degree of integration the transparency of the production pro-cesses increases. The possibility for a close-to-reality planning of inter-corporation order management grows accordingly. At the same time the de-mand of information as well as an exceeded planning horizon and a greater planning complexity increase. Figure 2 gives an example of a production network whose material flow is coordinated by different integration degrees.

The communication protocol for the exchange of bilateral constraints is modelled in the form of statechart diagrams. Message types of the international electronic data interchange standard (EDIFACT) are used and others are added in order to perform the whole range of constraint exchange. A statechart diagram models the business rules by defining possible sequences of states and actions as a result of the reaction to incoming messages.

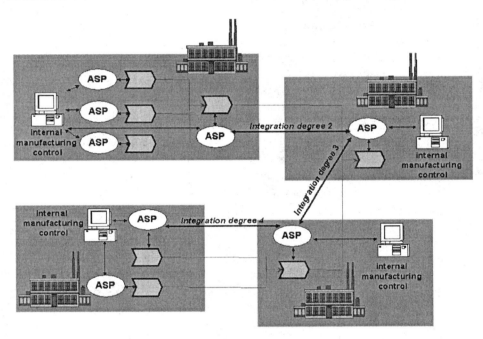

Figure 2. Production network (sample)

4.   AN INTRA-ORGANIZATIONAL FRAMEWORK

The next step is the integration of this software-system into an organizational framework. The authors carried out a study [6] to identify objectives used in companies for manufacturing control and to develop a ranking according to their importance. It turned out that it is primarily important to keep due dates for as many jobs as possible. A short mean flow time is less important. This latter objective is equivalent to other objectives like low inventory and high throughput.

Thus due dates and mean flow times seem to be the two most useful basic components for an objective function for schedule optimization. In order to calculate them the scheduling tool needs due dates for each job and each production area as input from the ERP-system. Sometimes this causes problems in companies with more than one production area where all areas need to have their separate scheduler. Many PPC-Systems calculate due dates with no regard to capacities. Others calculate virtually exact due dates for each operation but cannot distinguish where a production area ends and where the next one starts. Finally there are systems which do not keep records of the next job, for which one component is needed for. This makes it impossible for the scheduling system to derive a network of operations needed for the final product from the data that is kept within the ERP-/PPC-System. Therefore the rough capacity scheduling in most PPC-systems has to be changed to a certain extent to supply the data needed for manufacturing control.

During the installation process of the scheduling system, objective functions have to be adapted to the special needs of the company. The responsible person can choose from a set of objective constellations covering usual situations. When the scheduling system is started for the first time, all jobs currently in production or released for production are read from the ERP-system and distributed to different work stocks for the different schedulers - one for each production area. Then the schedulers are started. They are permitted to work for five minutes before their first solution is distributed to the PDA-terminals. The solution is set active.

The start of work has to be reported to the scheduler via the bi-directional PDA-system. Jobs which are currently being processed are no longer part of the scheduling process. Workers are allowed to choose the next job they want to work on from a number of n jobs. This is because up to now we are not able to build minimal setup time sequences due to the lack of data. Although the algorithm could easily handle sequence dependent setup times, there are only a few companies that record this data.

While the work process is running, the scheduler's EA is continuously searching for improved schedules. If a better schedule is found, it is set active and the planned work sequence is changed accordingly on all machines. This might cause a certain amount of disturbance within a manufacturing system because job floor workers usually prepare for their next job while the actual job is still in process. Therefore the n jobs appearing on the PDA-terminal are being "specified". This results in the fact that the sequence cannot be altered by the EA any more. Therefore the EA is given the initial five minutes to make sure that a satisfying solution is the base of the first specification of the jobs on a machine. If a job is completed on one machine or if some disturbance occurs, the worker has to report this via the PDA-terminal. This information is sent back to the ERP-system where it is ready for further use. The scheduling algorithm is also informed and performs its task on this changed database. The same thing happens if a supervisor decides to move a job from one machine which is currently overloaded to an alternative machine or if new jobs are released within the ERP-system. Thus the whole system is working continuously allowing to use the power of the EA but also relying on intuition and experience to handle exceptions.

5.   SUMMARY

It has been argued by the authors that modern shop floor scheduling tools based on Evolutionary Algorithms prove their potential for solving complex benchmark scheduling problems but failed to be used widely in real-world applications throughout the last ten years. One of the reasons identified is the need for organizational changes in manufacturing control caused by systems. The authors are proposing an organizational framework which includes bi-directional PDA-systems in which EA-based shop floor scheduling tools can be successfully implemented. A simulation study has shown the potential of the concept. A real-world implementation is currently being developed in co-operation with two German software companies and one mechanical engineering company.

In addition, such coordination patterns are proposed which make it possible to integrate interdependencies of manufacturing control between collaborating companies. An integration platform with ASP-functionality is being developed to link companies up to a non-hierarchical regional network. A model-based approach is chosen with which companies are able to configure their customer-supplier relations according to their specific organizational needs.

REFERENCES

[1]   Bierwirth, C., Kopfer, H., Mattfeld, D., Utecht, T.: Genetische Algorithmen und das Problem der Maschinenbelegung. Universität Bremen, Lehrstuhl für Logistik, Bremen (1993)

[2]   Bierwirth, C.: A Generalized Permutation Approach to Job Shop Scheduling with Genetic Algorithms. OR Spektrum vol. 17 (1995) pp. 87-92

[3]   Davis, L.: Job shop scheduling with Genetic Algorithms. In Grefenstette, J. (ed): Proceedings of the International Conference on Genetic Algorithms and their Applications, Hillsdale, Lawrence Erlbaum Associates (1985) pp. 136-140

[4]   ISO TC184/SC5/WG1: ISO Reference Model for Shop Floor Production Standards Part 1\&2. ISO, ANSI-NEMA, Washington DC (USA) 1989, (1990)

[5] Käschel, J., Teich, T., Zacher, B.: An empirical study of manual shop floor scheduling in job shop environments. Working Report 30/2000. University of Technology of Chemnitz, Department of Economics, Chemnitz (2000)

[6] Käschel, J; Teich, T.; Meier, B., Fischer, M.: Real-World Applications: Evolutionary Real-World Shop Floor Scheduling using Parallelization and Parameter Coevolution? In: Banzhaf, et. al. (publ.), Proceedings of the Genetic and Evolutionary Computation Conference, Las Vegas, Nevada, July 8-12, 2000, pp. 697-701, Morgan Kaufmann

[7] Kirkpatrick, S., Gelatt, C., Vecci, M.: Optimization by Simulated Annealing.Science, Vol.220 (1983) pp. 671-680

[8] Mattfeld, D.C.: Scalable Search Spaces for Scheduling Problems. In: Banzhaf, W. et al. (eds): Proceedings of the Genetic and Evolutionary Computation Conference (GECCO'99), Morgan Kaufman (1999) pp. 1616-1621

[9] Mertens, P.: Integrierte Informationsverarbeitung. Gabler, Wiesbaden (2000)

[10] Muth, J. F., Thompson, G.L.: Industrial Scheduling. Prentice Hall, Englewood Cliffs, New Jersey (1963)

[11] Syswerda, G.: Uniform Crossover in Genetic Algorithms. In: Schaffer, J.D. (ed):Proceedings of the Third International Conference on Genetic Algorithms and their Applications, San Mateo, 1989, Morgan Kaufman (1989) pp. 2-9

[12] Teich, T.: Optimierung von Maschinenbelegungsplänen unter Benutzung heuristischer Verfahren. Verlag Josef Eul, Lohmar, Köln (1998)

[13] Van Bael, P. et al.: The Job Shop Problem solved with simple, basic evolutionary search elements. In: Banzhaf. W., et al. (eds): Proceedings of the Genetic and Evolutionary Computation Conference (GECCO'99), Morgan Kaufman (1999) pp. 665-669

[14] Yamada, T., Nakano, R.: Scheduling by Genetic Local Search with Multi-Step Crossover. In: Proceedings of the Fourth International Conference on Parallel Problem Solving from Nature (PPSN IV), Berlin (1996) pp. 960-969

BIOGRAPHY

Prof. Dr.-Ing. H. Meier: Born 15. April 1951 in Pivitsheide, Germany. Mechanical Engineering studies at the TU Berlin; From 1977 to 1983 research assistant at the "Institut für Werkzeugmaschinen und Fertigungstechnik" in Berlin; 1981 Doctor of Mechanical Engineering by Prof. Dr.-Ing. G. Spur at the TU Berlin; From 1983 to 1995 leading position in the mid-size company "Schleicher & Co."; From 1995 to 1999 holder of the chair "Automatisierungstechnik" in Cottbus. Since 1999 holder of the chair "Production Systems" at the Ruhr-Universität Bochum.

Dr. T. Teich: Computer Science studies at the TU Chemnitz, Germany; Since 1994 research assistant at chair "Produktionswirtschaft und Industriebetriebslehre" (TU Chemnitz); 1998 Doctor "rer. pol." by Prof. Dr. J. Käschel.

H. Schallner: Born 22. September 1969 at Münster, Germany. Computer Science studies at the University Paderborn; Since October 1996 research assistant at the Automation Research Institute (Ruhr-Universität Bochum).

DATA ACCESS CONTROL IN VIRTUAL ORGANISATIONS
Role-Based Access Control Patterns

Peter Bertok, Saluka R. Kodituwakku
Department of Computer Science
RMIT University, Australia
{ pbertok, sakoditu}@cs.rmit.edu.au
Tel: 61 3 9925 6124
Fax: 61 3 9925 6139

Abstract: Virtual enterprises bring together different companies under one umbrella, and the organizational structure is tailored to the common project rather than reflecting the participating companies' structure. A virtual organization is also dynamic in nature, as jobs/positions can be created or abolished as the project progresses. Access to information needs to be very flexible in such environment. On one hand, people should have access to all information they need to perform their duties, and those duties may change as time progresses. On the other hand, providing access to data other than needed for a particular job can lead to information overload and also poses a security risk, as it can lead to information leak or accidental modification of data. Role Based Access Control (RBAC) offers a solution to this problem by associating roles with jobs and assigning access privileges to roles.

This paper examines Role Based Access Control and proposes some modifications to conventional RBAC to make it suitable for virtual enterprises.

1. INTRODUCTION

Controlling access to data is one of the core issues in system security. Within a single company access to data is very closely related to the immediate work environment, e.g. designers have full access to their own design data while can not modify other designs. It is quite possible, however, that all designs can be freely viewed by all designers, as this can promote interaction between designers. In a virtual organisation, however, any type of access to any data is a potential risk. While members in a virtual organisation are working on a common project, each member company has its own set of product, design etc. data, and only a small subset of these data, that belongs to the common project, should be accessible by partner organizations. [1]

Virtual organizations are also dynamic in structure. Responsibilities are defined and re-defined as the project progresses, necessitating different types of access to data at various stages of the project.

Access control in general is concerned with restricting the activity of legitimate users in a system. It relies on and coexists with other security services, such as authentication that establishes the correct identity of the user.

Several access control models have been developed, such as discretionary access control (DAC), mandatory access control (MAC) and role based access control (RBAC). In the following we give a short overview of the different access control methods, and then we describe RBAC patterns that are particularly suitable for the complex organisation of virtual enterprises.

2. ACCESS CONTROL METHODS

2.1 Approaches

Access control consists of two main components. Policies govern user access to information, and mechanisms are used to implement access control. The same set of mechanisms can be used in different ways with different access control policies. With discretionary access control the decision to allow or deny access to information is based on user identity. Every user or every user group has access privileges on individual objects. Each request is validated against those privileges and only requests passing this authorisation test are granted access. DAC can provide great flexibility via access policies, e.g. users can grant or revoke access rights on objects under their control, and this flexibility can be very useful in many cases. However, it may also pose a security risk, as users can open up confidential documents to other users; e.g. setting wrong access privileges enables access to product, design etc. data by partners in a virtual organisation.

Mandatory access control overcomes this by nominating a security administrator who solely has the responsibility of granting and revoking access rights. MAC policies are considerably less flexible than DAC policies, and can enforce security hierarchies straightforwardly. On the other hand, this rigidity can be a significant obstacle in a virtual enterprise, where partners need to respond to frequent changes swiftly.

Role based access control policies are based on activities of users. A role is a collection of tasks performed and responsibilities assumed in the organisation. A user acting in a certain role has to have all the privileges and access rights needed to perform the tasks attached to that role. With the increasingly widespread use of object-oriented technologies, the information accessed is often stored in the form of objects. [2][3][4]

When setting up the framework for RBAC, first roles are identified within the organisation, and then privileges are assigned to the roles. A user acting in a particular role is granted all privileges (also known as access rights) assigned to that role. Users may be able to assume several roles at the same time, or can be restricted to one role at a time. The structure of a RBAC system is shown in figure 1. RBAC implements two levels of association. First, access privileges are associated with roles, and second, roles are associated with users. A subject is a user acting in a particular role, and a user in multiple roles is represented by multiple subjects.

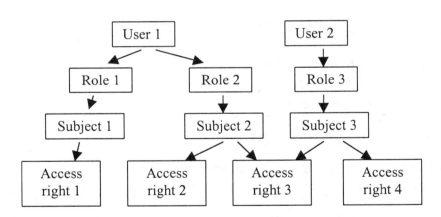

Figure 1. Associations between users and access rights

There can be mutually exclusive roles, which means that a user is not allowed to act in both roles; e.g. designers may not countersign their own designs. This is also known as separation of duties. Some roles have cardinality constraints; i.e. only a limited number of users can take that role at a particular point in time. [8]

The role-based approach has several important advantages, including the following.

- It implements separation and logical independence of user-role and role-privilege assignments.
- Role hierarchies can be used to support roles with overlapping responsibilities and privileges. A role hierarchy can contain non-overlapping sub-roles at a lower level, and upper-level roles can contain different, sometimes overlapping combinations of lower-level roles.
- It can easily implement the principle of least privileges, i.e. a subject acting in a role has no more privileges than those needed to act in that role.

2.2 Administration of privileges

Administrative policies control the allocation and modification of privileges. In DAC a security administrator assigns and withdraws user rights. Users then can control access to objects they own. In MAC privileges are based on security classification of subjects and objects. A security administrator assigns security levels to users, and the security level of an object is determined by the security level of the subject that owns it. It is the administrator only that can change the security level of subjects and objects. In RBAC the management of user privileges can be one of the roles. This administrative role, however, is different from the common roles it administers.

Delegation of responsibilities often makes administrative work easier. DAC allows this by letting users control access to objects they own. MAC is not really suited for delegation of responsibilities. RBAC can easily implement delegation of responsibilities by simply creating additional administrative roles with lower privileges.

3. ROLE BASED ACCESS CONTROL PATTERNS

3.1 Building blocks

RBAC is made up of several modules. [6] In the following we describe the building blocks that help to set up RBAC in a particular application environment.

Role-definition, privilege assignment

The first step is identification of job titles and responsibilities, and then a set of roles and privileges have to be assigned to each job title. Initially, roles are derived from job titles and privileges from responsibilities. When a user is assigned a particular role, it means that the user is given all the privileges and allowed all the operations that belong to that particular role.

Role administration

Administrative and common roles are similar in structure, but differ fundamentally on the objects they operate on: administrative roles operate on roles and privileges while common roles operate on ordinary data. To avoid any interference that may compromise security, administration roles are distinct from user roles i.e. only administrators can assign and revoke roles and permissions, and administrators are restricted to perform only these functions. Users are expected to keep a strict separation of administrative and common roles.

In a complex system there can be several administrators, with different, sometimes overlapping responsibilities. E.g. some administrators may assign, update and revoke roles and privileges, others may only manage user-role assignments; or certain administrators can be restricted to a particular domain and manage only a particular set of roles, etc. In fact, separation of authorization, i.e. role-privilege assignment from user-role assignment results in a more easily manageable system. This type of relationships between roles can be part of an administration role hierarchy, where upper-level roles can be combinations of lower-level roles and additional responsibilities.

Administrators may administer common roles and lower-level administration roles. Notwithstanding, administrators should never be able to increase their own rights under any circumstances.

Access control

After the administrator has assigned roles to users, the system must validate user requests against privileges assigned to roles: operations allowed for any of the user's roles can proceed while requests for other operations are denied. A straightforward approach to this problem is to provide a common interface to all operations, and individual operations are accessed via this common interface; in object-oriented terminology all operations inherit the common interface. The common interface implements access control, and the actual functionality of the operation is implemented in a derivative class.

This hierarchical implementation of operations ensures that all operations use the same access control mechanisms, and significantly simplifies the set-up and maintenance of role-privilege assignment by security administrators.

Object collections

In object-oriented systems the information is stored in form of objects. There can be any number of operations defined on these objects, and each operation can be permitted or forbidden for an individual role.

As the number of objects can be very large, it is very useful to combine smaller objects into larger aggregates, and define roles and privileges that affect the whole aggregate. This aggregation should be transparent to the user.

Roles as proxies

A consistent way of implementing RBAC is when roles access data and perform operations on behalf of the user, i.e. roles act as proxies for users. This simplifies the administration of privileges, as any role-privilege assignment is performed only once, and each instance of a role will have all privileges of the given role without any additional effort. A proxy role can be created only if the user has been assigned to that role.

A proxy role forwards any requests from the user to the actual operation. As actual operations are invoked via a common interface, an access request coming from a proxy role can easily be assessed within the common interface and allowed to proceed or denied, depending on the privileges of that role.

Role classification and access validation

The validation of a request consists of two steps. First, the privileges of the role requesting the operation are retrieved. Then the privileges are checked against the requested operation: permitted operations proceed, others are rejected.

Classifying roles and identifying common elements can be used for setting up role hierarchies and for rationalization of common functionality. [5]

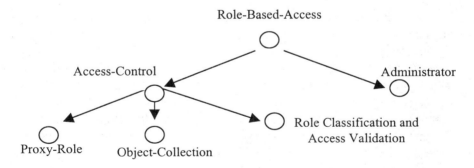

Figure 2. RBAC Components

3.2 Static and dynamic properties

Static properties form the skeleton of a RBAC system. Assignments that don't change very often, such as role-privilege assignment, and properties related to them are considered static. They are specified at role authorization time, and maintained throughout the role activation. There can be constraints of static nature as well, such as limiting the number of users in a particular role, also called cardinality constraint, and static separation of duty that describes roles that can not be taken by the same user.

Traditionally, properties referring to rules that apply while a role is active are called dynamic properties. Most of the static properties have a dynamic equivalent; e.g. dynamic separation of duty refers to roles that cannot be active by the same user concurrently.

We also introduced a new feature, which refers to the history of the user with regards to objects. If that is applied to separation of duty, it means that a user can not assume a certain role when dealing with a particular object. This can be easily understood by using an example in the area of design: a design may not be countersigned or authorized by the same person who actually produced it. However, without the history constraint, it would be easy to promote a designer to chief designer, who can then authorize all designs including his or her own.

4. A ROLE BASED ACCESS CONTROL IN PRACTICE

4.1 Roles and rules

RBAC can be implemented in several ways. In the following we present a pattern, in which two companies, company A and company B team up in a virtual enterprise to design and sell a particular product. Company A produces a piece of hardware that includes a common-off-the-shelf (COTS) computer, and company B produces software that will run on the computer. The team consists of the software and hardware designer groups, a sales representative and the manager of the manufacturing workshop and the manager of the software development group. Each designer group has a chief designer. Additionally, there is a project manager responsible for the whole project.

We consider the design phase of the project, which includes the following operations: (1) start a new design (2) delete an existing design (3) read (look at) a design (4) write (save) a design (5) modify an existing design (6) approve / reject a design.

A sample set of rules and constraints were applied, which consisted of the following:

(1) designers can write their own design

(2) designers can have a look at any design of their own group

(3) project manager and designer are mutually exclusive roles

(4) a person can be designer and chief designer but not at the same time

(5) the chief designer can have a look at any design, and can modify them.

(6) the chief designer has to sign off any design except those that he designed when he was an ordinary designer

(7) sales representative, manufacturing manager and software development manager have read only access to design data; if they want some modifications, e.g. for manufacturability reasons or to convey some request from the customer, they have to ask the designer to introduce those changes.

4.2 System architecture

In software development it is very common that data files are stored in a central repository, and users need to check out files if they want to work with them. This method offers an obvious point for implementing access control. When a user checks out a file, the access control system checks the user's role(s), and access privileges are attached to the file. In this way a user can get a read-only file or can get a file that can be read and written, or access may be denied altogether. When a designer wants to provide access to a particular design for other people, the file is checked in, i.e. a copy of the data file is put into the repository. This repository is shared with an existing Unix file versioning system, such as CVS, which also has the feature of maintaining a full design history. A versioning system like CVS can provide additional benefits, like checking for conflicts during check-in if two users had write access to the same file, but interpretation of data can be difficult. [7] From an architectural point of view the role based access control system constitutes a middle layer between the CAD systems used by the designers and the file system of the computer in use, as shown in figure 3.

Figure 3. System Architecture

4.3 System elements

The RBAC system maintains a history of accesses to each file, and also inserts this information into each data file in form of a header. This header includes file ownership, a history of modifying authors, and the authors' roles at the time of modification. A sample header is shown in figure 4.

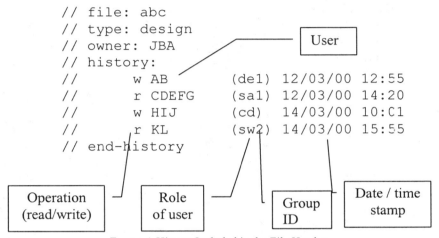

Figure 4. History Included in the File Header

The system stores the rules in the form of data, as opposed to code, which provides great flexibility as rules can easily be modified. Traditional RBAC systems implement rules in code, which means the source code needs to be modified, re-compiled etc. in order to change the rules.

The RBAC system maintains a simple user-role assignment table, and the operation – read or write – has to be specified at checkout. User ID and date/time stamp can easily be obtained from the operating system.

The rules are stored in a simple format. One group of rules describes access rights to objects, as shown in figure 5.

```
file-type: design
owner:    r w
group:    r
cdesign:  r w
sales:    r
manuf:    r
sw:       r
```

Figure 5. Access Rules for a Specific Object Group

Another group of rules describes role hierarchies from the prospect of privileges; i.e. a role has all the access privileges of all other roles that are on its 'includes' list. A third group describes 'separation-of-duties' rules. Examples of these are shown in figure 6.

When a user requests a file, RBAC establishes the user's identity. Then it ascertains the user's roles, and the rules relating to those roles. Subsequently it validates the request, and if no conflict is found, the file is checked out and passed on to the user with the appropriate read/write privileges. Requests that would violate existing rules are rejected. The system also keeps a log of checked-out files, mainly for error-detecting purposes.

```
cdesign  includes          design
manager  includes          cdesign, sales

manager  excludes(static)  cdesign
design   excludes(dyn)     cdesign
cdesign  excludes(hist)    design
```

Figure 6. Relationship between Roles.

4.4 Implementation

The RBAC system consists of three main modules, the Rule Management module, the User-role Management module and the Access Control module, as shown in figure 7.

RBAC maintains two databases. One contains the rules, the other contains user-role assignment data. When a request comes from a user to access a particular data file, the Access Control module (1) retrieves the relevant data from the databases, (2) evaluates the request, and (3) sends an appropriate answer to the user requesting the file. In step (1) the Access Control module determines the user's identity, retrieves the user's roles form the User-role database, then extracts the rules associated with that role from the rule data, and finally retrieves the history of the file. In step (2) it reads the rules associated with the roles, and checks them against the data in the history file. If the request passes the evaluation process, the Access Control system checks out a copy from the repository and passes it on to the user. The access privileges are set at this point, i.e. RBAC checks out a read-only copy or a read-write copy, as required. If the request fails the test, an error message is sent to the user.

When a user wants to check in a modified file into the repository, RBAC first checks if the file was checked out for writing. If it was, then it updates the file header from the history file kept in RBAC, i.e. inserts information about the user's name and role and puts in a time stamp. Then the file with the updated header is checked into the repository.

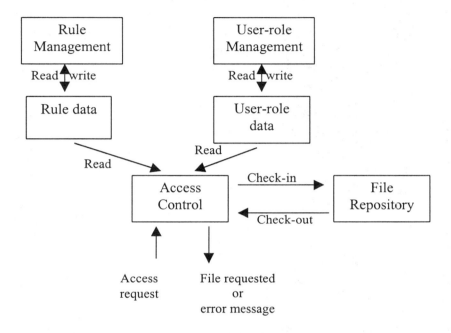

Figure 7. System Components

5. SUMMARY

This paper examined the use of role based access control in virtual enterprises. It was found that RBAC is suitable for such applications, and has some very distinctive advantages, as follows.

- There is a great flexibility in assigning roles to users, and user roles can be changed, users can be promoted or demoted, easily.
- RBAC provides a clear separation of role management and everyday task management. The roles of privilege assignment to roles and the role assignment to users are separate, and can be performed by different administrators.

- Implementing RBAC over a traditional operating system / file system is very straightforward and can easily be coupled with other, existing resources, such as CAD systems used in the enterprise.
- Features not used in conventional RBAC, such as rules relating to object history are needed in virtual enterprises.

An RBAC pattern was also examined. It was found that

- a RBAC system can be straightforwardly implemented as a middle layer between users and accessed files,
- in addition to user-role and role-rule databases a history of each file has to be maintained for assessing object-history-based rules.

6. REFERENCES

[1] W. Essmayr, E.Kapsammer, R.Wagner, G.Pernul, A.M.Tjoa: Enterprise-Wide Security Administration, Proc. of the 8th Int. Workshop on Database and Expert Systems Application, IEEE Computer Press, 1998,pp 267-272

[2] D.F.Ferraiolo, J.A.Cugini, D.R.Kuhn: Role Based Access Control: Features and Motivations, 1995, Proc. Of the 11th Annual Computer Security Applications Conference.

[3] S.I.Gavrila, J.F.Barkley: Formal Specification for Role Based Access Control User/Role and Role/Role Relationship Management, 1998, Third ACM Workshop on Role-Based Access Control

[4] W.A. Jansen: A Revised Model for Role Based Access Control, NIST-IR 6192, July 9, 1998

[5] W.A. Jansen: Inheritance Properties of Role Hierarchies, 21st National Information Systems Security Conference, October 6-9, 1998, Crystal City, Virginia

[6] S.R.Kodituwakku, P.Bertok, L.Zhao: A Pattern Language for Designing and Implementing Role-Based Access Control, KoalaPlop The Second Asian Pacific Conference on Pattern Languages of Programs, 29 May –1 June 2001, Melbourne, Australia

[7] A.Lam: Collaborative Design in a Distributed Environment and Object Versioning, Honours thesis, RMIT Australia,1996

[8] R.S.Sandhu, J.C.Edward, L.F.Hal, E.Y.Charles: Role-Based Access Control Models, IEEE Computer, Vol.29 February 1996, pp38-47

Knowledge-Based CAD/CAPP/CAM Integration System for Manufacturing

Kesheng Wang, Meng Tang, Yi Wang and Leif Estensen
Department of Production and Quality Engineering, NTNU,
N-7491 Trondheim, Norway
Kesheng.wang@ipk.ntnu.no

Pål A. Sollie and Mohsen Pourjavad
Xplnno AS, Trondheim, Norway

Abstract: Integration of CAD/CAPP/CAM is a natural result of scientific advance; such integration fulfills an overall automation and intelligence from product design to manufacturing in enterprises. In the paper, a new CAPP system is involved in and some new methods are introduced. The CAPP system is based on UG software, which is used as a platform of CAD/CAM system. A new way adopting the UG/API function to develop an interface between CAD and CAPP is employed to make feature extraction that can be transferred from CAD to CAPP. Then part information is sent to a knowledge base in the CAPP system. Process planning can be automatically generated through reasoning processes. The architecture of integration system and design methods are presented.

Key words: CAD/CAPP/CAM, API, feature extraction, Intelligence Manufacturing Systems.

1. INTRODUCTION

Integration of CAD/CAPP/CAM is a natural result of scientific advance; such integration fulfills an overall automation and intelligence from product design to manufacturing in enterprises. Productivity, quality, and manufacturing capacity are greatly improved by such integration. [Wang, 1991] Besides, the management methodology of an enterprise is also heavily influenced. However, the integration is a complex systematic project. The cost may be rather high, and if not implemented accurately, the benefits coming from CAD/CAPP/CAM integration will be compromised. Thus, one of the current trends in mechanical manufacturing area is to implement such an idea effectively with light investment, which is also our general rule when implementing the integration. CAD/CAPP/CAM integration is also required by the production and supported by the technical advances.

As separate systems, CAD, CAPP and CAM have been widely used in enterprises. However, due to the different historical backgrounds for these three systems, there are some limitations for the current application methods. Although these three systems are widely used in industries, there is limited information sharing among them, which reduces the efficiency of the system. Since these systems are separated in nature, resources in each system are also not fully utilized. With the advance of networking technique, information exchange among these three systems is highly desired. Processing information and knowledge should also be put on the first priory in development of integrated systems. In practice, different enterprises have different views of CAD/CAPP/CAM integration, but there are some common aspects related to system integration:

1. Parts information from CAD system has to be extracted and transferred to CAPP system completely and then CAPP system should generate the corresponding process plans automatically.
2. The CAPP system has to offers functions to manage information and knowledge. It means that the system provides mechanisms for users to acquire, share, utilize and create the knowledge in process planning.
3. In CAPP/CAM integration, the process planning information that generated by the CAPP system, should be transformed into NC codes in order to control NC machines.

It is very necessary to manage all data, information and knowledge related to manufacturing processes, such as, product data, CAD information, resource information, CAPP knowledge and so on.

Based on above principle, NTNU together with Xpalno developed an intelligent system for manufacturing companies to implement CAD/ CAPP/CAM integration. The system was based on the UG software as CAD/CAM platform. It successfully extracted the design feature with UG/API method. It can automatically generate process planning using a knowledge base which users are able to create and modify all the reasoning rules in the system. [Wang, 2000]

2.   THE AIM OF THE SYSTEM

2.1   Integration

The integration of CAD/CAM/CAPP is very important for improving efficiency of manufacturing and sharing production information. In order to take advantage of sharing information, the integration with PDM, ERP is also requested.

2.2   Intelligence reasoning

How to utilize the design and manufacturing information from CAD system to generate the process planning in CAPP system is a complex problem. Our aim is to generate the process planning based on feature information from CAD automatically. In the process of manufacturing planning, intelligence approaches are used to develop reasoning processes.

2.3   Information and knowledge management

The complex process planning is related to many data, factors, information and knowledge of product and production in manufacturing environment. Managing knowledge is very important, imperative and valuable in order to generate a perfect process planning. How to offer an efficient environment to user and how to organize, manage knowledge is also our tasks in the development of the integrated system.

3.   THE PRINCIPLE OF SYSTEM DESIGN

The integrated system is intended to provide practical solutions to the enterprise, but not a general system used for all enterprises. The following rules are followed during the system development:

3.1   Usefulness and practicability

The designed system should provide practical and useful functions to user, which can be used to solve practical problems.

3.2 Reliability and safety

Since implementation of the integrated system will be related with some low-level techniques of CAD/CAPP/CAM system and database system, reliability and safety are key factors for the success of the integrated system.

3.3 Expandability

Since the system is not a general solution to every enterprise, expandability will be considered during development in order to accommodate potential, specific requirements from different users.

3.4 Adaptability and Flexibility

Parameterized design concept and OOP technology are implemented in the system development. These techniques will ensure the adaptability and flexibility of the system when those items are actually changed during operations.

4. THE ARCHITECTURE OF INTEGRATION SYSTEM

To integrate CAD/CAPP/CAM, the key factor is to implement data sharing and exchange among individual system, which ensures sharing of design information, process information and manufacturing information. Figure 1 shows the CAD/CAM/CAPP integration architecture.

UG II software, from EDS, is applied in the system. Besides providing various supports for design (part design, 3D feature design, feature recognition, assembly and division etc.), UGII also provides support for automatic generation of NC codes and drives NC machines directly. Thus, UGII is chosen as CAD, CAM platform in the integrated system. The main aim for UG/CAD is to provide users with powerful part design functions. Besides, the system also provides 3D feature design functions, feature recognition functions and data exchange interface. [EDS Unigraphics]

The CAPP system are generating part process planning automatically through design information transformation interface and process reasoning procedure, and providing users with process editing and managing supports. At the meantime, it also manages related information and feeds the process planning back to CAM module to generate NC codes through manufacturing information interface and driving NC machines directly.

The interface between UG (CAD/CAM) and CAPP system is a key to the system.

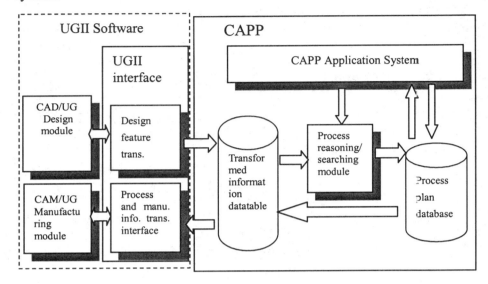

Figure 1. the CAD/CAM/CAPP integration architecture.

5.   THE APPROACH TO IMPLEMENT

5.1   CAD/CAPP Integration

In order to implement CAD/CAPP integration and transform part design information into process information for manufacturing the part, the following procedures are employed.

With the information exchange supports provided by UGII software, UG part design information (including geometric information and part feature information) is extracted from UG system through design information transformation interface, and stored (according to certain self-defined format) in transformed information database.

The transformed information is further processed by feature extraction, recognition and classification etc. to obtain original information for generating process plans.

On the basis of the original information, process plans are generated automatically by inference engine and are stored in CAPP process information database.

As we known, there are some criterions to use for information exchange, for example, STEP standard, XML and open XT format standard in UG. The UG/CAD also supports the standard. But the standard is too complex to use efficiently. In our system, a simple method called UG/API to extract information from CAD is applied.

Through the UG/API method provided by UG software for supporting the secondary development, we can develop the interface to extract the feature information and attributes. Based on analysis of information exchange, we need to get the following information from UG/CAD system to achieve the integration of CAD/CAPP. The most important information extracted from UG is listed in fig.2

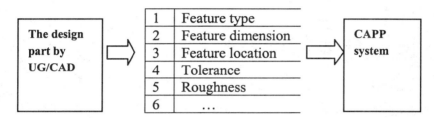

Figure 2. The most important information extracted from UG

5.2 CAPP System

Application system in CAPP can then be used to process planning (edit, manage, and dispatch etc.) The CAPP system provides functions for automatic process generation, manual generation, process editing, process management, information combination, process dispatching, resource management and process information sharing etc. providing user interface for PDM is desired. The system includes following major modules:

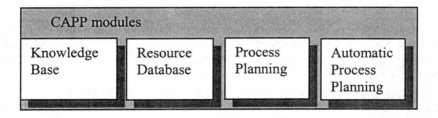

Figure 3. CAPP models

- ***Knowledge base:*** The reasoning rules for generate process planning is stored in the knowledge base. The main function of which is to collect the experience and knowledge from experts and operation engineers in the company. [Riley, 1989]

- **Resource database:** All resources of the manufacturing environment, including machines, cutting tools, fixtures, etc can be stored in it.
- **Process planning:** To make templates, edit the existed the process planning and monitor planning status.
- **Automatic Process planning:** To generate the process planning based on knowledge base and resource database; offer the manual function to edit process planning. Template function is offered to share process planning information, offer tools to get correlative knowledge of process planning. In this module, the integration of CAD/CAPP/CAM is implemented and realized. The reason process in the module is key to make the system intelligent.

In CAPP module, the intelligence plays an important role. It includes defining knowledge base, managing resource and automatically reasoning for process planning. The reason rule can be defined in knowledge base. The resource database is used for managing all manufacturing resources. In reasoning process, a reasoning engine generates process planning based on the reasoning rules and the information of resources. The reason mechanism is shown in figure 4.

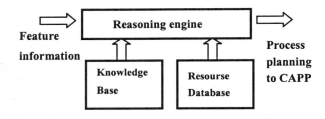

Figure 4. The reason mechanism

We define following steps for reasoning process planning:

- *Manufacture method selection*

According to some design and manufacturing features extracted from UG/CAD, including feature type, feature dimension, tolerance, roughness, the finial manufacture approaches can be selected. For example, if we know a cylinder and its dimension, roughness; then we are able to get the finial manufacturing method to make the feature. The rules about how to select the manufacturing approach can be designed in knowledge base by users.

- *Raw material selection*

The rules of raw material selection define how to select raw material according to overall dimension and shape identification of the part. So, it is necessary to calculate overall dimension and shape identification. We can get the overall dimension and shape based on dimension and location of each feature.

- *Machine selection*

The rules for selecting machine are based on manufacture methods, raw material dimension, machine parameter and performance, etc..

- *Fixture selection*

The rules of fixture selection can be defined according to overall dimension, raw material and machine interface. For example, using the rule of the chuck, chunk jaw can be selected.

- *Cutting tools selection*

Cutting tools selection refers to many options. In this phase, we not only select cutting tool material, shape and interface, but also select adapter and basic holder to fix the cutting tool. The rule is defined according to feature type, manufacture method, work piece, cutting tool parameter, adapter parameter and basic hole parameter.

- *Manufacture Sequence selection*

Based on the feature types, manufacture methods and location relation among each part feature, the sequence of process planning can be obtained. Some rules can be defined in knowledge base in order to adjust the sequence of process planning if the users want to modify the sequences generated.

- *Dimension calculation*

Dimension calculation is a complex process. It is based on the different feature type, manufacture method and model.

- *Cutting data selection*

Cutting data includes cutting speed and cutting feed. The rule is defined according to cutting method, cutting tool and work piece material.

In CAPP system, the Information management is also employed to manage all data, for example, part information, process-planning information, etc..

5.3 CAPP/CAM Integration

In UGII (CAM) system, the NC code for a manufacture part can be generated automatically. In the process of generating NC code, process planning plays an important role. It is necessary to transform the information of process planning into CAM system correctly and smoothly for generating the NC code automatically.

In the process of generating NC code, the information extracted from CAPP includes three types:

- *The manufactures sequence of parts*
This information will guide to generate NC code sequence for a part.

- *The method of manufacturing of parts*
This information will guide to generate different NC codes for different features.

- *The manufacturing dimension and precision*
This information will guide to generate different NC codes for different dimension and precision. It can help to match the manufacture tools automatically.

Those information must be transferred from CAPP system to CAM system for generating NC code. The interface of CAPP/UGII (CAM) plays an important role in the generating NC codes. According to the interface function, the CAPP/CAM interface mainly has two functions.

1. Extracting information from CAPP system
2. Treating these information according to data format and rules, and send some information to CAM system through API functions.

6.   THE FURTHER WORK

How to decide a process planning is a very complex process, there are many factors and Knowledge to be affected in the process. How to manage and utilize these knowledge becomes more and more important in manufacturing enterprises. Knowledge management about process planning is planed to involve in further work.

The integration is not only among CAD, CAPP, CAM, but also includes with PDM, ERP system. How to implement with PDM, ERP system is also considered in further work.

In order to improve the intelligence of system, the artificial intelligence, such as, Artificial Neural Networks, fuzzy logic systems and Genetic Algorithms, will be applied in futher development of the system.

7. CONCLUSIONS

In the paper, the CAD/CAPP/CAM integration system is introduced. A new approach with API to integrate CAD/CAM system is also presented. the system architecture is described in details. It is demonstrated that integrating CAD/CAPP/CAM based on the above architecture and technology is feasible, efficient and easy to implement for manufacturing enterprises. The integration, intelligence and knowledge management are an important development and application fileds in future manufacturing enterprises.

8. REFERENCES

1. EDS Unigraphics, Hybrid Modeling Fundamental student Guide. Version 13.
2. EDS Unigraphics, the reference of API Based on EDS Unigraphis.
3. Giarratano Riley, (1989), "Expert systems, Principle and Programming", PWS-KENT Publishing Company.
4. Hsu-Pin Wang, jian-kang Li, "Computer-Aided Process planning", Elsevier Science Publishers B. V. 1991,
5. Kesheng Wang,Tang Meng, Hilrpa L. Gelgele, (2000), Process Planing and its integration with design and machine process. ICME The 8th International Conference on Manufacturing Engineering, Sydney.
6. http://www.ugsolutions.com

The fabric feeding management for automatic clothing manufacture

G.M. Acaccia, A. Marelli, R.C. Michelini and A. Zuccotti
Industrial Robot Design Research Group - University of Genova - Italy

Abstract: The exploitation of "intelligent" factory set-ups enhances the competitiveness of the EU textile and clothes industries, by enabling collaborative design-and-manufacture options, while achieving economies of scope with effective exploitation of (strategic/tactical/execution) flexibility. Simulation turns to be reference aid for developing and acknowledging the appropriate set-ups and the adaptive schedules. The paper, besides reference concepts, summarises a case example related to the management of the fabric warehouse, in order to grant adaptive sorting and dispatching of bolts, by on-process optimisation of the actually used cloth, with account of fabric quality-data and of time-varying schedules. The combined-mode schedules show the benefits for qick response, leaving open middle/long horizons issues; checks on alternatives are provided by virtual reality tests.

Key words: Fabric warehouse automation, Quality manufacture, Computer simulation

1. INTRODUCTION

The *factory automation* concept bears little popularity among clothing manufacturers: the *information* flow (from ideation and strategic planning, to communication, flexibility ruling and quality assessments) runs separately from the *material* flow (from fabrics and supplies, to suits, dresses, apparels and underwear). Textile and clothes industries still prefer to preserve labour-intensive work-cycles, with fragmentation of material flow and *productive break-up* to low-wages countries; qualified businesses, at times, look after computer aids, aiming at using the information flow to oversee material procurement, marketing orders, etc. and, more specifically, to help devising innovative artefacts or, to a lower extent, to rule the shop floor schedules. Front end automation, in particular, lags behind; this concerns the flow-shop out-fits for mass-production through planning and scheduling organisation ruled by (off-process chosen) optimal control; this deals, as well, with flexible manufacturing of customised offers, by means of adaptive planning and scheduling driven by on-process govern. The *productive break-up* policy, thus, mainly aims at: «let do jobs by cheaper workers to improve effectiveness by better value/cost ratios», rather than at: «let do jobs by trained operators to embed extra-value from advanced technologies».

A different way to achieve return on investment leads to customer-driven product mixes, with quick-response frames and certified quality patterns. This brings to *intelligent* factory, with integrated control on, both, material and information flows; standard quality is steady issue of process visibility and up-keeping actions; client whims are readily satisfied by data transfer and on-line adaptivity. *Factory automation* will not cover the all *material* flow, rather a subset of it, but shall expand on the all *information* flow, to supply unified view of the production process, even when the material flow breaks into several segments. In the paper, a contribution in that direction is given, dealing with the manufacture of quality suits. The process description refers to a modular lay-out; the fabric storing-sorting section, only, is addressed to enable *intelligence*, with the joint exploitation of material-and-information, by warehouse automation and supervised fetching/dispatching with account of the pieces flaws or mismatches.

2. DISTRIBUTED INTELLIGENCE OPTIONS

With focus on the information frame, knowledge bases and knowledge processing shall be properly analysed. Qualifying bases embed: - fabrication agenda adaptivity: effectiveness aims at customers satisfaction, with pressure on fast changing fashion and quality preservation, by means of operation, tactical and strategic flexibility, in keeping with process monitoring; - fabric quality management: optimality looks at properly exploiting the available material, with account of nominal peculiarities (e.g. for fancy cloth) and of acknowledged faults (distinguished into classes and sizes).

An integrated information system qualifies by real-time adaptivity driven by on-process data; it supports economy of scope, say: flexible specialisation with lean manufacturing, company-wide quality, continuous betterment with simultaneous engineering, etc. Efficiency is built removing function/resource redundancy, as discontinuities and unexpected occurrences are overrun by prompting the technological versatility with the conditioning knowledge.

As a matter of facts, the 'creation' of high-standing clothes is still ascribed to handicraft; the innovation aims at enhancing the skilled labour at the ideation phases, improving the critical jobs with resort to intelligent automation and granting competitiveness with the build-up of items' *value* at buyer's satisfaction; rules require privileged knowledge sharing.

Combined control-and-management of flexibility is consistent with varying product mixes and with changing work-conditions. The strategic horizons comply with the business policy; quick-response replaces inventory by market-pull agendas. The tactical horizons deals with optimal schedule; versatility assures assorted batches with high productivity. The operation horizons tackle discontinuities; wild event tolerance and recovery help fabrication continuity. Intelligent manufacture grants flexible specialisation, but does not come to the same thing with the *unattended* 'factory-of-the-future', rather with *craftiness* helping steady high quality.

The fabric supply system is clearly critical to grant fabrication continuity, forwarding the items on a market-pull base. The information flow shall provide all pertinent data: warehouse general inventory, with cloth types, lengths, procurement policy, etc; individual roll special data, with handling rules, characteristic features, fault classification and mapping, etc.; these data, indeed, are actually background knowledge of front-end operators; they shall become foreground information to implement combined control-and-management logics. The goal is achieved, based on the factory automation software, a well developed branch of the computer technology, that simply needs include the peculiarities of the textile/clothing industry. A few hints are recalled in the following section, dealing the shop warehouse.

At the level of generalities, computer simulation and testing with virtual set-ups should be mentioned as powerful decision aids, for off- and on-line use to deal with flexibility. The tools are typical issue of the IT, providing: - at the facility *design-development* stage: resources setting needs comply with enterprise sale policies; agendas are stated for balanced throughput and due time; - at the facility *management-fitting* stage: production schedules are up-dated by on-process data, to face planned (e.g. itemisation) or unpredictable (e.g. failures) discontinuities. Simulation codes are, now, standard options, mainly, based on object languages and modular structures. Modularity is useful, to focus the attention on subsets of quantities, while leaving unaffected other parameters, drawn out from the facts to be assessed. The plant effectiveness, actually, depends on a large number of properties and the investigation should distinguish direct from cross-related effects, so that the knowledge frame is step-wise built up to the required level of completeness.

Extended tests on real facilities are clearly not possible. Then, simulation has to deal with a series of checks, each one providing the particular view of the studied factory. At the design stage, production facilities are compared as for enterprise policies; at the management-fitting stage, production plans are assessed as for delivery requests. The monitoring of value grow, by respect to cost build-up, is performed in virtual reality, to establish comparative enterprise forecasts and to anticipate benefits or drawbacks of the (finally) selected enterprise policy. The clothing industry benefits moves that way; the 'intelligent factory' concept is observed with caution: technology-driven additions, to a labour-intensive environment, cannot be accepted without fully acknowledging the return on investment. The throughout testing of achievements and drawbacks of virtual plants offers an affordable commitment, making easy to rank competing facilities and/or plans.

3.   THE FABRIC WAREHOUSE: DESIGN DATA

For competitiveness, the manufacture of quality clothes privileges styling and critical tasks (e.g., cutting), possibly, moving work-intensive phases (e.g., sewing) where labour wage is smaller. Return on investment establishes on condition that: - market demand takes over the amounts of ready-made suits or dresses, delivered by (large-enough) season's batches (to optimise the productivity on the tactical horizons); - flexibility is dealt with by 'quick response' opportunities, so that 'extra' items are managed on-process, to personalise size/details (as case arises, on the operation horizons).

FABRIC FEATURES: - Composition - Number of ends and picks - Weave - Colour fastness to light - Colour fastness to perspiration - Colour fastness to dry-cleaning (using perchloro-ethylene) - Colour fastness to washing - Colour fastness to dry pressing or ironing - Colour fastness to rubbing - Colour fastness to dry spotting water - Colour fastness to rubbing: organic solvents - Breaking strength and elongation: longitudinal, transversal - Breaking strength: grab method - Tear strength - Resistance to abrasion - Resistance to piling - Resistance to snagging - Resistance to dry ravelling - Seam slippage - Size stability to washing - Size stability to dry cleaning - Crease recovery

---

Figure 1. Typical characterising features of woven fabrics

Efficiency requires work-plans *leanness*, with visibility on cost build-up and quality transfer and removal of 'intangibles', with 'little' business benefit. Leanness entails benchmarking decisions, with purport on management (to embed administrative rules) and on technical issues (to plan out product and process innovation). Hereafter, the factory particular view tackles over the fabric warehouse. Due to flexibility, the fabrication agendas shall modify as for trade-driven (e.g. individual buyer whims) or leanness-bent (e.g. scrapes reduction) data. The latter option is outcome of fully exploiting available information by means of proper fixturing and processing equipment.

The cloth characteristics are object of proper standards and measurement techniques. It is possible to know: weigh, thickness, worsted number, warp and weft strength, bent and twist compliance, crease and crumple count, woof class, etc. and, in general, the different nominal properties associated to given commercial brands, *Fig. 1*. Fair trade, moreover, establishes a set of rules to specify the defects pertaining to individual cloth bolt.

A fault is defined as a non-conformity as compared to fixed agreements. The European Clothing Association, ECLA, suggests a reference set of specifications, covering: - yarn defects: broken thread, curling, twin top yarn, uneven glossiness, knot, etc.; - warp defects: barred weaving, furrow, stack, faulty warping, chopped off warp, etc.; - woof defects: lacking yarn, wrong weft, stretched/winding thread, fake reach, etc.; - flaws after dyeing, printing, finishing: unevenness, stains, blurs, creases, slobbers, etc.; - selvage flaws: folded listing, holed selvedge, damaged edges, etc.; - further deficiencies: skewness, drawing asymmetry, holes, inclusions, uneven width, non steady yarn, scrape marks, etc..

The process of deficiency sizing distinguishes: - *slight faultiness*: defects do not affect the cloth roll practical exploitation, due to location or low probability of being seen; - *heavy faultiness*: defects, if transferred into finished garments, are apparent and compromise quality; - *fatal faultiness*: the defective pieces of cloth, if used, make final artefacts unsafe or wrecked.

Business agreements include classing, sizing and mapping of defects. The cloth roll useful length is bounded between starting and ending strips; further strings mark the location of faults with, possibly, colour codes. Detection, location and marking are mainly accomplished on automatic unwrapping fixtures, so that classing, sizing and mapping are easily encoded on smart-cards or other chips, solid to the individual bolt (the addition of fillets or traps follows habits consistent with the manual processing of clothes).

The fabric cost is a non-negligible part of quality garments (from 40%, as a standard) and an enterprise will greatly benefits from waste decrease. The interest in special tools, such as **CAFE, C**omputer-**A**ided **F**abric **E**valuation, increases: - for textile shops, to bind product and process data and to establish setting and recovery actions assuring steady quality; - for garments shops, to devise efficient cutting policies, so that the (residual) defective stripes are moved off the used parts, properly exploiting the material and carefully avoiding the risk of faulty artefacts. The **CAFE** rig directly follows the loom and (usefully) enables corrective feedback; the monitoring is done by a camera (to perform pattern recognition) and restitution devices (to provide the length and width location). The mapping report is printed on tickets or encoded on read-only cards; the use of read/write cards would be helpful, to make easy the address of individual fabric bolts, during subsequent processing in and out the warehouse.

For an effective cloth management, the pieces to be cut on a given horizon should match the characterising quality/defect features of the stored material. Joint to the individual bolt, up-dated information shall provide: · fabric nominal identifier; · quality reference data; · rack and array location; ·

residual useful length; · defects mapping. Related to forecast work-plans, the current information will give: · requested nominal amount and delivery schedule; · pieces nesting programming and sequencing; · cutting policies, with due account of fault maps. The knowledge of faultiness degree and location (as for length and breadth) is used to compare the stored bolts by assessing the scrapes actually out-coming with the available nested cutting policies.

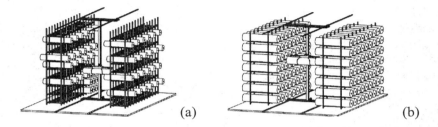

Figure 2. Warehouse lay-out for 350 mm (a) and 500 mm (b) diameter rolls

The betterment depends on many facts, such as: - delivery schedule horizon; - bolt placing and withdrawal capability; - cut pieces recognition and picking options; - defect sizing and description effectiveness; - fabric characteristics; - garments specifications; - faults removal policy; - data sharing and transmission devices. Efficiency requires balanced situations, when fully enabled hardware/software resources are managed with assessed return on investment. This quite naturally leads to increase automation, so that critical duties are moved off human reach into the domain of robot repetitiveness. For an effective cloth management, the pieces to be cut on a given horizon should match the characterising quality/defect features of the stored material. Joint to the individual bolt, up-dated information shall provide: · fabric nominal identifier; · quality reference data; · rack and array location; · residual useful length; · defects mapping. Related to forecast work-plans, the current information will give: · requested nominal amount and delivery schedule; · pieces nesting programming and sequencing; · cutting policies, with due account of fault maps. The knowledge of faultiness degree and location (as for length and breadth) is used to compare the stored bolts by assessing the scrapes actually out-coming with the available nested cutting policies.

The betterment depends on many facts, such as: - delivery schedule horizon; - bolt placing and withdrawal capability; - cut pieces recognition and picking options; - defect sizing and description effectiveness; - fabric characteristics; - garments specifications; - faults removal policy; - data sharing and transmission devices. Efficiency requires balanced situations, when fully enabled hardware/software resources are managed with assessed return on investment. This quite naturally leads to increase automation, so that critical duties are moved off human reach into the domain of robot repetitiveness.

The warehouse deserves special attention, since today manual operation is main resort. An overview of the prospected lay-out addresses to parallel scaffolds of superposed bolts, *Fig. 2*, each row is composed by frames with structural curved shapes, to grant roll steadiness. A translating tower accomplishes drawing in and out tasks; a cross-bar with latching/unlatching pliers assure the fetching and holding sequences. The tower is borne by top and bottom rails. The latching is done by pliers slight sink and quarter rotation, assuring safe cross-bar release with minimal operation space, *Fig. 3* **a**) *and* **b**). The tower feeds a series of AGVs, with racks, to stock the bolts, before dispatching to the proper laying and cutting station. The pertinent data are simultaneously transferred and, once the mattress is composed and the cutting accomplished, the picking of resulting pieces is automatically performed to establish assorted bunches exactly with the forecast order.

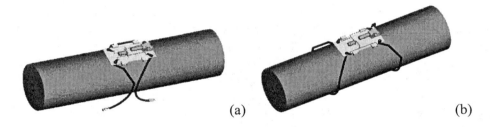

(a)   (b)

Figure 3. Grasping of rolls (a: approach phase; b: rotation phase for safe latching)

4.   WAREHOUSE MATERIAL AND DATA FLOWS

The advantages of process-data visibility need be assessed for actual running conditions. To comparison of competing solutions is beforehand ran by simulation: The **RSS**-SIFIP code is developed for the central warehouse feeding the subsequent laying and cutting sections; bolts are selected with up-dated account of, both, requested delivery order and cloth cutting plans optimised as for faultiness maps. The code avails of the MODSIM **II** object-oriented software and develops on a modular architecture to make easy resources changes or additions and/or control logic up-grading. Aiming at high-standing garments with market-driven customised satisfaction, quick-response and total quality are demanding requisites to win the competition in a world-wide market. Cost monitoring has to deal with material logistics and piece cutting programmes, with attention on data communication and on automation appropriateness.

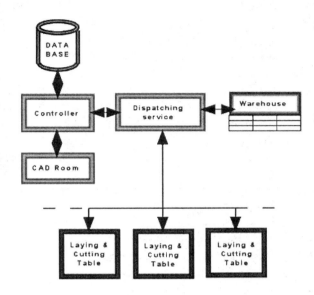

Figure 4. Overall architecture of the RSS-SIFIP simulation code

The **RSS**-SIFIP code, *Fig. 4*, distinguishes: the controller, with related data-base and CAD tool; the dispatching service; the addressable warehouse. The controller enables task schedule and co-ordination; given the fabrication agenda, it selects, from the cloth data base, the bolts allowing minimal waste and oversees the material and information transfer to the laying and cutting section. Dispatching includes the alternatives for roll latching/unlatching, withdrawal, buffering, transfer, and delivery to the individual laying/cutting station stores; the description of the all equipment is afforded at different levels: - for design, the translating towers (one for each bolt rank) and related cross-bar and pliers, the interfacing buffer, the shop-logistic facilities and the related releasing rigs are individually specified; - for operation testing, the overall service is combined into equivalent unified elements, each one with job-dependent properties. The cloth warehouse size is related to the enterprise strategic plans: e.g., with buyers-driven manufacture, total inventory is critical figure when quick-response is strict request.

Intelligent flow-shop should grant combined material and data processing. The **RSS**-SIFIP code turns to be a reference means for developing and acknowledging appropriate lay-outs and effective schedules. The fabrication agenda appears at the input; it is a file, with business-driven cutting schedules, specifying: - order identifier; - batch size; - fabric types; - clothes size mix; - mattress data (number of layers, number of windows, etc.); - laying data (due to quality specification); - cutting data (due to defect maps). The agenda specialisation is automatically accomplished by the shop controller, by using the cloth data-base. This up-dated information of the existing bolts: physical location (row, rank and layer), fabric nominal attributes (type, etc.); bolt properties (residual length, defect mapping, etc.); other (e.g., pointers for expensive, temporarily not present cloth, retrievable by the dispatching fixtures).

A simulation code module selects the fabrication agendas: these are either pre-established to test and compare the enterprise delivery strategies, either generated mixing structured files with random orders to assess the adaptive flexibility effects. The cloth data-base needs be up-dated and, possibly, expanded with temporary information (e.g., on-duty bolt work-cycle, so that a single item could be engaged by subsequent agendas with no in between storing). The study has, conveniently, to assess the warehouse lay-out: - arrays spatial built-up (row, rank and layer); - cradle size (and distribution in view of bolt specialisation); - storing up policy (handling priority, modular allotment, etc.); - input/output buffering capacity; - and so on. In the present case, for instance, the intermediate buffering is simply provided by a set of AGVs or shuttles, with a rack to be either loaded with the bolts to be forwarded, either unloaded of the coming back ones.

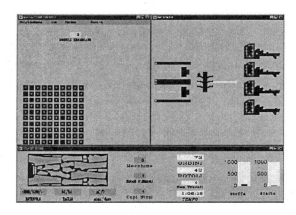

Figure 5. Pop-up windows of RSS-SIFIP for monitoring simulation progression

The facility shall permanently be monitored. The **RSS**-SIFIP code, similarly, displays the current state by a multiple window arrangement, *Fig. 5:* · the left window gives the addressed warehouse module, showing the occupied or empty cradles; · the right window specifies the considered buffering section, with the related work-status and set of required cutting stations; · the lower window provides the series of cut pieces lay-outs; · a set of slits supplies current reference data; · another set of slits presents cumulated process figures (utilised fabric, etc.). The knowledge of the defect maps may suggest special pieces nesting; alternatively, a subset of pieces could out-come with fatal/heavy faultiness: these appear marked-up and replacing pieces are separately cut from an extra length of the laid down cloth, *Fig. 6*.

The controller has to oversee the sequences of fabrication agendas; for efficiency, batch orders are forwarded with, mixed in, one-of-a-kind items, to satisfy given purchasers. This is always possible if the laying tables accept *extra-length* for out-of-batch articles (or for duplication, to replace faulted pieces, as just mentioned). The planning flexibility makes easy to include extra items or to push some others at a later batch, e.g., when the selected fabric in temporarily not available. The adaptive programming can report to on-line operators or to an *expert* module, with a rule-based decision logic. When this kind of investigation has to be pursued, the shop logistics is summed up into an AGVs equivalent dispatching service. In the practice, the bolts are looked for at first in the intermediate buffering section and only subsequently in the all warehouse: cloth recovery the re-routing of given AGVs for in-between work-station service.

Summing up, the simulator is task driven, each task split into a series of occurrence: a cycle starts with the new task (from the fabrication agenda); the cloth bolt selection is done, for the feeding-run rack-mix; the loaded AGV service follows, before the return-run or, possibly, an inter-station service; the information up-dating gives account of every in-between occurrence. The number of AGVs, having given in charge the dispatching, depends on the through-put; the simulation helps assessing the need of the intermediate buffering and alternatives can be devised with shuttles on fixed rails or other transfer means. The control logic, as well, can embed several options, e.g., rack selective loading-unloading, to keep in the foreground the bolts of fabric known to be requested within a given horizon span.

The simulation study is pushed up to compare facilities with different level of integration, as plants are conditioned by contingencies which might frustrate expensive changes, when actual advantages are left in-exploited. The results, given by the **RSS**-SIFIP code, provide hints on actual returns for different factory situations. Let refer to the following assumptions: · the available warehouse section composes of four scaffolds with 10x10 frames, for 400 total bolts, say, 20 equally-distributed cloth types; · the initial roll length ranges between 18 to 20 m, with 5 to 15 duly mapped heavy/fatal faultiness; · the artefacts mix has varying assortment, requiring a proper number of *cut span* on the laying table. Depending on the batch (number of products), the fabric laying with account of the defects mapping allows a considerable reduction of wastes, as compared to the plain afterwards recognition of non-usable pieces. The on-process use of cloth data requires the handling of a larger number of bolts, but this is compensated as the later handling of the (same) bolts for cutting the replacing pieces is avoided. The bolt optimal search, in this case, turns to be simple due to the large availability of alternatives.

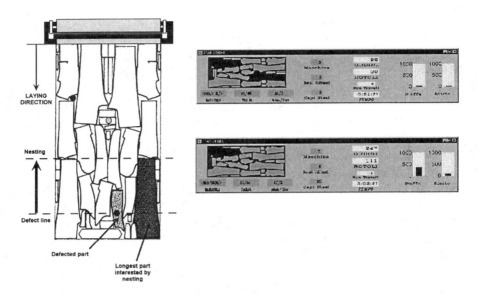

Figure 6. Windows for management of defected parts

Let, thereafter, modify the previous assumptions so that the number of requested items almost matches the available material and the fabrication agenda should comply a strict cloth type order. The bounded inventory keeps benefits, with higher effects for larger batches. The available bolt number is varied, to assess inventory influence with constant batches,. The study goes on with focus on the faultiness ratios; let assume: · to have a large inventory (20 bolts each cloth type) and to request a mix of 20 products (15 items each), for a total of 300 items batches; · the defects might vary between 2 and 20. The benefit of the early report of faultiness is quite evident. The situation modifies, with an inventory bounded at the current batch size. Basic results, obtained assuming full exploitation of the handled bolts, are shown, as explanatory examples, at the presentation. For high standing, one-of-a-kind production, the bolts are progressively consumed turning the useful length down to less to the minimum *cut span* of a complete suit. The influence of roll-ends is needing attention, specially for expensive materials. The previously found benefits modify and the two situations of high inventory, or batch limited inventory might be compared.

5. CONCLUSIONS

The paper introduces to the economy of scope, by intelligent clothing manufacture. The competition between enterprises resorts to the process-added value of *actually sold* apparel, rather than to large products batches, requiring to run after buyers, with advertising or lower sale prices. High-standing clothes are noteworthy, as clients require personalised quality and quick service. The discussion offers hints to look after the integrated manufacturing approach and the fabric warehouse management is specifically dealt with. The process description is based on a modular lay-out, to separate the effect of influence quantities and to *investigate details*, preserving the overall view of the business. In this study, fabric faultiness is directly considered, showing the benefits out-coming from the before-hand exploitation of the defect maps to lower the material wastes.

One should emphasise the fact that, to-day, the clothing industries are work-intensive set-ups and extensively resort to the on-line operators versatility of to modify production, while the process progresses; this possibly hinders the benefit of *intelligent* manufacturing, based on the concurrent run of the material and the information flows for adaptive flexibility: - at the organisational range (process-attuned managers): to select the fabrication agendas; - at the co-ordination range (decentralised controllers): to optimise the cloth bolts choice; - at the execution range (real-time supervisors): to adapt the material dispatching service.

The example discussion shows that flexible automation can deal with foregoing information (cloth defect maps) on self-sufficient bases; actually, the benefits depends on a large number of cross-related facts and actual implementations, hard to be fixed, remain out of the reach of front-end operators. The area of high-standing garments, satisfying varying market requests, is exemplary case where automation provides critical support for quality certification. A complex innovation of, both, material and information resources shall be promoted, e.g.: · cloth warehouse with high dexterity, efficiency, flexibility and versatility fixture and with effective data-bases storing and management capabilities; · sophisticated shop logistic service for the material dispatching ruled by a decision logic incorporating on-process information by plausibility reasoning. The changes, however, need be investigated in terms of the expected returns; simulation studies are dominant help, to compare competing alternatives referring to actual production contexts and, moreover, to provide explanatory examples with training purport immediately related to sets of feasible implementations.

HIERARCHICAL KNOWLEDGE-BASED PROCESS PLANNING IN MANUFACTURING

Ferenc Deák
Budapest University of Technology and Economics
dfery@bigmac.eik.bme.hu

András Kovács
Budapest University of Technology and Economics
csandris@sch.bme.hu

József Váncza
Computer and Automation Research Institute, Hungarian Academy of Sciences
vancza@sztaki.hu

Tadeusz Dobrowiecki
Budapest University of Technology and Economics
dobrowiecki@mit.bme.hu

Abstract   Artificial intelligence planning methods haven't been used until recently to address the problem of computer-aided process planning (CAPP) in manufacturing in its entirety. They were simply not developed enough to tackle real-world problems of that complexity. In the paper we show that with so-called Hierarchical Task Networks, a recently matured general-purpose domain-independent planning method, we could model the planning process itself, represent and utilize different kinds of technological knowledge and keep in check the complexity of the plan generation process. To this aim the planner was extended with search methods for finding the best plans and supporting mixed-initiative, interactive planning. The proposed CAPP system deals with geometry analysis, setup planning, selection and ordering of machining operations and the assignment of resources. The first experiments with prismatic and rotational parts show considerable merit of the approach.

Keywords: Computer-aided process planning, artificial intelligence, hierarchical task network

1. INTRODUCTION

Planning of manufacturing processes provides the link between design and production. Its task is to determine a plan of discrete manufacturing operations that, when executed in an actual production environment, will produce the part as required by its design description. Computer-Aided Process Planning (CAPP) may result in better designs, lower production costs, larger flexibility, improved quality and higher productivity.

Over the last three decades vast efforts have been made in developing novel methods and architectures for CAPP systems. The last twenty years of research has been dominated by the application of artificial intelligence (AI) methods and tools. Oddly enough, results of planning - a mainstream AI research area - found only scattered and rather simple applications in the domain of manufacturing. The reasons of this mismatch are twofold:

- CAPP is a complex problem that includes part analysis, selection of operations and resources, operation sequencing, setup planning, fixture design, and the determination of process parameters. Hence, the domain knowledge of a process planner has to cover geometry and tolerances, material properties, manufacturing processes and tools, fixtures, as well as machine tools. Besides generating executable plans, the optimal allocation of resources is the main concern of planning.
- General-purpose AI planning systems provided clear-cut logic-based representation formalisms and more and more efficient solution methods. However, the restricted representation formalisms did not allow to capture all of the relevant domain knowledge and to define planning strategies. Solvers could not handle optimization objectives and support mixed-initiative, interactive problem solving. Hence, they could not fit the real-world problems like the CAPP problem.

The goal of our research was to show that despite all the above difficulties the CAPP problem could be approached and solved as a planning problem. With the application of a recently matured general-purpose domain-independent planning method, the so-called Hierarchical Task Networks (HTN, [10]), we could model the planning process itself, represent and utilize various kinds of technological knowledge and keep also in check the complexity of the plan generation process.

In the paper a short analysis of the CAPP problem is presented. The adopted perspective is the complexity of the information management, the variety and the volume of information involved and the demands it presents for the automation and interactivity of the planning process. Next we show how the general-purpose, public domain planning methods should and can be extended to tailor them into efficient CAPP tools. The newly developed algorithms are presented in detail. The proposed approach is experimentally verified for a set of prismatic and rotational parts manufactured on vertical machine with a configurable set of tools. Finally, we discuss future extensions and the integration of the proposed methods.

2.   THE CAPP PROBLEM

The *problem of CAPP* in the domain of manufacturing can be stated as follows: given (1) the descriptions of the blank part and of the finished part (in terms of geometry, dimensions, tolerances, material, and quantity), (2) the available production resources (machine tools, fixtures and tools), (3) the technological knowledge and (4) some optimization objective, find an executable and close-to-optimal plan [4,16].

There are two main approaches to process planning, the variant and the generative one. *Variant* methods are based on the retrieval and the (manual) adaptation of previous plans; they can be supported by database retrieval, and as recently, by case-based reasoning. *Generative* planners synthesize process plans. They almost unanimously depart from the geometrical CAD model of the part and work with descriptions enhanced by manufacturing *features*. Features (like e.g. holes, slots, pockets, etc.) tie together frequently occurring sub-problems with their corresponding solution patterns, i.e., geometry and tolerances with particular production methods and resources [16]. Features realize micro-worlds with both design and manufacturing related information [13]. They decompose the problem and make it amenable to efficient and automated problem solving. However, decomposing the problem is not sufficient for conquering it. Features - or rather the operations that produce them - often interact, hence the selection and merging of the appropriate plan fragments is not that straightforward. Further on, planning has to account for global technological requirements (concerning, e.g., fixturing), and overall optimization objectives as well.

The *process plan* describes how to produce the part by using the available resources. It specifies the *operations* and their *resources* (tools, fixtures and machines), the *sequence* of operations and the groups of operations - so-called *setups* - that will be performed together, by using some common resource(s). Every setup begins with mounting the part to the machine tool. The orientation of the part on the machine tool determines which operations can be executed. For some operations the orientation of the part must be changed. Hence, the setup is changed: the part is mounted to the machine tool (or to an other machine tool) in a new position.

The *planning process* can be usually considered as the hierarchy of:
1) *Setup planning*: The determination and sequencing of setups and the selection of machine tools and fixtures.
2) *Operation sequencing*: The determination and sequencing of operations and the selection of tools.
3) *Operation planning*: Determination of the machining parameters (cutting speed, feed rate etc.) and the trajectories of the tools.

There are many commercial tools to aid the last step. However, the first two steps are very hard because the knowledge provided by the experts is usually fragmentary and inconsistent. In computer aided systems such contradictions must be solved by the human experts. Nowadays the manufacturing process is more or less automatic, but process planning still requires much work of qualified personnel. Hence, process planning is the "bottleneck" of the production [8].

3.   ARTIFICIAL INTELLIGENCE AND PLANNING

Although artificial intelligence is by definition involved in the design of intelligent systems, the question of intelligence is in process planning not that important. The reason why the working knowledge and the application of the AI methods is nevertheless essential, even if no intelligent system is actually being built, is the fact that the scaling up of the real-life manufacturing problems leads to a vast body of heterogeneous, uncertain and inherently inconsistent information. In the traditional system design approach these are the major obstacles, which prohibit the usual formal specification and verification of designs.

In artificial intelligence, on the contrary, exponential complexity, uncertainty, inconsistency, interaction of various kinds of knowledge are treated as inherent attributes of complex real-life problems and numerous methods of representation and manipulation are designed to handle them [13]. Furthermore, no other methodology provides tools to tackle complicated sub-problems and to integrate the solutions into a whole.

Planning, i.e., generating sequences of actions to perform tasks and achieve objectives belongs to the core paradigms of AI [19,20]. It embraces a considerable body of abstractions, methods, and tools applicable at, and even spanning (as in hierarchical planning) many levels of abstractions. Although the paradigm of planning is one of the oldest in the AI, it is also that paradigm where the creative mixing of the ideas took and continuously takes place (consider e.g., partial ordering, least commitment strategy, using logic within a non-logical framework, anytime planning, monitoring and re-planning, planning with imperfect information, probabilistic planning, hierarchical planning, etc.).

Planning is also considered one of the most important paradigms in the future AI, serving as a kernel method for information gathering and sharing, setting up co-operation, maintaining dialogue protocols, etc. in loosely coupled intelligent systems. As a consequence, we can expect a constant flux of new developments to be tried also in the process planning domain.

General purpose, symbolic *planning methods* of mainstream AI offer in some sense a wider, in another sense narrower conceptual background than actually needed for process planning. So-called classical planners [13,19] provide a formal language for describing the initial and goal states as well as the possible actions that may change the states. They have a plain and single measure of plan performance (plan length) and give no means for expressing control knowledge in an explicit way. They are weak in geometric reasoning and require a consistent world model [9]. The efficient generation of plans calls either for specialized methods, or for the translation of the problem case into a general combinatorial problem. Hence, such planners can hardly support mixed-initiative problem solving. This approach is also called primitive-action planning because plans are constructed solely by making use of the descriptions of the actions that may appear in the final plan [20]. Until recently, classical planners could be applied only to simple CAPP problems.

A powerful tool to overcome problems stemming from heterogeneity and complexity is the hierarchy of abstractions. Hierarchical approach, by processing information confined to the hierarchy levels and by passing information between them, makes it possible to maintain descriptions of much more complicated real systems and activities. Viewing intelligent activity as a hierarchy of tasks performed by agents and requiring expertise and resources at several levels served as the basis of the recently developed KADS knowledge modeling approach and similar efforts worldwide [15]. It came up also in planning, in various forms of so-called Hierarchical Task Network planners [10], which are now considered the most appropriate vehicle for solving real-world planning problems [20].

HTN planners generate plans that accomplish so-called *task networks*. A task network is a set of tasks to be carried out, together with constraints on the ordering of the tasks and the possible assignments of the task variables. The variables can typically represent resources assigned to and/or demanded by the tasks. Further constraints can express conditions on the states of the world before and after executing the tasks. Tasks can be decomposed into subtasks. Planning progresses then through the stages of top-down task decomposition, constraint satisfaction and conflict resolution.

4.   PROCESS PLANNING WITH HIERARCHICAL TASK NETWORKS

In the proposed approach both the representation formalism and the planning engine of SHOP, a domain-independent HTN planner were used [11,12]. This planner was provided with domain-specific knowledge concerning parts, features, tools, setups and machines, and extended with search methods aimed at finding the best (i.e., the cheapest) plans and supporting mixed-initiative, interactive planning. All in all, our CAPP system deals with geometry analysis, setup planning, selection and ordering of machining operations and the assignment of the resources.

By applying SHOP we combined the domain-specific and the domain-independent aspects of planning. Domain-specific planner – our CAPP system – is encoded in the domain description language of SHOP. Then the planning engine of SHOP compiles and runs this planner.

4.1. The Model

The planning problem is specified in the problem domain description language of SHOP. This input contains the feature-based model of the part, as well as the enumeration and geometrical models of the applicable tools.

The raw material can either be a prismatic or rotational stock or a pre-product with some features in an advanced state. Its fixturing can be done using a vise or a classical 6-points holding device. The part orientation, the holding device and the positioning surfaces determine together the fixtures.

Technological instructions on when and how to use the tools and fixtures are also part of the domain model. A process plan is a totally ordered sequence of machining and fixturing operations. The cost of a plan is determined by the costs of the individual machining operations plus the costs of resource (tool and fixture) changes.

The task is to generate alternatives of executable and close-to-optimal process plans.

4.2. System Structure

The planning system has several modules. A pre-processor unit analyses the part and generates the operations needed to manufacture the part, as well as the constraints on their ordering. Different modules deal with the selection of appropriate fixtures and tools, and handle the resource assignment and ordering of the operations. Plan correctness is guarded by a body of technological constraints that are expressed as axioms of the CAPP domain. A machine-tool specific module performs simple geometric reasoning. We intend to extend this component to a general-purpose geometric reasoner (see Section 5).

The task of the core, generic planning engine is to solve the planning problem by harmonizing the work of the modules: it has to assign fixtures and tools to the operations and to organize them into an executable, correct sequential plan in the most economical way.

4.3. Hierarchical Planning

Given the feature-based model, the pre-processor composes the list of required operations in compliance with the initial and required states and geometrical parameters of the features. It also recognizes *precedence constraints* between operations and translates strict tolerances between features into statements asserting that manufacturing of those features should be done in the same setup. Such constraints express *setup coincidence.*

Since the pre-processing determines the operations and their execution costs, which are supposed not to depend on the tool selection, the optimization can be performed by finding an adequate plan and resource assignment with a minimal number of setup and tool changes. According to the hierarchical structure of the manufacturing process, planning is executed at the levels of setup planning, tool selection and operation sequencing.

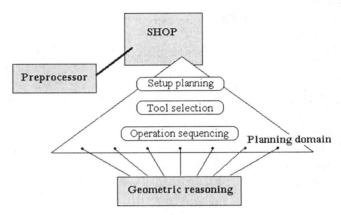

Fig. 1. Architecture of the CAPP system

At the highest, setup planning level the aim of finding a minimal grouping of operations is achieved by an iterative deepening search that minimizes the number of setups. For each group of operations a fixture is selected. While investigating the feasibility of the partial plan, we have to consider the following factors:

- *Precedence constraints*: Groups of operations are subject to constraints inherited from the contained operations;
- *Setup coincidence constraints*;
- *Resource constraints*: There should to be at least one tool available for each operation, able to execute it and being potentially able to access the feature realized by the operation. 'Potentially' means here that the workpiece orientation is convenient, and that the operation can only be obstructed by the removal volume of another feature.

Similarly, the tool selection is also performed by an iterative deepening search according to the number of tools, separately for each setup. Operations that can share tools are collected in sub-groups. However, the decision here is less complicated, since we have only to examine whether there is a tool to execute all operations belonging to a sub-group.

At the operation sequencing level the resource assignment is completed. The operations using the same resources are organized into operation clusters. Hence, only their total ordering is to be established, in accordance with the precedence constraints. There are two types of such constraints. *Explicit precedences* can be fetched from the statement list of the planner (either recognized by the pre-processing module or specified by the user). *Implicit precedences* can only be obtained by an on-line test, assuming full knowledge of the state of the part at the moment of the operation in question. This information is provided by the forward-chaining search mechanism of SHOP. An example of such hard-to-recognize constraints is shown in Fig. 2.

Consider a part with an overlapping hole and a slot, where the hole is deeper than the length of our twist drill. The hole can be drilled only after machining the slot, assuming that the latter is wide enough to admit the chuck. Geometric reasoning is dedicated to investigate such questions and to verify whether an operation can be inserted into the plan as the next step.

This kind of precedence constraint, if realized only during operation sequencing, can make it necessary to backtrack, and in the worst cases, even to reconsider the tooling or the setup plan. However, the experience shows that such situations occur quite rarely and the decomposition achieved by the hierarchical planning offers efficient means to cope with the enormous search space and results in reasonable technological plans.

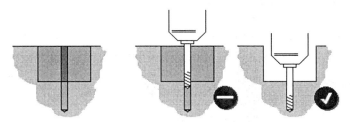

Fig.2. Implicit precedence constraint

5. GEOMETRIC REASONING

Geometric reasoning in CAPP systems has to be accomplished by performing collision tests between solids that represent the part, its features, the machine tool and the fixtures, as well as the tool path. Our planner handles now special cases that are characteristic to the problem instances, but we intend to perform geometric reasoning in a more principled way. Below we describe shortly the basic concepts of our model and algorithms.

Solid models are built of *primitive solids* of simple shapes (box, cylinder). They represent the stock, the removal volume of a feature, or a component of a tool or fixture.

Compound solids are constructed by applying regularized Boolean set operators to primitive solids. Using De Morgan's laws, the resulting CSG tree can be converted to a special tree, in which the operator *complement* is applied only on primitives. Hence, a compound solid is (1) a primitive solid, (2) the complement of a primitive solid, (3) the union, or (4) the intersection of two compound solids.

We make a reasonable restriction on primitives: they can either be independent or overlapped, but not tangent. Thus, the intersection of any two of them is a finite (maybe empty) set of continuous curves, while that of three primitives is a finite (maybe empty) set of discrete points.

Consider two compound solids A and B, where neither of them contains the other (i.e., $A \not\subset B$, and $B \not\subset A$). They are *colliding*, if there exists a point p such that $p \in A \cap B$. Let us call these points *common points*.

The planner has to determine whether two compound solids are colliding or not. Our algorithm accomplishes it according to the *generate-and-test* paradigm [2]: First, candidate points of interaction are searched for; then each candidate is tested whether it is inside both solids.

If the two solids collide, there also exists an intersection of their boundaries: $p' \in b(A) \cap b(B)$. Furthermore, there is also a common point in the finite set of discrete points that is generated as follows:

- For each pair of primitives (Ta, Tb), such that Ta is a constructing primitive of A, and Tb is that of B, an arbitrary point is chosen on each continuous curve segment of $b(Ta) \cap b(Tb)$.
- For each triplet of primitives (Ta, Tb, Tc), in which there are primitives of both A and B, all the intersection points are generated.

Finally, the so-found points are checked if they are inside the solids A and B. The test is first performed on the ponated or negated primitives of the solids. Then, the results are propagated along the arcs of the CSG tree. The following rules are applied:

Let X and Y be two compound solids, and p the point to be tested. Then

1) $p \in X \cap^* Y \Leftrightarrow p \in X \wedge p \in Y$,
2) $p \in X \cup^* Y \Leftrightarrow p \in X \vee p \in Y$.

Note that the restriction we made on primitives is necessary to justify these apparently trivial rules for regular operators.

The collision test succeeds once a common point is found. The negative result means that the two compound solids do not interact, and there is no geometrical hitch to execute the operation on issue.

6. EXPERIMENTS

The introduced approach was validated on a set of rotational and prismatic test parts. As an example, see our result on a vertical, 3-axis machining center that had to produce a prismatic part with 9 features. Features are referred to by their type and normal vector. This experiment was run on a 233 MHz PC with 32 Mbs of memory. It took 11.25 seconds for our CAPP system to generate the following manufacturing operational plan (see Fig. 4).

Noticeably, the hierarchical decomposition, i.e. the tooling being done separately for setups, results in a surplus tool change. In the second setup, there is no reason not to execute the operations with the endmill first, and those with the sidemill after it. However, on machines with an automatic tool change, it means no extra time, since the tool and setup changes can be done simultaneously, and this handicap is a reasonable price for the achieved saving of the search time.

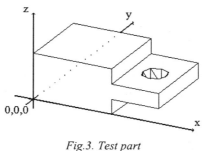

Fig.3. Test part

```
Fit in a vise, z- up, on surfaces x+, x-
              Change tool: 0 → sidemill
                        Machine face_y-
                        Machine face_y+
              Change tool: sidemill → endmill
                        Machine face_z-
                        Rough step_z-
                        Finish step_z-
Release
Fit in a vise, z+ up, on surfaces y+, y-
              Change tool: endmill → sidemill
                        Machine face_x-
                        Machine face_x+
              Change tool: sidemill → endmill
                        Machine face_z+
                        Rough step_z+
                        Finish step_z+
              Change tool: endmill → center-drill
                        Center hole
              Change tool: center-drill → drill
                        Drill hole
Release
```

Fig.4. Manufacturing plan for the test part

7.   DISCUSSION AND CONCLUSIONS

In this work we followed the traditional CAPP developments that took the hierarchical approach to planning and optimisation. Such planners first attempted to form optimal setups, and then to sequence the operations optimally relative to this so-called setup plan (e.g., [1,3,14,22]). In contrary to earlier results, we expressed *all* available domain knowledge in the declarative knowledge representation scheme of a general-purpose planner.

The idea of a planning engine that could synthesize the activities and results of several domain-specific reasoning modules appeared in [6]. This process planner worked also with a kind of hierarchical task network. However, neither the HTN engine nor any of the attached reasoning modules concerned with the optimisation criteria. A similar modular architecture appeared in [17] while [21] suggested a blackboard-based platform for integrating the advice of various modules and the decisions of the human planner.

In our previous works we had a two-phased approach to CAPP: In the first phase the search space of solutions were set up by knowledge-based reasoning, and then close-to-optimal solutions were extracted either by genetic algorithms [5] or by case-based reasoning [18]. Recently, we proposed a constraint-based planner engine that could interleave steps of inference, conflict resolution and search [7].

Reasoning and search performed at several levels of abstraction is what makes our hierarchical planner an efficient CAPP system. The first experiments with a limited set of prismatic and rotational parts are encouraging and show the following merits of the proposed approach:

- HTN planning provides a declarative representation scheme with well-defined semantics for capturing all principal kinds of the necessary domain knowledge, including planning strategies, as well.
- The knowledge representation and the plan generation (i.e. search) are separated. Plans can be generated by a sound and complete planning engine. Plan generation can be interleaved with domain-specific (e.g. geometric) reasoning and user interaction which are essential in supporting engineering problem solving.

ACKNOWLEDGEMENT

We are grateful to the creators of the SHOP system for making it freely available.

REFERENCES

[1]   Britanik, J., Marefat, M.: Hierarchical Plan Merging with Application to Process Planning. In: *Proc. of the IJCAI'95*, Montreal, 1677—11683, (1995)

[2]   Cameron, S.: Efficient Intersection Tests for Objects Defined Constructively. *International Journal of Robotics Research*, **8**(1) 3—25, (1989)

[3]   Gupta, S.K., Nau, D.S., Regli, W.C.: IMACS: A Case Study in Real-World Planning. *IEEE Intelligent Systems*, **13**(3), 49—60, (1998)

[4]   Halevi, G., Weill, R.D.: *Principles of Process Planning*. Chapman& Hall, 1995.

[5]   Horváth, M., Márkus, A., Váncza, J.: Process Planning with Genetic Algorithms on Results of Knowledge-Based Reasoning. *Int. J. of Computer Integrated Manufacturing*, **9**, 145—166, (1996)

[6]   Kambhampati, S., Cutkosky, M.R., Tenenbaum, J.M., Lee, S.H.: Integrating General Purpose Planners and Specialized Reasoners: Case Study of a Hybrid Planning Architecture. *IEEE Trans. on Systems, Man, and Cybernetics*, **23**(6), 1503—1518, (1993)

[7]   Márkus A., Váncza J.: Process Planning with Conditional and Conflicting Advice. *Annals of the CIRP*, **50**(1), (2001), in print

[8]   Marri, H.B., Gunasekaran, A., Grieve, R.J.: Computer-Aided Process Planning: A State of the Art. *Int. J. of Advanced Manufacturing Technology*, **14**, 261—268, (1998)

[9]   Nau, D. S., Gupta, S. K., Regli, W. C.: AI Planning Versus Manufacturing-Operation Planning: A Case Study. In *Proc. of the IJCAI-95*, Montreal, 1670—1676, (1995)

[10] Nau, D. S., Smith, S. J. J.,Erol, K.: Control Strategies in HTN Planning: Theory versus Practice. In *Proc. of AAAI-98/IAAI-98*, Madison, WI, AAAI Press, 1127—1133, 1998

[11] Nau, D. S.: Documentation for SHOP 1.6.1 and M-SHOP 1.1.1. http://www.cs.umd.edu/projects/shop/, 2000

[12] Nau, D.S, Y. Cao, A. Lotem, and H. Muñoz-Avila. "SHOP: Simple Hierarchical Ordered Planner." In IJCAI-99, pp. 968-973, 1999.

[13] Russel, S., P. Norvig: *Artificial Intelligence. The Modern Approach*, Prentice Hall, 1995

[14] Sarma, S.E., Wright, P.K.: Algorithms for the Minimization of Setups and Tool Changes in "Simply Fixturable" Components in Milling. *Journal of Manufacturing Systems*, **15**(2), 95—112, (1996)

[15] Schreiber, G, et al.: *Knowledge Engineering and Management. The Common KADS Methodology*, The MIT Press, 1999

[16] Shah, J.J., Mantyla, M.: *Parametric and Feature-Based CAD/CAM*. Wiley, 1995.

[17] Teramoto, K., Onosato, M., Iwata, K.: Coordinative Generation of Machining and Fixturing Plans by a Modularized Problem Solver. *Annals of the CIRP*, **47**(1), 437—440, (1998)

[18] Váncza, J., Horváth, M., Stankóczi, Z.: Robotic Inspection Plan Optimization by Case-Based Reasoning. *Journal of Intelligent Manufacturing*, **9**(2), 181-188, (1998)

[19] Weld, D.S: Recent Advances in AI Planning. *AI Magazine*, **20**(2), 93—123, (1999).

[20] Wilkins, D.E., desJardins, M.: A Call for Knowledge-Based Planning. *AI Magazine*, **22**(1), 99—115, (2001).

[21] Zeir, G. van, Kruth, J.-P., Detand, J.: A Conceptual Framework for Interactive and Blackboard-Based CAPP. *Int. J. of Production Research*, **36**(6), 1453—1473, (1998)

[22] Zhang, H.-C., Lin, E.: A Hybrid-Graph Approach for Automated Setup Planning in CAPP. *Robotics and Computer-Integrated Manufacturing*, **15**, 89—100, (1999)

Supply Chain Management in Agile Manufacturing Environment

Toshiya Kaihara
Dept. of Computer and Systems Engineering, Faculty of Engineering, Kobe University
E-mail: kaihara@ms.cs.kobe-u.ac.jp

Abstract: Supply chain management (SCM) is now recognised as one of the best means by which enterprises can make instant improvements to their business strategies and operations. Computational virtual market based supply chain operation solves the product allocation problem by distributing the scheduled resources based on the agent interactions in the market. To explore the use of market mechanism for the coordination of distributed planning module in supply chain management, we have developed a prototype supply chain model in dynamic environment for specifying and simulating computational markets with circulative structure. It has been clarified that the proposed methodology facilitates sophisticated SCM under dynamic conditions.

Key words: supply chain management, market-oriented programming, dynamic system, multi-agent

1. INTRODUCTION

Many manufacturing operations are designed to maximize throughput and lower costs with little consideration for the impact on inventory levels and distribution capabilities. Purchasing contracts are often negotiated with very little information beyond historical buying patterns. The result of these factors is that there is not a single, integrated plan for the organization - there were as many plans as businesses. Clearly, there is a need for a mechanism through which these different functions can be integrated together. Supply chain management (SCM) is a strategy through which such an integration can be achieved (Goldratt, 1992) (Fisher, 1994).

Companies, both manufacturing and service, are creators of value, not simply makers of products. SCM focuses on globalisation and information management tools which integrate procurement, operations, and logistics from raw materials to customer satisfaction. Future managers are prepared to add product value, increase quality, reduce costs, and increase profits by addressing the needs and performance of supplier relations, supplier selection, purchasing negotiations, and so on.

Solving product distribution problem in SCM presents particular challenges attributable to the distributed nature of the computation. Each business unit in SCM represents independent entities with conflicting and competing product requirements and may possess localised information relevant to their interests. To recognise this independence, we treat the business units as agents, ascribing each of them to autonomy to decide how to deploy resources under their control in service of their interests.

In this paper, a distributed product distribution method can be analysed according to how well it exhibits the following properties:
− Self-interested agents can make effective decisions with local information without knowing the private information and strategies of other agents.
− The method requires minimal communication overhead.
− Solutions don't waste resources. If there is some way to make some agents better off without harming others, it should be done. A solution that cannot be improved in this way is called Pareto optimal.
− The environment surrounding agents has dynamic nature.
Conventional straightforward distributed policies do not possess these properties. Assuming that a product distribution problem in SCM must be decentralised, market concept can provide several advantages as follows:
− Markets are naturally distributed and agents make their own decisions about how to bid based on the prices and their own utilities of the goods.
− Communication is limited to the exchange of bids and process between agents and the market mechanism.

In this paper we formulate SCM as a discrete resource allocation operation under dynamic environment, and demonstrate the applicability of economic approach to this framework by simulation experiments.

2. BASIC CONCEPT

2.1 Product distribution problem

Product distribution in SCM is generally proceeded by distributed autonomous dealings amongst business units. There generally exist several criteria in the business dealings, and we focus on principle elements, price and quantity, as a basic study in this paper.

Product distribution belongs to multi-objective optimisation problems, and severe conflicts between suppliers and demanders would occur due to the tradeoffs of their utilities. Market mechanism is expected to solve the problems by presenting a Pareto optimal solution for all of the business units in the market.

2.2   Market metaphor and SCM

Virtual market is a kind of computational market which consists of two types of heterogeneous agents, supplier and demander. Agent activities in terms of products required and supplied are defined so as to reduce an agent's decision problem to evaluate the tradeoffs of acquiring different products in market-oriented programming (Wellman, 1993). These tradeoffs are represented in terms of market prices, which define common scale of value across the various products. The problem for designers of computational markets is to specify the mechanism by which agent interactions determine prices.

In this paper the framework of general equilibrium theory (Okuno, 1985), which is proposed in microeconomics research field, has been adopted. In economics, the concept of a set of interrelated goods in balance is called general equilibrium. The general equilibrium theory guarantees a Pareto optimal solution in perfect competitive market. The connection between computation and general equilibrium is not all foreign to economists, who often appeal to the metaphor of market systems computing the activities of the agents involved. Some apply the concept more directly, employing computable general-equilibrium models to analyse the effects of policy options on a given economic system (Shoven, 1992). Obviously SCM model is well-structured for market-oriented programming, and that means the proposed concept takes advantage of the theory, and a Pareto optimal solution, which is conducted by microeconomics, is attainable in product distribution problem in SCM (Kaihara, 1999a).

2.3   Market-Oriented programming

Market-oriented programming is the general approach of deriving solutions to distributed resource allocation problems by computing the competitive equilibrium of an artificial economy (Kaihara, 1999b). It involves an iterative adjustment of prices based on the reactions of the agent in the market. Negotiation mechanism in market-oriented programming is shown in figure 1.

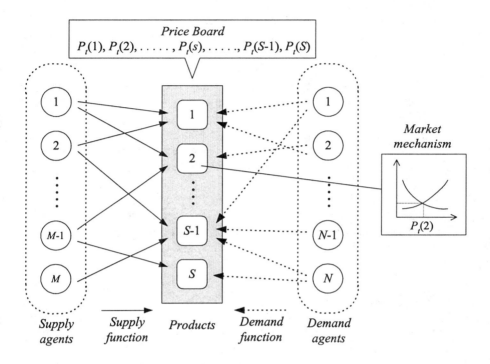

Figure 1. Virtual market structure

Let $P_t(s)$ be the price of resource s at time t. Market mechanism computes an equilibrium price in each separate market. It involves an iterative adjustment of prices based on reactions of agents in the market. Product s receives supply and demand functions and the market mechanism involved in the product adjusts individual prices to clear, rather than adjusting the entire price vector by some increment. The mechanism associates an auction with each distinct resource. Agents act in the market by submitting bids to auctions. In this paper bids specify a correspondence between prices and quantities of the resource that the agent offers to demand or supply as a basic study. Given bids from all interested agents, the auction derives a market-clearing price.

The algorithm of the proposed market-oriented programming in SCM is shown as follows:

Step 1: A supply agent m sends bids to the market to indicate its willingness to sell the product s according to its current price $P_t(s)$

in time t. The supply agent willingness is defined as a supply function in the bid message. The agent can send bids to the market within the limits of its current inventory level.

Step 2: A demand agent n sends bids to a market to indicate its willingness to buy the product s according to its current price $P_t(s)$ in time t. The demand agent willingness is defined as a demand function in the bid message. The agent can send bids to the market within the limits of its domestic budget. Each product has its own market, and they construct a competitive market mechanism as a whole.

Step 3: The market in product s sums up supply functions and demand functions, then revises balanced price $P_t'(s)$ of product s in time t. All the market must revise their balanced price via the same process.

Step 4: Check the balanced prices of all the products and if all the prices are fully converged, the acquired set of the prices is regarded as equilibrium price, then go to *Step5*. If not, go to *Step 1*.

Step 5: If dealing time is up, then stop. And if not, $t = t+1$ and go to *Step 1*.

Each agent maintains an agenda of bid tasks, specifying in which it must update its bid or compute a new one. The bidding process is highly distributed, in that each agent needs to communicate directly only with the auctions for the resources of interest. Each of these interaction concerns only a single resource; the auctions never coordinate with each other. Agents need not to negotiate directly with other agents, nor even know of each other's existence. As new bids are received at the auctions, the previously computed clearing price becomes obsolete. Periodically, each auction computes a new clearing price if any new or updated bids have been received, and posts it on the tote board. When a price is updated, this may invalidate some of an agent's outstanding bids, since these were computed under the assumption that price for remaining resources were fixed at previous value. On finding out about a price change, an agent argues its task agenda to include the potentially affected bids. At all times, the market-oriented mechanism maintains a vector of going prices and quantities that would be exchanged at those prices. While the agents have nonempty bid agendas or the auctions new bids, some or all resources may be in disequilibrium. When all auctions clear and all agendas are exhausted, however, the economy is in competitive equilibrium.

In the following chapter we clarify the heterogeneous agents definitions in dynamic environment, where trading conditions change dynamically during the task processing.

3. ECONOMIC AGENT

3.1 Production function

Suppose supply agent k has a production function f_k. In this paper we adopt a simple linear function as the production function described in equation (1).

$$y_k = f_k(x_k) = a_k x_k \qquad (1)$$

where x_k and y_k are the amount of input / output resources of agent k, respectively, and a_k is a constant value that represents productivity.

The relation between the amount of input materials i (x_{ki}) and the cost (c_k) is the other important factor in SCM, and we define the following function based on logistic curve. The reason of applying the logistic curve is that we have a limitation of production due to resource constraints, in other words, diminishing returns.

$$\frac{dx_{ki}}{dt} = h_k(1 - \frac{c_k}{r_k}) \times c_k \qquad (2)$$

Then we have

$$x_{kit} = \int^{t+1} dx_{ki} = h_k(1 - \frac{c_{kt}}{r_k}) \times c_{kt} \qquad (3)$$

where x_{kit} and c_{kt} are the amount of input material i and its cost of agent k at time t, respectively.

3.2 Cost function

The turnover of agent k (w_k) by selling product j is shown in equation (4).

$$w_k = q_j s_{kj} \qquad (4)$$

where q_j and s_{kj} are the selling price and the total amount of supplying product j in agent k, respectively.

The total cost of agent k (v_k) is

$$v_k = p_i d_{ki} + c_k \qquad (5)$$

where p_i and d_{ki} are the buying price and the total amount of demand product i in agent k, respectively. Then finally cost function is defined as follows:

$$w_k = q_j s_{kj} - p_i d_{ki} - c_k \qquad (6)$$

3.3   Commerce in virtual market

All the agents in the virtual market try to get the maximised profit in their trading. The basic principle of the agent is that it requires just the amount of materials for its production at this moment.

It requires processing time for manufacturing, and the trading conditions should be changed during the manufacturing process in dynamic environment. The objective function (u_{kdt}) for demand of agent k at time t is

$$u_{kdt} = (a_k q_{jt} - p_{it})h_k(1 - \frac{c_{kt}}{r_k})c_{kt} - c_{kt} \tag{7}$$

where p_{it} and q_{jt} are the demand price and supply price at time t, respectively. Then the demand (d_{kit}) of agent k at time t is attainable as the maximum value of u_{kdt}. Since the logistic curve is a concave down function, u_{kdt} is maximised to substitute the following c_{kt} for equation (3):

$$\begin{cases} (a_k q_{jt} - p_{it})h_k(1 - \dfrac{2c_{kt}}{r_k}) - 1 = 0 & (\text{ for } (a_k q_{jt} - p_{it}) > 0) \\ c_{kt} = 0 & (\text{ for } (a_k q_{jt} - p_{it}) \le 0) \end{cases} \tag{8}$$

The objective function (u_{kst}) for supply of agent k at time t is

$$u_{kst} = q_{jt}s_{kjt} - p_{i(t-1)}\frac{1}{a_k}s_{kjt} - c_{kt} \tag{9}$$

where $$s_{kjt} = a_k h_k(1 - \frac{c_{kt}}{r_k})c_{kt} \tag{10}$$

and the amount of supply (s_{kit}) of agent k at time t is attainable as the maximum value of u_{kst}. Then u_{kst} is maximised to substitute the following c_{kt} for equation (10):

$$\begin{cases} (a_k q_{jt} - p_{i(t-1)})h_k(1 - \dfrac{2c_{kt}}{r_k}) - 1 = 0 & (\text{ for } (a_k q_{jt} - p_{i(t-1)}) > 0) \\ c_{kt} = 0 & (\text{ for } (a_k q_{jt} - p_{i(t-1)}) \le 0) \end{cases} \tag{11}$$

3.4   Agent optimisation strategy under budget constraint

All the agents have budget constraint so as to realise our market model in SCM. They naturally try to maximise their profit in their economic behaviours under the constraint. Suppose the total expense for resource set l to produce j from i in agent k at time t is $c_{kl,t}$, and the maximum budget is e_k, then we have

$$\sum_l c_{kl,t} \le e_k \tag{12}$$

Agents naturally try to maximise their individual utility and we can adopt Profit Maximise Theorem (kaihara, 1999c) as the agents' strategy. The Profit Maximise Theorem requires two types of preconditions, such as:

i) f_k (production function) is differentiable

ii) $\forall l \ F\dfrac{\partial f_{kl}}{\partial x_{kl}}\bigg|_{x_{kl}=x} > \dfrac{\partial f_{kl}}{\partial x_{kl}}\bigg|_{x_{kl}=x+\Delta}$

and these conditions are obviously satisfied by equation (1),(3) in this paper. The theorem endorses that agent k maximises its profit to calculate the minimised m_k, which satisfies the following equation (13):

$$\forall l \ F\frac{\partial u_{kd}}{\partial v_{kl}} = m_k \left(m_k \ge 0\right) \ \cap \ v_k \le e_k \tag{13}$$

Finally we can get x_{klt}, which maximises the function u_{kdl}, by substituting c_{kl} acquired from equation (14) for equation (3).

$$\frac{\partial u_{kdl}}{\partial c_{kl}} = (a_{kl}q_i - p_i)h_{kl}(1 - \frac{2c_{kl}}{r_{kl}}) - 1 = m_k \tag{14}$$

4. COMPUTER SIMULATION

4.1 Experimental circulative SCM model

A primitive product life cycle SCM model, shown in figure 2, was developed to investigate the validity of the proposed approach. As a basic research, the experimental model is quite simple enough to evaluate the basic dynamism of the agent interactions, defined in the previous chapter.

There exist two agents, which construct a circulatory resource flow via two markets, and this model represents the simplest structure of real supply flows. The agents would affect themselves with some time delay in the model. Each market has outside supply / demand, which are defined as primitive Cobb-Douglas function (Okuno, 1985) shown in equation (15).

$$y_i = a_j p^{b_j} \tag{15}$$

The experimental parameter settings are as follows:

Agent: a_k=1.2, h_k=5.0, r_k=100, $e_{k,0}$=10, $z_{kj,0}$=10
Outside supply function: a_j=0.1, b_j=1.2
Outside demand function: aj=10.0, b_j=-1.0
Market: $p_{j,0}$=1.0, $q_{j,0}$=1.0, k=1,2

where $*_{k,0}$ and $*_{j,0}$ mean initial value of agent k and resource j, respectively.

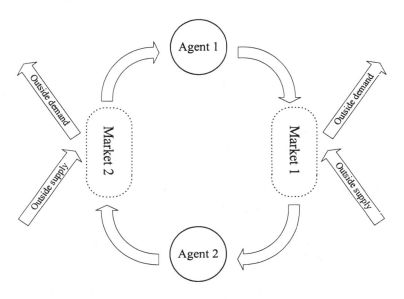

Figure 2. Primitive product life cycle SCM model

4.2   Basic dynamism

Basic dynamism of resource allocations in terms of financial items (budget and price) and stock (product) is shown in figure 3.

Initially (t<4) agents try to hold down their production and sell their stocks due to cheap selling prices. Then the over demand situation appears, and that causes the increase of the selling prices (t<6). Finally the markets are gradually converged into the equilibrium prices endorsed by microeconomics.

Since the model has circulatory structure, initial resource prices should affect the system dynamism obviously. Figure 4 illustrates the simulation results in ($p_{j,0}$=1.0, $q_{j,0}$=2.5). The result with the vibrated curves is completely different from figure 3. It is quite interesting to confirm that the small initial price difference causes strong influence to the markets due to a dynamic interaction and two-way enrichment process between micro and macro processes of change.

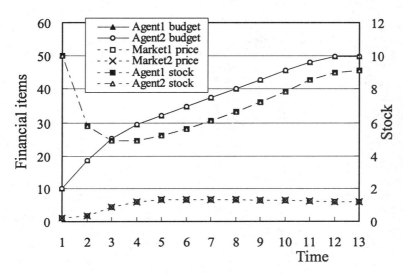

Figure 3. Basic dynamism ($p_{j,0}$=1.0, $q_{j,0}$=1.0)

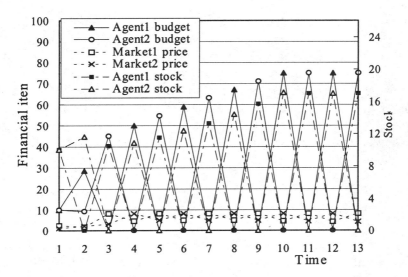

Figure 4. Basic dynamism ($p_{j,0}$=1.0, $q_{j,0}$=2.5)

4.3   Productivity effects

Constant parameter a_k in equation (1) is regarded as productivity in the agent definition. The experimental results on the productivity are shown in figure 5. The agents productivity (a_k) was set to 2.0 in this figure.

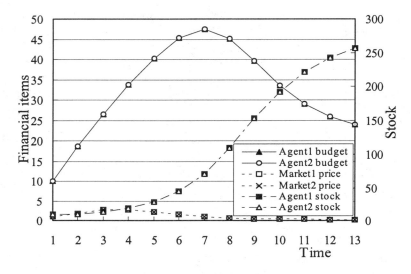

Figure 5. Productivity effects

Generally it is well known that the rapid increase of total productivity leads to excessive competitions in the market. Product prices decrease there, and that causes financial difficulties in all the business units. We can often observe it in semiconductor production industry as an example.

It is quite interesting to have confirmed that the experimental results in figure 5 have some analogy to the general market phenomenon. It is obvious that the rapid productivity increase is short-sighted and makeshift in the resource-oriented ceiling market. Formulating the agent strategy with dynamic utilities enables us to handle the transitional trends with reality.

5.   CONCLUSIONS

In this paper SCM was formulated as a discrete resource allocation problem with dynamic environment, and the several experimental results demonstrated the applicability of economic analogy to this framework. It has been confirmed that careful constructions of the decision process according to economic principles can lead to efficient distributed resource allocation in

SCM, and the dynamic behaviour of the system can be analysed in economic terms.

Acknowledgements
This research has been supported by IMS international research program in Japan, under contract No.0019 (HUTOP project).

REFERENCES

1. Fisher ML, Hammond JH, Making supply meet demand in uncertain world, Harvard Business Review, May/Jun, 1994.
2. Goldratt EM, The GOAL, North River Press, 1992.
3. Kaihara T, Supply Chain Management with Multi-agent Paradigm, Proceedings of the 8th IEEE International Workshop on Robot and Human Interaction, pp394-399, 1999a.
4. Kaihara T., Supply Chain Management based on Market Mechanism in Virtual Enterprise, Infrastructures for Virtual Enterprises, Kluwer Academic Publishers, Boston, pp.399-408, 1999b.
5. Kaihara T., Supply Chain Management with Market Economics, Manufacturing for a global market, M. T. Hillery & H. J. Lewis Eds., Vol. 1, pp.659-662, 1999c.
6. Okuno H, Suzumura K, Micro economics I, Iwanami, 1985.
7. Shoven JB, Whalley J, Applying General Equilibrium, Cambridge University Press, 1992.
8. Wellman MP, A Market-Oriented Programming Environment and its Application to Distributed Multi- commodity Flow Problems, ICMAS-96, 385-392, 1996.

BIOGRAPHY

Toshiya Kaihara received the B.E. and M.E. degrees in precision engineering from Kyoto University, Kyoto, Japan, in 1983 and 1985, respectively. He received the Ph.D. degree in mechanical engineering from Imperial College, University of London, London, UK, in 1994. He is currently an associate professor of Computer and Systems Engineering at Kobe University. He is a member of Japan Society for Precision Engineering, IEEE, IFIP and many others.

Communicating to a Manufacturing Device using MMS/CORBA

T.Ariza, F.J. Fernández and F.R.Rubio

matere@trajano.us.es fjfj@trajano.us.es rubio@cartuja.us.es
Dept. Ingeniería de Sistemas y Automática.
Escuela Superior de Ingenieros. Univ. de Sevilla. Spain

Abstract: The advantages of distributed systems are also applied to manufacturing systems due to their inherent distribution. However, the heterogeneity found in the current hardware and software and the underlying communication between the different components of the system make the development of these systems a difficult task. MMS allows a uniform communication with different hardware systems. On the other hand, CORBA makes the communication between several objects easier, reducing the implementation cost to a minimum. By joining both technologies a communication method between devices can be achieved. This method is location transparent and specific physical features independent. The MMS implementation over CORBA allows new devices to be added in a natural way, making use of the inheritance available in object oriented programming and the facility supplied by CORBA in the communication. In this work, this implementation has been used in order to communicate to a real device.

Keywords: Distributed Manufacturing System, Object Oriented Programming, CORBA, MMS, JAVA.

1.   INTRODUCTION

Distributed systems have been widely introduced in manufacturing due to their flexibility, reliability, incremental growth and better price/performance rate. But the main problem that arises is the communication between several interconnected elements, which are very different.

MMS is an application layer protocol that homogenises the use of the devices that compound the system. MMS makes use of this approach to specify the services that a user can invoke to communicate with these devices. On the other hand, distributed object methodologies, where CORBA is framed, provide all the object oriented methodology advantages in distributed systems. CORBA can make the MMS service user application to request the service easier, as it allows to structure the system in objects and at the same time to have these objects distributed along the several components of the system, making the distribution transparent to the user. JavaIDL has been chosen to implement CORBA objects. It automatically creates the skeleton and the stub starting from the IDL specification.

This work intends to show how using the MMS services and CORBA for the communication between devices makes the construction of distributed manufacturing systems easier. The creation of new VMD objects for new devices can be based on the existent objects using inheritance. An application that uses these objects does not have to worry about the communication, nor the location of the rest of elements. This has been used in order to build the VMD object for a specific manufacturing device, the RX-90 Robot by Stäubli Unimation.

2.   BACKGROUNDS

In this section a brief revision of the MMS protocol and the CORBA architecture is introduced. They are the bases of the work that is mentioned in this paper.

2.1   MMS

In distributed heterogeneous systems, where each device has different features, carries out different tasks and is probably owned by a different manufacturer, the need to interconnect all devices that compound the system rises in order to achieve the integration of each one in the whole system.

Manufacturing Message Specification (MMS) is a communication language to aid the interconnection of devices in a heterogeneous environment. It is a protocol that falls in the application level standardised by ISO (International Standard Organisation).

Although MMS is not a complete Object-Oriented language, as it does not support all the features of the Object-Oriented programming, it splits the system into objects and presents a well defined interface for each one. The most important object in the system is the Virtual Manufacturing Device (VMD). The VMD hides the real manufacturing device to the programmer. The system is structured in a set of VMDs.

MMS is also based on the client/server model. The VMD acts as a server, and so carries out a set of services that are described in the MMS protocol. The clients are the requesters of the services provided by the VMD.

A particular implementation of the MMS server must provide the mapping between the VMD model, which is an abstraction, and the functionality of the real manufacturing device.

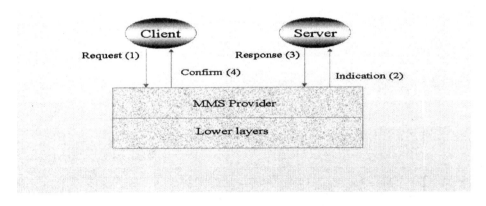

Figure 1. System based on MMS

The scheme of a system based on the communication protocol MMS is showed in figure 1.

Each MMS server manages a set of objects associated to the VMD. Some of the most important are the following:

– **The Domain Object:** It represents a subset of the capabilities of the VMD which is used for a specific purpose.
– **The Program Invocation Object:** It is a dynamic element which most closely corresponds to an execution thread in a multi-tasking environment.
– **The variable Object:** It is used to model real variables of the VMD.

Other important objects that are managed by the VMD are events. The event management services provide facilities that allow a client MMS-user to define and manage event objects at a VMD and to obtain notifications of event occurrences.

A revision of MMS can be found in [5,13] and a complete description of all these services and the protocol specification in [6,7].

2.2   CORBA

The Common Object Request Broker Architecture (CORBA) is a distributed object architecture that allows software objects to interact across networks. CORBA was first introduced in 1991 by the Object Management Group (OMG). It is an international consortium of over 800 software vendors, developers and end users.

The aim of this group is to specify an open software bus on which object components written by different vendors can interoperate regardless of the implementation language, location or host platform.

The object bus provides an Object Request Broker (ORB) that lets clients invoke methods on remote objects either statically or dynamically.

As a result of their work, OMG approved a set of specifications -called CORBA 2.0- in late 1994.

The Object Management Architecture (OMA) [10] specified by OMG, is shown in figure 2 and consists of the following elements:

–   Object Request Broker (ORB): It provides the mechanisms by which objects transparently make and receive requests and responses. In doing this, the ORB provides interoperability between applications on different machines in heterogeneous distributed environments and seamlessly connects multiple object systems.
–   Object Services: It provides basic operations for the logical modelling and physical storage of objects. The operations provided by Object Services are made available through the ORB.
–   Domain Interfaces: They represent vertical areas that provide functionality of direct interest to end-users in particular application domains.

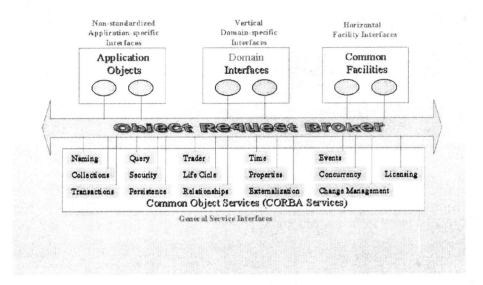

Figure 2. The Object Management Architecture (OMA) specified by OMG.

– Common Facilities: Services of direct use to application objects. It provides a set of generic application functions that can be configured to the specific requirements of a particular configuration.
– Application Objects (AO): They are specific components to end-user applications. It corresponds to the traditional notion of an application. AOs represent individual related sets of functionality.

The object specification is carried out using the Interface Definition Language (IDL). It is a descriptive language used to define the interface through which a client may access a server. IDL provides operating system and programming language independent interfaces to all the services and components that reside on a CORBA bus.

The method invocation used by the CORBA's implementation is shown in figure 3. The following components are necessary to carry out the invocation:

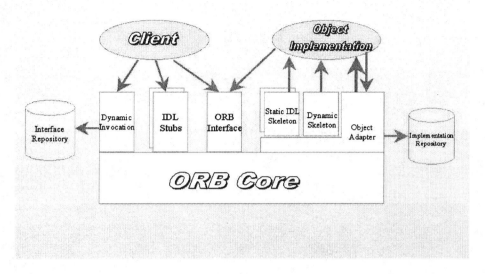

Figure 3. Object's method's invocation.

– Client IDL Stub: Client code used by an object to encode invocations in a form which can be handled by the ORB, and to decode replies received via the ORB.
– Skeleton: Server code up-called by the ORB, capable of decoding requests transmitted by the ORB, converting it into an invocation of the implementation object, and encoding the results to be sent back to the client via the ORB.
– Object Adapter: It defines how an object is activated. It can do this by creating a new process, creating a new thread within an existing process, or by reusing an existing thread or process.

More information regarding CORBA can be found in [10,11,12] and about CORBA applied to manufacturing systems in [4,9].

3. AIM OF THE WORK

The objective of this work is to implement the VMD object for the RX-90 robot. However, the development has consisted in two phases. In the first one, a generic VMD has been carried out conforming to ISO/IEC 9506 part 1 [6] and part 2 [7]. This generic VMD (presented in [2,3]) can be inherited by other classes in a natural way. Doing this, the construction of a specific

VMD is more straightforward and it is suitable not only to build the VMD for the RX-90 but for whatever manufacturing device that it is necessary.

Figure 4. RX-90 Robot

The specific VMD for the RX-90 is built starting with the generic VMD, conforming to ISO/IEC 9506 part 3 standard [8], taking into account that this implementation must be used as the base to build other specific VMD's for robots making the least number of changes possible.

The prototype includes the parts concerning the VMD, Domains, Program Invocations, variables and events (this last one, only in the generic VMD). It does not include Semaphore, Operator Station, or Journal objects, but the system could be extended to them.

The main objective is to build classes that can be used as the base for the implementation of VMD objects that represents other different devices. In addition to this, a MMS client has been developed, which can be used in the future in order to build control distributed systems.

4.   SYSTEM DESCRIPTION

The scheme of the system architecture is shown in figure 5. As the standard describes, the MMS client also carries out MMS services, its behaviour is as a server from this point of view.

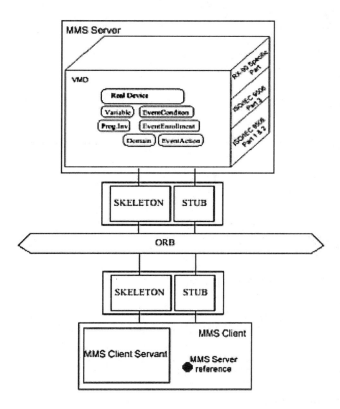

Figure 5. System Architecture

On the server side the components are the following:

— MMS Server: CORBA object that can be accessed through the object bus. The services provided by this object are the services specified in the MMS standard for the server. The description of this object is specified using the Interface Definition Language (IDL).
— VMD: the MMS Server delegates in the VMD to resolve the MMS services.
— Skeleton: It allows the clients to call the MMS Server methods using the ORB.
— Stub: It is used in order to carry out the remote invocation to the methods of the MMS Client Servant object so that the server can notify the events to the client when they happen and carry out requests of MMS services that the client can accomplish.
— Real Device: It is the physical device that is hidden by the MMS Server object. In this work the RX-90 [1] device is used.

On the client side, the following components can be identified:

- MMS Client: It is responsible for calling the services using CORBA. Its task is to get the server references and to register the client servant in the ORB.
- MMS Client Servant: It carries out the client services and receives requests from the MMS servers.
- Skeleton: It allows the MMS Client Servant object methods to be called through the ORB.
- Stub: It undertakes the responsibility for calling the remote methods of the MMS Server object.

The VMD is divided in three different layers, which are the following:

- Generic VMD: It corresponds to ISO/IEC 9506 parts 1 & 2 that include the general services of the VMD.
- Robot VMD: It corresponds to ISO/IEC 9506 part 3 that include the companion standard for robots.
- RX-90 specific part: This is the layer that depends on the specific device.

5.   IMPLEMENTATION

A prototype for this work has been built, where the programming language JAVA has been used. JAVA language has the following features:

- Object Oriented programming language: It has the benefits of this methodology (reusability, facility in the integration, debugging and maintenance). Throughout, e.g. no coding outside of class definitions, including main(). An extensive built-in class library.
- Familiarity: It is similar enough to C and C++ that experienced programmers can get going quickly.
- Simpler than C & C++: Because it has no pointers, no preprocessor and automatic garbage collection
- Portability: Code is compiled to bytecodes, which are interpreted by the JAVA virtual machine.
- Robustness: Exception handling built-in, strong type checking (i.e. all variables must be given explicit type), local variables must be initialised.
- Threading: Lightweight processes, called threads, can easily be spun off to perform multi-processing.
- Dynamic Binding: Even if libraries are recompiled, there is no need to recompile the code that calls classes in those libraries since binding, i.e.

the linking of variables and methods to where they are located, is done at runtime.

- Platform Independence: The JAVA Virtual Machine is available in many types of computers and OS's. The Code that can be exchanged without requiring rewrites and recompilation would save time and effort.
- Security: No memory pointers exist. Programs run inside the virtual machine sandbox. The code is checked for pathologies by the bytecode verifier, the class loader and the security manager.

In order to communicate with the real device the javax.comm library has been used. The prototype can be run in any computer and operating system where the Java interpreter can be run. The VMD used is a CORBA object developed in [2] that attends to MMS requirements with extensions for the robot.

CORBA has been used as the communication architecture and the ORB chosen is JAVAIDL because previous work had been done with it.

A client with the basic functionality has also been implemented. It allows the operator to load and run programs over the robot, show variable values, etc.

6. CONCLUSION

In this work, the VMD for the RX-90 robot over CORBA and a basic client that allows access to the MMS services have been implemented. With these components, a set of functions can be carried out over the robot, as load and run programs in a remote way, create, read and write variables, and so on.

Reusability and modularity are features achieved in this development. It allows having a base to implement new clients using inheritance. On the other hand, using CORBA, as the communication platform to accomplish the interaction between clients and servers, facilitates the creation of control distributed systems.

7. ACKNOWLEDGEMENT

This work is supported in part buy the CICYT under grant num. TAP-98-0541

8.   REFERENCES

[1] Adept Technology, Inc, "V+ Language Reference Guide", Part Number 00961-00100,Rev. B. Version 11.3T. July 1996.

[2] T. Ariza, F.R. Rubio, "Communicating MMS Events in a Distributed Manufacturing System using CORBA", Preprints DCCS'98,1998.

[3] T. Ariza, F.R. Rubio, "MMS-Manager: Device Management in Heterogeneous Environment Based on CORBA", Preprints Controlo'98,1998.

[4] Carvalho, A.S. and M.J. De Sousa. "Development of an ORB for Distributed Manufacturing Applications", WFCS'97 Workshop (1997).

[5] CCE-CNMA, ESPRIT Consortium. "MMS: A Communication Language for Manufacturing", Springer, 1995.

[6] ISO/IEC 9506-1. "Industrial Automation Systems- Manufacturing Message Specification", Part 1: Sevice Definition. International Standars Organization, 1990.

[7] ISO/IEC 9506-2. "Industrial Automation Systems- Manufacturing Message Specification", Part 2: Protocol Specification. International Standards Organization, 1993.

[8] ISO/IEC 9506-3. "Industrial Automation Systems- Manufacturing Message Specification", Part 3: Companion Standard for Robotics. International Standards Organization, 1993.

[9] P. Newmann, F. Iwanitz. "Integration of Fieldbus Systems into Distributed Object-Oriented Systems", WFCS'97 Workshop (1997).

[10] Object Management Group. "CORBA: Common Object Request Broker Architecture and Specification", Published by the Object Management Group (OMG), Framingham, MA. 1995.

[11] Object Management Group. "CORBA services: Common Object Services Specification", Published by the Object Management Group (OMG), Framingham, MA. 1995.

[12] R. Orfali, D. Harkey, J. Edwards. "The Essential Distributed Objects. Survival Guide", Wiley, 1996.

[13] Pimentel, Juan R. "Communication Networks for Manufacturing". Prentice Hall, 1990.

ERP Interfaces for Enterprise Networks
An XML approach

Szilveszter Drozdik
drozdik@sztaki.hu

Abstract: Although Enterprise Networks may be formed for different purposes, the need of integration always pops up. Integration requires interfaces, clear and consistent constructions of the communication. *FLUENT (Flow Oriented Logistics Upgrade for Enterprise Networks)* is a joint European project (Esprit IV) in which – among others - *ERP (Enterprise Resource Planning)* interfaces are defined and implemented for systems, like SAP and Baan. ERP systems provide proprietary interfaces, so FLUENT founded a common core interface specification easy to fit for any ERP interface. FLUENT ERP interface architecture defines XML schemas and transformations to convert a proprietary ERP data set to the format of another one and vice versa. This architecture is based on standardised XML solutions so exploits the advantages of the emerging XML trend

Key words: Enterprise Network Integration, ERP, XML, EDI, FLUENT

1. INTRODUCTION

FLUENT [FLU] is an international, multi-sectorial attempt to develop methods and tools for managing complex logistic flows, occurring in a distributed manufacturing network with multiple plants and co-operating firms. This kind of organisation has been attracting great interest from the industry community world-wide, under the impulse of:

- *Emerging trends in logistics management*: Virtual/extended enterprise paradigms and Integrated Supply-Chain Models
- *Evolving market conditions*: Decentralisation of facilities, Strategic alliances with suppliers and Complex distribution networks

– *Enabling technologies available on the IT market*:Networking and
   Workflow solutions, Standards for ERP integration and Supply Chain
   Planning solutions.

Appealed by this scenario, companies of every size strive to integrate
their logistics processes, but a coordinated IT solution is not yet available.
So, traditional logistics functions like sales and purchase are left alone to
face problems far beyond their intended role.

2.   ERP INTEGRATION

The main purpose of the FLUENT ERP interface software is to achieve a
seamless integration between the supply logistic chain that will be managed
by FLUENT and the ERP product.

Supply chain management does not mean exposing ERP transactions via
an Internet interface. FLUENT aim is to implement new functions
supporting network-level processes, complementary to the standard
functionality of ERP at the local level.

Theoretically, FLUENT allows integration with any ERP system,
including legacy systems. ERP products of SAP, Baan/Singular, Diapason,
system were considered and an EDI interface were planned to support EDI-
linked business.

Hence, the integration software is based on the principle that no local
function or ERP transaction shall be replicated in the new system. Hence,
only data really necessary to node interactions in the supply network will be
exposed at the network level.

This requirement can appear straightforward, but it is very common to
find "web-transposed ERP functions" when looking at supply chain
solutions on the market. It is thus necessary to mark clearly the difference
between network transactions, to be supported by FLUENT, and local
transactions supported by the ERP:

– Network transactions are those requiring the co-operation between nodes
   in order to take decisions, and for FLUENT the only data to be exposed
   at network level are those supporting these transactions. For example, a
   network process is the planning and negotiation of a supply flow through
   a link between two nodes. A data item relevant to this network-level
   transaction is the price applying for the transfer of product.
– Local transactions are those performed within the scope of a single node,
   and not requiring interactions at the supply network level. For example,
   local transactions are those of creation, definition and approval of a price
   list for a supplier or class of suppliers. The "price" data is obviously the

same as in the previous case, and a proper connection will be needed to update the price in the "Link" FLUENT object on the basis of listings changes. What is not required is to expose the "Listing" document (with all its approval, accounting and other details) at the network level. This would mean translating ERP functions into the new system.

Connections between two systems are realised through asynchronous alignment of data and events:
- Each system (FLUENT or the ERP) imports from/exports to the other system according to its own schedule. No synchronised update or common schedule between the two systems is required.
- Both systems are able to operate in their respective domains even in absence of up-to-date data from the other system. This means, for example, that FLUENT uses last availability data published from the ERP, while "real" availability can have changed in the meantime in the ERP domain.
- From a technological point of view, alignment of data can take place indifferently via LAN, WAN, or any other file transfer mechanism.

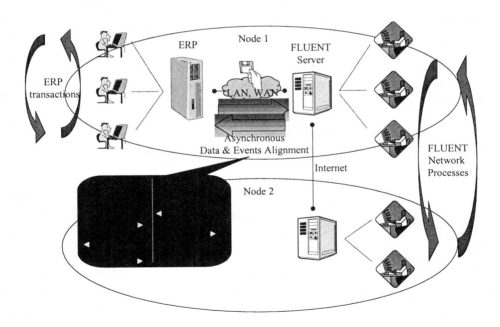

Figure 1. ERP Integration Model

To achieve this seamless integration a long-term strategic approach was chosen satisfying these goals:

- Usage of common available technology.
- Usage of a common language for exchanging data.
- Usage of simple and standard formats for data exchange.

The answer to these requirements is represented by XML.

3.   XML IN A NUTSHELL

Here in this paper there is no place to introduce XML [XML], we just hope the reader has already met it. For a newcomer, very briefly, **XML** (*Extensible Markup Language*) is a widely adopted standard, is available on a large number of platforms, is pretty simple and straightforward and, because is based upon exchange of formatted ASCII files, can be easily implemented also if the platform upon which the ERP product is based doesn't support it. All is needed is simply to write, read and transfer well formed text files. In some features:

- XML is a way of adding intelligence to your documents. It lets you identify each element using meaningful tags and it lets you add information ("meta-data") about each element.
- XML is very much a part of the future of Web, and part of the future for all electronic information.
- XML is syntax for marking up data and it works with many other technologies to display and process information. It looks and feels very much like HTML.

In the following we apply some XML companion standards like XML Schema [XSCH] and Extensible Stylesheet Language Transformations [XSLT]. For further reading, see the references.

4.   FLUENT ERP-XML INTERFACE

For understanding the interface, let we consider a typical communication scenario. The flow of communication will proceed according to the following steps, shown in Figure 2 (the described scenario is export of data from ERP to FLUENT, but the same considerations apply also to the reverse case):

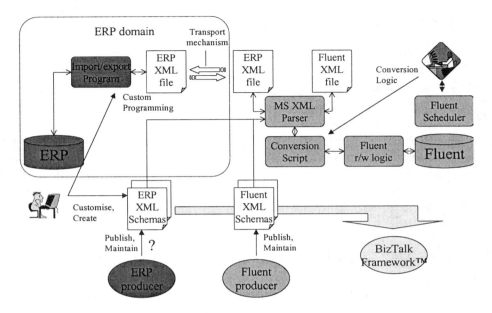

Figure 2. Communication Scenario

1. The ERP system will describe an XML Schema of the data it will make available for FLUENT, each entity that will be transferred to FLUENT will need to have a schema defined. Based upon this schema the ERP system will generate the XML files upon some arranged scheduling (ie: each time a new file is needed, each night, each week), the XML schema will be also made available on the FLUENT machine. This schema will be referred as "Starting schema" in the following.

2. For each entity that FLUENT will need to process a schema will be defined by the FLUENT Consortium. This schema will be referred as the "FLUENT schema" in the rest of the paper.

3. The Starting schema and the FLUENT schema can be the same or can be different, it depends on the ERP system. This choice (having two possible schemas) was made because there is a increasing interest in XML by the ERP producer and so there can be a situation where an existing schema can be already available for the ERP product the customer has available, in this case there is no need to create a new schema but the existing one can be used.

4. Recently Microsoft has been supporting a common framework for these operations named "BizTalk" that is already endorsed by the major ERP producer (SAP, Baan, PeopleSoft and JD Edwards just to name few). FLUENT and is also part of this initiative and so all the schema will be defined using the BizTalk format.

5. The XML files will be transferred on the FLUENT machine using some kind of common transfer method (e.g.: FTP, Samba, CFS or some other kind of file transfer mechanism), at this point the ERP system has finished its work. As far as the ERP is concerned, this approach seems to be the less demanding in terms of adoption of new technologies and changes to the local environment. The EDP personnel has to know:
– How to create and manage an XML Schema.
– How to write XML text files according to this schema.
– How to transfer files to the FLUENT machine.

The last two of the operations above would be needed even if XML schemas and XML files were not involved in the process.

5.   AN INTERFACE EXAMINATION METHOD

Our examination on the interfacing issues lays on some obvious considerations. Let us have two systems (A and B) and two transformations (T_{AB} and T_{BA}) to exchange some structured data (x) between them (Figure 3).

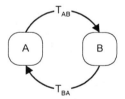

Figure 3. Simple Transformations

Let we define a test transformation:
$$T = T_{BA} \circ T_{AB}, \text{ so}$$
$$T(x) = (T_{BA} \circ T_{AB})(x) = T_{BA}(T_{AB}(x))$$
The test transformation simply transfers data from A to B, and backward. We would like to get back the same data we sent, but our wishes rarely come true. Repeating T again and again on the result of the previous T we can make a sequence of x_i :
$$x_{i+1} = T(x_i)$$
And if T satisfies some criteria it should do, the limit value of the sequence signed χ must exist:
$$\lim_{i \to \infty} x_i = \chi$$

Obviously χ is the fix point of T as

$$T(\chi) = \chi$$

Here comes the sense of the thing as here we get back the same data we sent. Unfortunately T is not supposed to behave so nice in case of arbitrary data. χ is the only one T likes, and χ may be completely useless for us. In order to be able to gauge the usefulness of χ or any other x_i, we should introduce an evaluating function.

The evaluating function E(x) measures the efficiency of the utilisation of the structure, that is E compares the actual data to the ideal amount of data the structure could carry. Function E produces a value between 0 and 1 where 0 means data good-for-nothing and 1 means data the structure is completely utilised by. The construction of the value is hidden and does not matter. Now we may examine the amount of the information χ contains.

In the best case $E(\chi) = 1$. It means T is perfect, consequently T_{AB} and T_{BA} are both perfect, so the interface works fine in both direction and we may use it as it is.

If $E(\chi) = 0$, the interface worth nothing. It does not mean that T_{AB} and T_{BA} are both inaccurate, both of them may be good enough and the interface can work quite fine without the loopback.

In any other case, we may consider whether χ provides data reasonably enough or not. Anyway $E(\chi) < 1$, there must be some interface insufficiency or excessive requirements occur. If χ is good enough, we should compare x_0 to χ and decide to eliminate data loosing substructures, because those needlessly engage resources.

A good approach is to define an X data structure first as necessary and sufficient for our systems, then we design and interface of T_{AB} and T_{BA} where

$$X = \chi$$

6. INTERMEDIATED TRANSFORMATIONS BY FLUENT ERP INTERFACES

Our case study begins with three systems (Figure 4). Let system A the FLUENT one, system C is the ERP one and system B is the intermediate one.

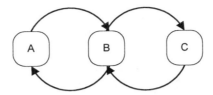

Figure 4. Intermediated Transformations

Why intermediation necessary? In case of n systems without an intermediating one there are $(n^2- n)$ transformations. The intermediating solution requires $(2 * n)$ transformations only. The role of the intermediate system is to provide a common base for any system to join and participate in the communication.

Unfortunately ERP systems do not really support a common way to exchange information between them. In most case an ERP product declares some interface for the world outside its own way. Recently XML became an interface language of the communication and ERP systems are announced to support XML interfaces. It does not mean a common interface with anybody using XML, since the XML Schemas ERP systems use are different ones. XML just makes the interfacing easier by a standardised data format. FLUENT recognised and exploited the advantages of the XML-based interfacing and declared its own XML interface.

In a typical case system S specifies an XML subsystem S' and export/import transformations where $E_{SS'}(\chi) \approx 1$. System A (FLUENT) and C (an ERP) do provide such A' and C' XML spaces and transformations. The only thing to do is to implement transformations between A' and C' (Figure 5).

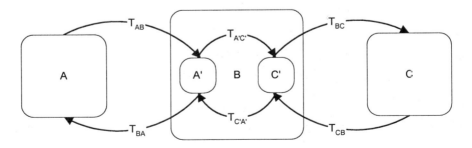

Figure 5. Intermediation Detailed

The solution of Figure 5 should not be called really intermediated as system B is phantom one. System B is a collection of XML Schemas, transformations and tools supporting the interface between A and C. Furthermore B is unclosed since when a new ERP system (D) joins the network B should be expanded with D' and corresponding transformations to A'.

Although there is no real ERP intermediating specification, FLUENT itself could be seen as a kind of ERP integration solution since ERPs could be connected to each other through FLUENT (Figure 6).

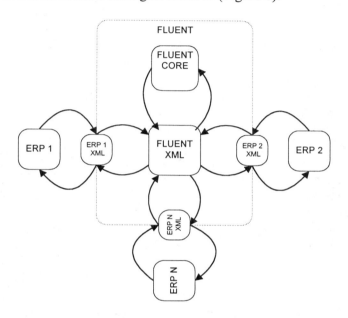

Figure 6. Intermediation by FLUENT XML

Our considerations on interface examination can be extended for the case of using FLUENT to integrate different ERP systems. As any combination of transformation routes may occur, we should test chains of transformations. According to a well-known saying a chain is as strong as the weakest link does. Consequently the chain can be tested link by link. A link test produces χ and $E(\chi)$ values for a link. We also can produce a value for the chain as the minimal value of the links:

$$Min\ (E_i(\chi_i)\ )$$

This way we can simply point out that one transformation loosing data might infect the whole system. If there is a chain going through each ERP system and through FLUENT between ERPs, the chain uses each of the transformations at least once. Obviously, if a transformation loose data somewhere, there is no way to reproduce it, so the following links get less information.

7.   FLUENT SPECIFIC INTERFACING PROBLEMS

Although FLUENT does not specify requirements for the ERP integration, the nature of the integration assumes some features the ERP should satisfy.

First we consider the main problematic point of the integration. As FLUENT itself works with Business Objects (BO), the work around is BO centric. Each import operation results one or more BOs and each export operation is fed from BOs. FLUENT objects represent Product, Aggregate BOM, Available Inventory, Planned Inventory, Independent Demand Order, Materials Requirement, Shipment, Receipt, Supply Item, Supply, Forecast and Unit Of Measure descriptions.

The FLUENT ERP Interface software realises a given number of *Connections*. Each connection represents an automated process for importing, exporting or aligning data between the ERP and FLUENT Business Objects.

Import connections update FLUENT objects based on ERP data; *Export* connections update ERP tables based on FLUENT objects; *Align* connections operate in both directions, keeping data aligned with respect to latest changes in both systems.

In the case of an import operation, we should create FLUENT Business Objects from the ERP side. The import procedure should solve the following problems:
- Semantically mapping ERP information into BO attributes can be difficult or even impossible and require ERP queries.
- One result may be calculated from other ones, so a dependency tree grows up.
- Some BO attributes reference to other ones (in FLUENT syntax of course), so the procedure should check the existence and coherence of the references at FLUENT side.

FLUENT having recognised the problems above, classified BO attributes as:
- *Mandatory*: The field value is required by FLUENT in order to define a proper instance of the object (defining attribute).
- *Default available*: If the ERP has it, the field value is useful on the FLUENT side. Event if the field is not exported by the ERP, new instances can be defined since FLUENT can use a default value.
- *Optional*:   The field is optional on the FLUENT side, and it will be instantiated only if the ERP has it.

Unfortunately even if we consider the mandatory attributes only (limiting quite much the usability of the interface), we could not ward off the great mass of the complications.

Export operations face similar difficulties, and in the align case, conflict may arise if a data element is updated in both systems. In such cases, as a general rule, the winner will be the latest update.

8. FLUENT - EDI INTERFACING PROBLEMS

Special cases of ERP integration when we are to build interface for another interface like **EDI**. EDI (Electronic Data Interchange) is a computer-to-computer transmission of (business) data in a standard format. Parties using EDI exchange messages. EDI allows specifying specialised message types (so called subsets) over the general syntax and semantics. Companies specified countless subsets for their own needs as ordering or offering products, notifying price changes and so one. In many cases, EDI is used to exchange ERP like information, so at this point we may introduce FLUENT as a solution of ERP integration.

When we build FLUENT Business Objects from EDI messages, we face some deeper difficulties as if we would work upon an ERP interface:
— There is no standardised subset each company use, so we should produce an interface for each subset.
— We just get messages often reflecting completely different approaches of the FLUENT use.
— There is no way to query the ERP for filling up a BO.
— When we export a FLUENT BO, we should build up messages filling up with EDI and subset specific data fields FLUENT does not provide.

Considering the problems above, we suggest specifying FLUENT compatible EDI subsets. Unfortunately there is no guarantee anybody accepts new subsets.

9. CONCLUSION

FLUENT provides software tools and techniques for the integration of distributed logistic and business flows.

The integration is based on standardised XML documents and transformations.

The integration can be difficult in case of special ERP systems. Although we may solve the integration is some way, we lose data during the interfacing procedures.

More problems arise when we integrate FLUENT with EDI.

10.   REFERENCES

Table 1. References Table

Ref.	See Here
FLU	http://fluent.gformula.com
XML	http://www.w3.org/XML
XSCH	http://www.w3.org/XML/Schema
XSLT	http://www.w3.org/TR/xslt

Multi Agent Systems for Multi Enterprise Scheduling

Dieter Spath, Gisela Lanza, Markus Herm
Institute for Machine Tools and Production Science, Technical University Karlsruhe, Germany
gisela.lanza@mach.uni-karlsruhe.de

Abstract: 'Multi Enterprise Scheduling' considers the temporal allocation of activities to resources. As soon as the specific targets and additional conditions have been achieved, the most suitable enterprise to execute the activity must be chosen. In the following article, the activity model presents the connections and logical sequence of the activities within the company overlapping business processes. The Organization units are identified in an Organizational and Resource Model. In the concept they are represented through Multi Agents, namely the Activity Agent and the Resource Agent. To meet the requirements of communication and cooperation across companies, it is attempted to implement this agent technology with web parts.

Key words: Manufacturing control and scheduling, Multi Agent System, Multi-Enterprise Scheduling, Business Process Modelling

1. INTRODUCTION

The demand for distributed value added nets [1] requires the conversion of dynamic process chains across several autonomous companies. Nowadays these so called Supply Chains are no 'chains' anymore, they occur as supply networks (see *Figure 1*).

In these dynamic supply networks, new optimization potentials must be developed through considering the involvement of a company with its customers and suppliers [2].

With virtual value added networks, or rather 'virtual companies', these partnerships are subjected to a relatively fast change. Thus, it is necessary for optimization to occur throughout the entire value added chain, that is to surpass the company's boundaries.

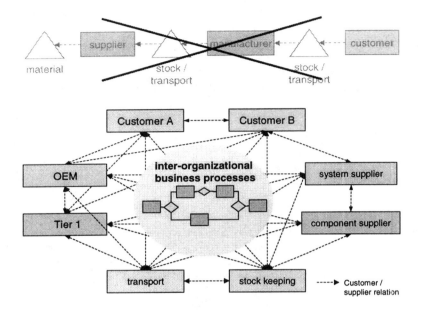

Figure 1: From Supply Chain towards Supply Network

1.1   Motivation

If one considers traditional solutions, the planning of the supply chain is initially strongly oriented towards production planning. Taking a rough plan as a starting point, which is then systematically dismantled into more refined plans up to the final production planning.

For this, Enterprise Resource Planning (ERP) Systems for materials management and production planning are introduced into companies. They are, however, not designed to optimize company overlapping supply chains in a decentralized way. They take fixed capacity quantities and production times as a basis.

It can be observed that stock transparency does not exist throughout the entire supply network, since standardized mapping of operative business processes is not available [3].

1.2   Scheduling Approach

The focal point of this article is that production processes are event-orientated and subject to certain interferences. This event-orientation is intensified when stock keeping and transport have to be regarded as the determining factors of schedule reliability (see *Figure 2*).

Consequently, only decentralized planning approaches can achieve a certain schedule stability, because local solvable deviations must no longer be taken into account in a global planning process.

This requires an event-oriented and decentralized scheduling approach of the company overlapping business processes. The presented solution with Multi Agent Systems meets these requirements. An uncomplicated adaptability at structural changes is achieved through the autonomous negotiation of the multi agents. Moreover a parallel scheduling is possible on distributed systems. To meet the requirements of communication and cooperation across companies, it is attempted to implement agent technology via web parts.

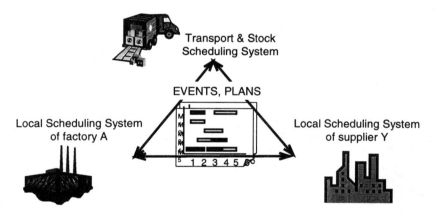

Figure 2: Multi-Enterprise Scheduling Scenario

2.   MODELING OF BUSINESS OVERLAPPING PROCESSES – ACTIVITY MODEL

The objective of the project is to generate an optimized and realizable operation plan, in which the diverse influential factors and restrictions are taken into consideration. An abstract model of the interorganizational processes is required to develop solutions for the scheduling problem. The basic activity model [4] presents the connections and the logical sequence of the activities in the regarded process. The modeling of activities is done through the generation of activity components (Enterprise Activities), which contains a status model as well as through the construction of rules (Business Rules), which are situated between these components [5].

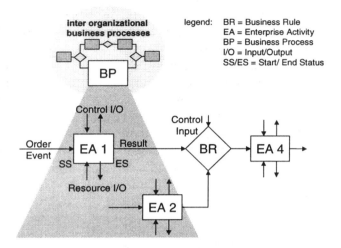

Figure 3:   The basic Activity Model of the Multi Enterprise Scheduling

For the Multi Enterprise Scheduling the activity model is characterized by the following objects and relations:

Orders for the product manufacturing: For products, various manufacturing alternatives with different production steps (work schedules) and alternative usable resources can exist. Orders are mapped in the model through Event-Input, which activates an Enterprise Activity. Here orders are defined through order number, start status, variant, batch number, input date, output date, prioritization and target costs.

Resources such as machines, raw materials, personnel, as well as warehouse and transport type, which are generally available for several products. The current capacity situation and future capacity changes must be considered in the optimization. Mapped in the model through Resource Input (Information from the resource pool, whether the required resources are available) and Resource Output (Resource de-allocation after termination or interruption); described through resource number, resource type (internal, external resource), resource class, availability, allocation and cost rate per unit of time.

Interferences of a technical or an organizational type - for example the loss of resources leading to modifications in the planning environment, which must be responded to appropriately. Mapped through: Control Input (interference signal or signal at resource reclaim) and Control Output (activation of alternative processes, for example reworking with interferences).

Constraints, that have to be (un)conditionally maintained and which are mostly technical, such as manufacturing specifications or the restriction of a double reservation of resources or a warehouse safety stock, a minimum batch size, etc. Mapped in the model through Business Rules, which describe the conditions under which the various activities are started.

Objectives such as strict schedule compliance, cost optimization, maximum plant utilization with maximum flexibility (robustness versus order variations), constant capacity utilization, minimum processing time, etc.

Optimization of scheduling is controlled through these constraints and objectives. The restrictions define the solution space. Via the objectives, the quality of the solution can be generated .

3. ORGANIZATIONAL AND RESOURCE MODEL

Why an organizational model? The structuring of business processes, as well as the allocation of the resources, can be clearly presented through the internal and intermediate organizational structures across collaborating companies.

In abstract terms, the core elements of company organizations are organizational units as well as relationships between them.

3.1 Organizational Model

The organizational model described here [6], considers the internal hierarchical structure as well as the intermediate net-like dimension. The latter gains increasingly in importance through the above-mentioned demands.

Hierarchical structures are the usual form of internal organizational structure, since they are ideal for the unambiguous definition of: power of decisions, instructions and control as well as supervisory duties. Hierarchical structures form the basis of this model. For this purpose, the organizational units are defined according to resource-oriented aspects and sorted into a hierarchy. Possible organizational units are for example, the entire company, suppliers, transport companies, sites, manufacturing areas, warehouse areas or machine groups.

Complex organizational forms are not, however, exclusively organized hierarchically. This applies in particular to legitimate and economically independent companies. The relationships between these include neither the

480   *Dieter Spath, Gisela Lanza, Markus Herm*

'power of decision' nor the 'authority to instruct', but they place the aspect of coordination and communication in the foreground. Thus the network-like, tendentiously dynamic dimension of the coordinating, logistic relationships are also considered on all levels in the hierarchy, as well as the hierarchical, or rather static dimension of the internal main administration system in the organizational model.

In the organizational model (see *Figure 4*), this overlay of the hierarchical and net-like structures is represented through two different types of relationship. Both types are orientated (Arrow) and define a potential use of an organizational unit by another one. The first can be described in the following as a customer and the other as a supplier. For example, a company (Arrow-source) loads an assignment (Relationship-arrow) in a specific plant (Arrow-destination).

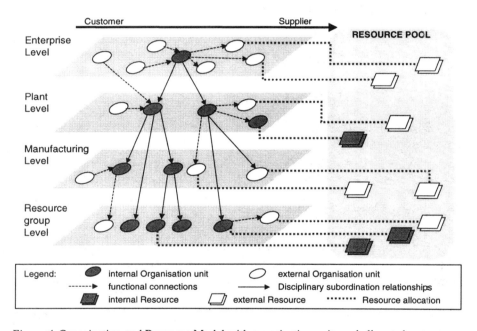

Figure 4: Organization and Resource Model with organization units and allocated resources.

3.2   Resource Allocation - Generation of a Resource Pool

The available resources are now allocated to the organizational units [7]. Resources for example, are employees, production systems, test beds, ...

In figure 1 the resources are coupled through connecting lines to the respective organizational units.

Coupling of the Activity Model with the Organization and Resource Model occurs because each individual activity out of the activity model is only processed through **one** organizational unit. If this organizational unit is an external one, for example a supplier, then the required resources will be administered by this organizational unit and must be inquired about there. The more detailed these activities are modeled, the more exactly the allocation of the executive organizational units and resources can occur.

The static capacity of a superordinate organizational unit is the sum of the respective resources of the disciplinary subordinate units. The sum is referred to as internal resources.

Resources of other organizational units can be used in the same way, for example the use of a test bed or the formation of a worker team. These combinations are marked through the functional connections in the organizational model. The subsequent useable resources are allocated as external resources of the superordinate organizational unit.

The Organizational and Resource Model administers the entire supply of the resources, which are available in a resource pool. Within this pool, one distinguishes between the company internal resources (sum of all internal resources) and company external resources, that represent the partial quantity of the external resources, which do not refer to other organizational units in the same company, but to external sources. The external resource allocation needs a specific communication and coordination, since arranged parameters (Priorities of the assignment, costs, etc.) have to be negotiated, in order to avoid, for example, double reservations

4.   MULTI AGENTS FOR SCHEDULING

For the scheduling of activities which need these external resources the requirements like 'information exchange', 'Negotiation' and 'Task' must be satisfied. These demands shall be achieved via three characteristics of the multi agent technology: The agent architecture, the agent interface (especially the agent language) and the agent platform (see *Figure 5*)

4.1   Agent Architecture

The agent architecture enables the autonomous, or rather partly autonomous, implementation of decisions and tasks. The agent makes assumptions about its environment (belief), it possess an objective function (desire) and an intention module (intention), in which it makes plans of action or decisions as to how it wants to adapt its environment to the target status (corresponds to BDI-Architecture [8]).

Based on the coupling of the two models every identified Organization unit is represented by two agent types with specific assignments.

The **Activity Agent** (or Order Agent) attempts to make activities that require external resources, capable of starting. Visualized through the functional connection in the Organization Model. Therefore the activity agent negotiates with the suppliers' agent (from the resource agent type) about the resource offers, under specific target restrictions. The most important assignments are dispatching enquiries, evaluating offers and eventually assigning orders.

The **Resource Agent** of an organization unit is interested in a constant use of its resource and pursues this objective, this means that it provides free capacities of its resource to the agents of the customers (from the activity agent type) and negotiates with them. It must take current allocation dates as a basis.

Additionally there are the agents of the customers and suppliers, which are registered (known) on the agent platform.

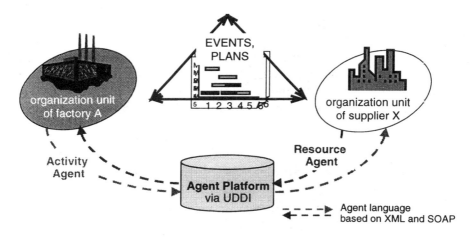

Figure 5: Multi Agents for Scheduling

4.2   Agent Interface

The agent interface permits the exchange of the agents' messages between systems on the same level, via an application independent agent language. For this, the agent uses so-called 'Speech act' with which it can express its objectives to other agents. For supply chain management, five categories of speech act are defined [9] (see *Table 1*)

The content of the message has an application independent semantic, due to it being defined as XML schemata. The agent language is developed on the basis of the internet standards XML and SOAP [10]. Through the Remote Procedure Calls (RPC), the Simple Open Access Protocol (SOAP) that is also based on XML, enables applications or modules to be contacted and executed via the internet. When using both protocols together they enable distributed and platform-independent communication and releasing of tasks.

Classes of speech acts	Meaning
Information transfer	The company-external sender delivers information to the recipient. He does not combine the execution of a task. - Resource availability: the enquired resource is reported as being available and then booked by the resource manager. - Interferences are directly reported to the event manager.
Information inquiry	The receiver requests specific information about a specific state of affairs. The resource input of an activity is enquired about.
Negotiation	Sender and recipient negotiate about variables (for example price, delivery date, quantity to be delivered)
Task	The sender combines with the message the execution of a task at the recipient (for example contracting)
Error handling	routines, if no speech act is accomplished

Table 1: Classification of speech acts for the Supply Chain Management

4.3 Agent Platform

The agent platform presents the environment, on which agents can be executed. The approach is currently followed to realize the agent platform according to the experience with distributed database application and to the UDDI efforts based on XML and SOAP.

The Universal Description, Discovery and Integration (UDDI) standard creates a platform-independent, open framework for describing services, discovering businesses, and integrating business services using the Internet [11].

The principle of telephone directories vividly illustrates the procedure: the section of the UDDI-index corresponding to the normal "white" telephone directory contains the names of the companies, the addresses and standard contact information (business entity element). The "yellow pages" provide information about the products and services, which the company offers (business service element). Finally, the new "green pages" inform about the protocols that the companies use in the electronic business commerce (binding template element).

In our scenario the call of a service or order, which is registered in the database, operates approximately as follows (see *Figure 6*):

An UDDI client, an activity agent of the customer Y traces the business entity information of the desired service via an inquiry in the business repository. The agent requests detailed information regarding the business service element. It negotiates with the UDDI client of the addressed company, for example the resource agent of supplier X, through standardized speech act according to its objectives, for example target costs and delivery date. If the objectives are achieved, the assignment is processed with the help of the binding template element, for example Purchase Order Confirmation (POC) and delivery schedule. Furthermore, a special authentication mechanism guarantees that only authorized clients may apply the web services.

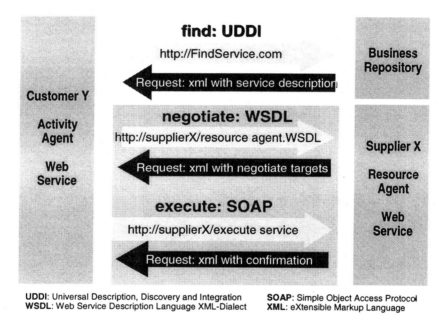

UDDI: Universal Description, Discovery and Integration   SOAP: Simple Object Access Protocol
WSDL: Web Service Description Language XML-Dialect   XML: eXtensible Markup Language

Figure 6: Agent negotiation via web services

5.   SUMMARY

The report presents an activity model that is connected with the responsible organizational units. It presents the company overlapping business processes, which considers, in addition to production processes, stock keeping and transport as determining factors of schedule reliability.

Distributed production and logistic networks generate a higher level of coordination requirement. The presented Organizational and Resource

Model maps these problems through the introduction of explicit communication relationships.

In the joint project 'Distributed modeling, simulation and control of business processes with generic Petri nets' supported by the ministry for economic affairs of 'Baden Württemberg', very positive results were experienced during the modeling and simulation of the Activity Model as well as the Organizational and Resource Model.

With the help of multi agents, the multi enterprise scheduling that is presented in this article, is now in the development stage: To produce a prototype there must be a specification of the negotiation protocols. Current research focuses on the development of the UDDI standards and investigates whether the above-performed scenario is realizable.

6. REFERENCES

[1] Scheer, A.-W.; Borowsky, R., 1999, Supply Chain Management: Die Antwort auf neue Logistikanforderungen, LM'99-Intelligente I+K Technologien, Springer Verlag, Bremen.

[2] Schönsleben, P., 1998, Integrales Logistikmanagement: Planung und Steuerung von umfassenden Geschäftsprozessen, Springer, Berlin, Heidelberg.

[3] Tiemeyer, E., 2000, Supply Chain Management - Konzeptentwicklung und Softwareauswahl in der Praxis, atp 42, Heft 10: 24-32.

[4] Kühnle, H., Sternemann, K.-H., Harz, K., 1998, Herausforderung Geschäftsprozesse – Den Wandel organisatorisch und technisch gestalten, Logis Verlag GmbH, Stuttgart.

[5] CIMOSA – Open System Architecture for CIM; ESPRIT Consortium AMICE, Springer Verlag 1993, ISBN3-540-56256-7, ISBN 0-387-56256-7, Heidelberg.

[6] Appelrath, H.-J., Sauer, J., Fresse, t., Teschke, T., 2000, Strukturelle Abbildung von Produktionsnetzwerken auf Multiagentensysteme, KI Künstliche Intelligenz, Heft 3/200: 64-70, arenDTaP Verlag, Bremen.

[7] Spath, D.; Sternemann; K.-H.; Lanza, G.: Supply Network Simulation. CIRP Proceeding of the 34th International Seminar for Manufacturing Systems, Athens 2001.

[8] Rao, Anand S.; Georgeff, Michael P.; Sonenberg; 1995, BDI Agents: from Theory to Practice, Proceeding of the First International Conference on Multi-Agent Systems, pp 312-319, MIT Press, MA.

[9] Stiefbold, O., 1998, Konzeption eines reaktionsschnellen Planungssystem für Logistikketten auf Basis von Software-Agenten, Forschungsberichte aus dem Institut für Werkzeugmaschinen und Betriebstechnik der Universität Karlsruhe, Dissertation, ISSN 0724-4967, Karlsruhe.

[10] W3C- specification, 2001, SOAP Simple Object Access Protocol, http://www.w3.org/TR/SOAP/, DATE 2001/03/10.

[11] UDDI.org, 2001, Universal Description, Discovery and Integration, http://www.uddi.org, DATE 2001/03/10.

Valu.e-networking

Taking the e-conomy from vision to value!

Dr.-Ing. Wilfried Sihn, Dipl.-Ing. Joachim Klink
Fraunhofer-Institute for Manufacturing Engineering and Automation
Nobelstr. 12 D-70569 Stuttgart / Germany
Email: whs@ipa.fhg.de, jfk@ipa.fhg.de

Abstract   The modern corporate management is still in a state of permanent change, whose dynamics of development is rather increasing than decreasing. This change takes place especially in a linking up of value added activities (that occurs more and more in a cross-company way) and the computerisation of business processes. It occurs under various approaches and keywords (e-business, supply chain management, virtual business, e-procurement,..). This change on the one hand includes a successive adjustment to known trends, and on the other hand it includes new challenges for corporate management. In this paper, the change in corporate management is discussed using the concept of the 'fractal business' as an example. Solutions for the future are sketched, too.

Changes in Management – Management of Change

For ten years, we have been living in a time that has been called a 'time of change' everywhere. Markets, products, and technologies are changing at an unprecedented speed. Since the beginning of the 90ies, when Europe was in the worst economic crisis of the post-war era, businesses have tried to keep up with that change. Market shares are being regained with various approaches, concepts, and strategies, as well as a reduction of costs and processing time and– last bust not least – the ability to react to changes quickly is being increased. It is interesting to watch how management concepts have developed during the past years.

Among all of the consultants', research institutes', and practical persons' trends, three fundamental developments can be extracted:

1. The development of 'adaptable' businesses
2. The development of business networks
3. The computerisation of added value (e-business)

When looking at these developments on an abstract level, the conclusion can be drawn that it is the development in the same state, seen from different perspectives. The first case is about transforming a rigid, centralistically organised structure into a dynamic shape with decentralised organisation units. In the second case, an attempt is made to co-ordinate and optimise several legally autonomous businesses together. In the third case, a technology revolutionises the possibilities of interaction between business and customer ("business to customer", B2C) and between business and business ("business to business", B2B). All of the three developments have a common objective: an added value, produced together and co-operatively with making full use of high degrees of freedom and the single added value partner's ability to react. Obviously, modern top-level organisations are a composite of extensively autonomous organisation units with common objectives and defined rules of co-operation. It is a form of co-operation we have propagated in the form of the 'fractal business' concept for nearly ten years and which has by now been practised in many businesses.

The Fractal Company – Origin of a Trend Reversal!?

In 1992, we published the book "The Fractal Company: A revolution in corporate culture". Connected to this book is a turn in the thinking of modern business management. The concept of the 'fractal business', as it has been called since the transmission to non-industrial branches, stands for a break with the paradigms that could then be found in theory and practice. New approaches, methods, and concepts have been developed and introduced in companies. For many, the 'fractal business' has become the 'European answer to lean production'. The management trends that were set then are today more relevant than ever. The direction in which we have been heading proved to be successful and is a signpost for modern corporate management. The features of fractal units of organisation are established in many businesses as a model.

Features of Fractal Units of Organisation

"A fractal is an autonomous unit whose objectives and performance can be described unambiguously" [1]. This is how the fractal was defined originally. With that, nothing is said about a fractal's size or organisational structure. A fractal is generally a unit of a business which has objectives, achieves results and acts autonomously. In this context, the description of objectives and achievements for business units is important. This was new at that time. What was new was that these objectives and achievements were not only fixed on financial scales. In contrast to the so-called Center-concepts, which generally include financial objectives, the objectives of a fractal are orientated on its contribution in the entire added value of an enterprise. This does not necessarily have to be a financial contribution.

Self-similarity

A fractal has objectives, achieves results, can be described through connections (both within the fractal and with other fractals), elements, and qualities. Therefore, from the single employee to the entire business, everything can be described in fractals. Using systems theory, a fractal can be thought of as an open system which itself includes (subordinate) fractals and is also part of a superordinate fractal. All units of organisation are businesses within businesses; employees are entrepreneurs in their company.

Self-organisation

Fractals run by self-organisation. They are allowed a defined amount of autonomy to reach their objectives and accomplish their results. Efficiency requires the ability to act. Fractals are able to act because they can command their resources which are necessary for the supply of performance on their own authority and own adequate freedom for acting and making decisions. Self-organisation also means market economy principles within a company. The fractals organise themselves and are not controlled from outside. The fractal's limits (see above) describe how far this self-organisation stretches, which scope of development it has. The rules of the game between fractals are set by their relationships.

Self-optimisation

Fractals do not persist on their status quo, but are part of the change. They revise their objectives, performance, and qualities permanently and adjust them accordingly to the market's requirements. In this, there are evolutionary changes, e.g. through adaptation of qualities, relationships, etc. However, revolutionary changes, e.g. in the form of a partition, separation, or even dissolution of fractals are possible.

Dynamics, Vitality, Adaptability

The features described above allow fractals not only to adapt to changes, but to act anticipatingly and on their own initiative. Fractals are, to a certain extent, "autonomously viable". For an extensive definition of the term "adaptability" we refer to the works of the Sonderforschungsbereich „Wandlungsfähige Produktionssysteme" (special research field "adaptive production systems")[1]. For a better understanding of what is described by adaptability, the following sentences may suffice: Adaptability is the ability to reach lasting changes out of one's own strength. These changes lead to conditions that were unknown before the changes occurred. Entirety – The Six Level Concept

The fractal business sees itself as a holistic concept. It is based on the insight that a company is only fully efficient as a whole, as a unit. Local or one-sided optimisations are not very helpful; the unit is as weak as its weakest part. The approaches of cost accounting, of compensation and of PPC must fit, the entire philosophy must be harmonious, and the concept must be consistent. A fractal (and with it, the entire company) is described using the levels 'culture', 'strategy', 'finances', 'socio-informational level', 'information', and 'material flow and process level'.

New Developments, New Business Environments, New Challenges

Since the original development of the 'fractal company' concept, several influential factors and business environments which are relevant for business management have evolved or changed. These changes are manifold; they concern among other things:

- A new generation of businesses and entrepreneurs. Along with the buzzword "new economy" an unprecedented founding wave can be noticed which questions many "laws of success" that were valid until now.
- New business models, new markets. The term "new business models" stands for innovative ways to achieving economic success. The added value itself changes. Many businesses make money by producing "traffic" and creating a "community".
- A change in the understanding of what a product is. Whereas in the past the term "product" usually meant a physical product, today the virtual share of a product has increased. Instead of buying a car, you merely purchase mobility or the sensation that is combined with the brand's image.
- New technologies. Especially in the information and communication technology sector massive changes take place with a noteworthy influence on industrial products and production.

---

[1] Westkämper, E.; Zahn, E.; Balve, P.; Zilebein, M.: Ansätze zur Wandlungsfähigkeit von Produktionsunternehmen – ein Bezugsrahmen für die Unternehmensentwicklung im turbulenten Umfeld. In wt Werkstattechnik 89 (1999) 9

- New methods and procedures. The industrial added value is more and more supported by efficient methods and is realised both better and faster.

New forms and models of organisation. The developments and new models of organisation focus on cross-company organisation. Be that in the form of virtual enterprises, business networks, or supply chains – the added value that is linked up across companies is established step by step.

Despite these multifaceted changes of the past years, no fundamental break concerning the basic orientation for successful business management can be noticed. Rather, the significant paradigms of a "fractal company" (the "turbulent environment", adaptive business structures and processes, holistic management,...) have been confirmed and applied in various concepts. Especially the mega-trend of "linked-up added value" with its many specific characteristics follows the basic philosophy of fractal organisation – the common added value of extensively autonomous performance units with the objective to react very flexibly and quickly to the market's requirements.

These modified but still valid business environments offer the modern business management freedom to act. They are a reason for testing the validity and potentials of the fractal company concept. Also, new adaptations and further developments should be identified. An extension of the fractal model accordingly to the latest market developments – namely the trend towards linked-up added value and "e-business" – is at issue. Those two developments are currently the most prominent drivers for concepts and methods of business management.

"Fraktal+" – Adaptation and Further Development of a Successful Concept

The fractal company concept is generally practical for taking up the developments sketched and for offering suitable solutions. As mentioned in the beginning, the basic principles within a fractal company can be compared to those of a business network or a virtual enterprise. The linked-up added value is in this case achieved through legally autonomous fractals, i.e. the companies involved. The self-similarity principle thus makes the mapping of this linked-up added value possible and thereby offers an approach to view this form of added value in the framework of the known theory. The definition "A fractal is a business unit that acts autonomously" must simply be modified to that effect that the company itself and systems comprising several companies are to be regarded as fractals, too. It could be something like "A Fraktal+ is an autonomous performance or business unit consisting of parts of companies, the company itself, or several companies,...".

However, the adaptation of the fractal company concept is slightly more difficult regarding the e-business phenomenon, which is marked by changes in the business model area. Yet, at second view, it becomes clear that the generic fractal model offers adequate application possibilities. A fractal is, among other things, described through its objectives and accomplishments. These objectives can, as mentioned above, be manifold; they are not restricted to the financial dimension. Neither are the achievements restricted to producing material goods, but they comprise any form of producing performance. Thus, the models for describing the performance of so-called "not directly productive fractals" (as, e.g. in a personnel or quality department) are, at least in principle, conferrable to the "new economy" business models. Here, too, we find performance units (the start-ups) which do not (yet) generate any financial yield, but are appreciated (in the capital market) because of other performance features ("community" size, degree of being known,...).

The general applicability of the "fractal company" concept, however, must not lead us to ignore that the concrete solutions and concepts to be developed are new. Yet, this fact does not pose a real problem, as each fractal company is unique. It is just the individual specifications of fractals – depending on the context of the entire economic unit and their contribution to the entire added value – that make this concept so multifaceted and thereby efficient. Seven rules for successful corporate management in valu.e networking

There are seven rules that companies must obey if they want to achieve lasting valu.e networking success:

1. Businesses must balance the interactive process of value creation. A product is not defined as the result of interaction between "supplier" and "customer", the product is interaction itself! The manufacturer becomes a service provider and catalyses the networked, interactive process of value creation. The linear process of value creation gives way to the linking-up of the value-adding community.
2. Businesses must be able to constantly innovate on their own initiative.
   Adaptability is crucial to being competitive in a turbulent environment. It allows companies to react quickly, efficiently and comprehensively to changes. Adaptability itself is not sufficient. Only the one who impacts these changes constantly and on its own accord will be the one to gain a competitive edge. Lasting success requires the ability to innovate products, processes, structures and methods constantly.
3. Businesses must understand value creation as added value creation.
   Added value occurs if a transaction among at least two value-adding units is successfully creating an added value for both units (e.g. money for product). This is why each value-adding unit

(from the single employee to the entire economy) must actively and permanently have an interest in offering, selling, communicating and realising added value. Added value is defined from the recipient's point of view and not from the added value creator's point of view!

4. Businesses must emphasise the communicative and emotive aspects of their products.
    Added value is not only a product, it is the random combination of physics/hardware, software, service rendering, knowledge, time, law and emotion. Added value is not restricted to an immediate or potential benefit, it might also include emotion. Added value is a subjective "additional value" from the recipient's point of view. A vital aspect is whether both recipients recognise the transaction as a personal added value.

5. Businesses must place greater emphasis on "partner" and "competitor" markets.
    The potential of vertical networking, i.e. along the value stream is currently being tapped into by all agents. It is not that these potentials hold any competitive advantages worth mentioning, they are merely necessary preconditions for being competitive. New potentials can be extracted from the axis "partner⟷ competitor". Along this axis, new dimensions of added value are waiting to be discovered, ultimately leading to a significant improvement of competitiveness.

6. Businesses must command "processes overlaying processes"
    The automation of manufacturing processes comes along with the consistent automation and standardisation of business processes. Human work shifts from executing single processes to managing structural and meta-data, and continuous process optimisation. It is not the direct added value activities that mark competitiveness but the meta-processes i.e. "processes overlaying processes" (CIP processes, change management, development of organisational structures).

7. Businesses must act in time
    Beware of falling a prey to the acceleration trap: Failed product launches, quality defects, misjudged market inertia and burn-out syndromes are the first signs of wear in the high-speed community. Success requires proper timing. That's why enterprises discover the "new slowness". Mature, "decelerated" processes ensure lasting success.

As a conclusion, you should bear in mind that the general approaches of the fractal business will be valid in the future and will offer the business management sufficient potentials. Furthermore, there are enough challenges and questions to be answered in this modified environment. We are looking forward to meeting these challenges.

Literature

[1]   Warnecke, Hans-Jürgen: Revolution der Unternehmenskultur: Das Fraktale Unternehmen. 2. Aufl. Berlin u. a.: Springer, 1993

[2]   Warnecke, Hans-Jürgen: Aufbruch zum Fraktalen Unternehmen: Praxisbeispiele für neues Denken und Handeln. Berlin u.a.: Springer, 1995

[3]   Westkämper, Engelbert u.a.: Ansätze zur Wandlungsfähigkeit von Produktionsunternehmen: Ein Bezugsrahmen für die Unternehmensentwicklung im turbulenten Umfeld. In: Wt Werkstattstechnik 90 (2000) 1/2, S. 22-26

[4]   Sihn, Wilfried: Service aus Kundensicht. In: Westkämper, Engelbert (Hrsg.); Schraft, Rolf Dieter (Hrsg.); Fraunhofer-Institut für Produktionstechnik und Automatisierung IPA: Neue Servicekonzepte im 21. Jahrhundert: Fraunhofer IPA Seminar F 52, 11. und 12. Mai 2000, Stuttgart, Getr. Z.

[5]   Sihn, Wilfried; Hägele, Thomas: Innovative company networks – Organization, technology and successful examples. In: National Defense Industrial Association (Hrsg.): CALS Expo International & 21st Century Commerce 1998: Global Business Solutions for the New Millenium, Long Beach/California, 26.-29. Oktober 1998. Arlington/Virginia, 1998

[6]   Sihn, Wilfried; Klink, Joachim: Added Value in Production Networks In: Roller, Dieter (Hrsg.): Advances in Automotive and Transportation Technology and Practice for the 21st Century - Abstracts: Complete Symposium Abstract Volume 32nd ISATA, 14th-18th June 1999, Vienna, Austria, Croydon, UK, 1999, S. 128

[7]   Sihn, Wilfried; Hägele, Thomas; Deutsch, Oliver: Competitive production networks through software-based reengineering and added value networks. In: Mertins, Kai (Hrsg.); Krause, Oliver (Hrsg.); Schallock, Burkhard (Hrsg.); IFIP WG5.7: Global Production Management: IFIP WG5.7 International Conference on Advances in Production Management Systems, Berlin, 6.-10. September 1999. Dordrecht u.a.: Kluwer, (International Federation for Information Processing (IFIP) 24), 1999, S. 432-439

[8]   Warnecke, H.-J.; Braun, J.: Vom Fraktal zum Produktionsnetzwerk; Berlin u.a.; Springer 1999.

The authors

Dr.-Ing., Dipl.-Wirtsch. Ing. Wilfried Sihn is director and head of the business management sector at the Fraunhofer Institute for Manufacturing Engineering and Automation (IPA), Stuttgart/ Germany.
Dipl.-Ing. Joachim Klink is project leader in the business management.

The Operating Models of Tomorrow Require New Control Concepts Today

Dieter Spath, Karl-Heinz Sternemann
University of Karlsruhe (TH)

Abstract: The growing number of product variation, unstable sales figures and the increasingly reduced lifecycle of products has lead to increasing flexibility requirements for the whole production process. Additionally, mechanical engineers pursue, against the background of the growing importance of services in the industrial sphere, the improvement of the existing supply of services in the direction of customer service. From an OEM viewpoint, it would be desirable to purchase the production as a service and to pay per produced unit. Beyond maintenance and upkeep, services emerge that grow into the core business of the system supplier. For the user, service, reliability and quality losses can be the consequence when the system supplier has a lacking or limited business experience. For the system supplier, as the operator of their system, the unrestricted availability and complete control over the reuse are the primary objectives. Once the net concepts and structures correspond to the modern requirements, then future operating models with suitable monitoring functions can be supported. It is necessary, that:

- The system, not the control system, is considered as the logical unit.
- A high flexibility through consequent use of TCP/IP and HTTP as a transportation and transfer protocol on the basis of Ethernet as net technology can be guaranteed.
- The communication of individual control units, sensors and actuators in a network based on switches with a quasi deterministic behaviour is allowed.
- Through internet-similar structures in the system, no additional programming costs for data communication and data synchronisation arise and secure, switched network hierarchies merge to a logical net layer, as a result the bottlenecks SPS-CPUs are cancelled.

Key words: control concepts, information and knowledge management, life cycle management, internet technologies, XML, SOAP

1. MOTIVATION

The growing number of product variation, unstable sales figures and the increasingly reduced lifecycle of products has lead to increasing flexibility requirements for the whole production process /GAB-94/. Additionally, mechanical engineers pursue, against the background of the growing importance of services in the industrial sphere, the improvement of the existing supply of services in the direction of customer service [LUC-97]. The path leads from the supplier model to the different participation models to the operation model. As a result, one increasingly ends up with a transfer of the risk from the user back to the system supplier. At present, the conditions are lacking to financially bear this risk.

Research must develop future comprehensive approaches that are economically optimal for the manufacturer, operator and society [STE-87]. The economic potential is particularly located in the area of reusing systems and components for new products or for other customers [SCH-00]. Deficits are caused by insufficient knowledge about the load situation of system components and their remaining lifespan. This has various causes:
– Insufficient modular design of machines and systems
– Lack of experience with using used components and accompanying risk estimation;
– The insufficient knowledge regarding component condition, i.e. insufficient statistical data for the financial and technical potentials of components;
– Lack of technical documentation of the system;
– Lack of concepts for the organisational integration of the new business field into the existing company.

Figure 1 portrays the development of a production system with all the connected development processes; preparation of the market or development of production processes accompany, or rather influence, the planning of the features of a new product, its realization as well as its application and sorting out. The starting point is an operating scenario for a production system, which contains a life cycle monitoring and prognosis including appropriate maintenance strategies and which considers an extended system responsibility. This scenario pursues the solution from the existing principle of the to date unidirectional business relation between system supplier and his customer, as far as an integrated business model. System adaptation and retreating work gain importance [KOR-99]. The objective is for a production system to be able to "produce" over several product generations; this means that the system should be able to flexibly adapt to product further developments, after the actual production commencement. The reuse for new products or for other customers is prerequisite to use the considerably longer lifetime of components in comparison to the product lifecycle.

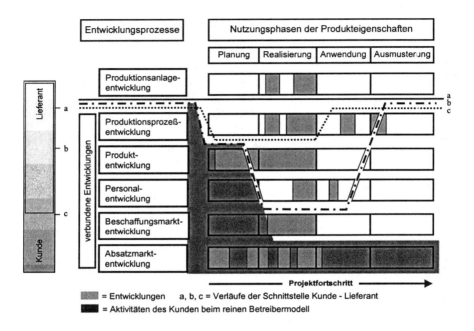

Figure 1. The change of the interface customer - supplier [GER-99]

2.   NEW SERVICES

From an OEM viewpoint, it would be desirable to purchase the production as a service and to pay per produced unit. In the automobile industry, corresponding examples of complete operating models already exist: MCC (Hambach) or VW do Brasil (Resende) [Koe-00]. Beyond maintenance and upkeep, services emerge that grow into the core business of the system supplier. For the user, service, reliability and quality losses can be the consequence when the system supplier has a lacking or limited business experience. For the system supplier, as the operator of their system, the unrestricted availability and complete control over the reuse are the primary objectives. From this the following core questions arise:
- How does one modularly design the system correctly?
- Which maintenance and upkeep strategies should be pursued?

– What must a control technology concept accomplish in the future, with regard to monitoring and lifecycle prognosis?

Loads over the whole lifetime of the component and the resulting service, maintenance and failure probabilities are the basis of a new type of calculation models, which will have a decisive influence on the development of systems or components. This requires a complete knowledge base concerning every single module, in order to determine, via the lifecycle or prognosis models, the possible failure probabilities and the remaining lifespan of the modules. This means that a comprehensive monitoring of system relevant information objects must be the input for diverse models for analysis, status description and prognosis of failure probabilities or remaining operation times. Figure 2 illustrates the available communication levels of the machine levels over the company internal network up to connection via the internet.

Conventional control concepts of systems are often not capable to fulfil the increasing requirements of complex production systems and they are moreover part of the monitoring and prognosis problems. One current hurdle that is difficult to clear is the classical PLC, which is currently still the central control element for automated production. New control concepts, which for mechanical modular design analogously contain a decentralized control periphery and a decentralized control intelligence, are therefore an indispensable condition of future operating models. An additional independent network of sensors and actuators without a media break to the production controlling nets (continuous TCP/IP) can achieve this under the condition of acceptable costs.

3.   WHY A NEW CONTROL TECHNOLOGY?

Lets begin with an excursion into nature. Walking means more than merely to swing the legs –the movement seems to provide common principles: the model of central pattern generators (CPG) can be understood as such a principle [GMD-00]. When we walk the muscles in our thighs are tensed and relaxed in a specific chronological connection (phase) to one another. This pattern is produced through the central pattern generators (CPG). They are so to speak small brains, which take over this control for us. Consequently, our real brain is relieved of this burden and is able to concentrate on other things, according to the principle: divide and conquer! This means distributed control. In nature there doesn't exist a nervous system that takes over all assignments on all levels. Instead, the tasks are shared out, this means that certain routine processes are out sourced to subordinate structures, which accordingly take over certain control processes autonomously.

The transfer to the automation technology means: future controls should be divided up into smaller local units, which only concentrate on certain processes. While a specified processor specialises in the analysis of sensory impressions, another takes over the central control of the whole system on a higher level. The local processors are structured similar to the central pattern generators (CPG).

They have a URI (uniform resource identifier) with methods and parameters; this means that information vectors enable a worldwide unambiguous addressing. The data to be exchanged is implemented in SOAP Objects (Simple Object Access Protocol) and uses standardised transport protocols such as TCP/IP on the basis of Ethernet. The control systems of the production have to be integrated into the administrative IT systems. The objective of these developments must be to continuously realize the data and information flow of the visualisation as far as the individual drive, the control unit or the relevant sensors of a system.

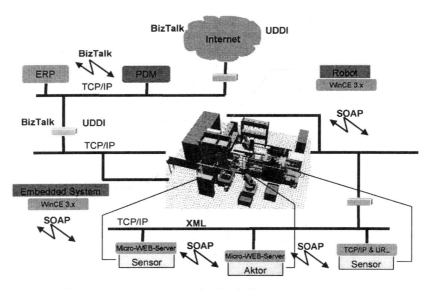

Figure 2: Levels of communication in the production field.

Decentral controls are the fundamentals of modern industrial control concepts and are based on:
- the decentral periphery (distributed installation) and
- the decentral intelligence (distributed functionality).

The objective of distributed installation is to reduce the wiring expense; minimal cable lengths and open and maintainable installations are thus achieved. The advantage of less wiring expenses is however confronted with the serious disadvantage of,

The structures of classical PLCs are not created for this.

Data must pass through Systems with the help of different drivers in order to reach the main level from the sensors. Less speed and a lack of transparency are the consequences. Not every control can communicate at any time with every other control. In many cases, the real time of the application and high flow rate of data can not be spoken of.

As a result, the required principle of decentral intelligence, which should distribute a control task among a number of intelligent control nodes, is not realizable with the classical PLC systems. Moreover, trends such as OPC (OLE for Process Control) only produce marginal improvements, as long as they are imbedded into the old structures.

Consequence: In order to achieve a drastic simplification, a reduction in programming expenses and a strengthening of interoperability, a paradigm change is necessary in control technology.

The future lies in distributed intelligence, which combines the advantages of central and decentral concepts. In the individual components of a system, specific function blocks, which for example contain high-integrated drive functions, must be implemented as well as the mechanical modularity. At the same time the openness of the systems is particularly worth paying attention to, this means that these components must be freely programmable, or rather parametrizable and operate as independent modules. The intelligent systems are no longer only driving systems or actuators, which are operated from a central control, but rather systems that independently and self-sufficiently take over the control of sub-systems.

Consequently data and information emerge that on the one hand, are directly integrated into the operating and monitoring process of the operator's IT system and which, on the other hand, present the basis for a target orientated monitoring and data processing of future operating models. This means, for example, that direct access from the ERP system to the operation unit or a specific sensor for foresighted maintenance assignments or for necessary recording of current components with reference to future taking back obligations becomes necessary, this means with this the pursuit of the current manufacturing process will attain a high priority in the future.

The internet enables an intensive and rapid collaboration between operator, system supplier and component manufacturer. The connection of current system information with the service departments of participating partners over the internet, accelerates error diagnosis and as a result the entire response time. Spare part orders can automatically and directly occur via corresponding business-to-business mechanisms such as BizTalk and UDDI (Universal Description Discovery Integration, an international repository for web services). When problems arise and their cause cannot be immediately and clearly classified, the internet enables a cooperative collaboration of all participators on a standardized database.

4.   DEMAND THAT ARE MADE ON FUTURE NET CONCEPTS:

A net technology, which has a capability of at least ten times more than that of introduced field buses, is the prerequisite for a decentralized, but simultaneously real time capable and deterministic concept. A suitable basis seems to be **Ethernet**, which must be made usable for applications in the production systems. A switched Ethernet meets all the necessary conditions:
-   High transfer capacity (up to 1Gbits/s)
-   Quasi deterministic behaviour with use of the switch technology;
-   In principle all the Net topologies are realizable via different media;
-   Based on an established transport protocol TCP/IP, which is equipped with a high availability of components for connection to classic information technology and as a result access to the entire network with low costs for the whole system.

The core of switch technology is a unit for the distribution of an Ethernet network into the various sub-networks. Switching has two main advantages: the possibility of application oriented scaling of collision areas (up to collision free Ethernet) and the extremely fast packet exchange between the collision areas. Figure 3 illustrates the connection between communication paths of a switched network with redundant communication paths. A switch switches through the (data) packet that arrives at an input port, due to a destination address being contained within the packet, unchanged to the correct output port. The switch analyses the packet on arrival, decides, according to a saved address table, to which output port the packet appertains and despatches it as soon as possible, this means the data packets are thus no longer released in the whole net and the network is relieved of the load.

An important function of a switch is the collision avoidance through switching a separate collision free channel with the full Ethernet bandwidth between the input and output port. Nonetheless, the switch allows direct data circulation control overlapping from module to module, without the data being led via a bottleneck of a CPU. What is even more important, however: through the consequential usage of switches up to the lowest level, the Ethernet becomes consequently deterministic and real time capable. As a logical result of the consequent decentralized structure, every intelligent module can also work in a standalone way in the net.

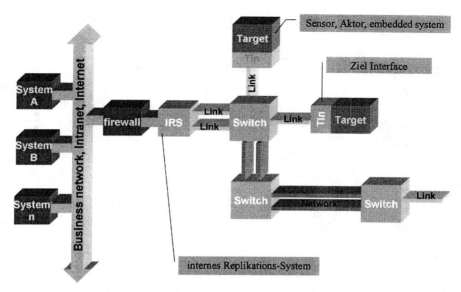

Figure 3: Network topology based on switches

5. OBJECT ORIENTATION AND MODELLING OF THE CONCEPT

A function is only defined once within a system and used for all the same functions within the system (inheritance). If this object orientation of the elements is connected with the modelling of the concept, it is possible to describe the system with information, action and description objects. As a result the simulation of the entire functionalities and the working process chains become reality. The long-distance maintenance and further telematic functions are thus, in principle, plausible from anyone location in the world via the internet with these control concepts .This means a new dimension of services.

The course to implement secured operating models is to supply qualified data and to offer digital services based on internet technology. In this connection SOAP (Simple Object Access Protocol) [SOAP-00; W3C-00] is a possibility to implement the data exchange; platform independent and from business-to-business.

6.   WHICH PROSPECTS DOES SOAP OFFER?

SOAP based on XML (Extensible Mark-up Language); with a high probability, this combination will represent, in future, the construction of internet-capable application solutions. XML is very simple to understand and use, quasi unrestricted in possible use and is one of the few recognised standards in the broad sphere of the industry. XML, as a common platform, is supported by various company consortia, in order to standardize data exchange between systems and different environments.

For a new generation of man-machine interfaces XML together with HTTP is already in use as a base technology. Nevertheless, even an extensively widespread and accepted data format does not alone accomplish the Paradigm change. In order to achieve this, XML requires a strong partner: SOAP as mechanism for the call of functions in computers that are distributed in the internet. SOAP combines the advantages of XML with protocols such as HTTP, SMTP or Message Queues, in order to obtain a new, powerful mechanism that has been especially designed for the internet: the integration of remote functions. Figure 4 portrays the principal elements of a SOAP Envelope.

SOAP is a very simple packaging and routing mechanism for XML data. Figuratively speaking, the XML data are a letter, which obtains its correct Envelope with sender and receiver information through SOAP. The SOAP Envelope (official term of the specification) contains descriptive information regarding data that are to be transported between two end points. Their format, coding as well as references, regarding how and from whom this data is to be processed, are also contained within this Envelope.

SOAP was designed especially for use in the internet with high latency times and frequently available, unreliable availability. This became necessary since the fundamental specifications of XML enable important enlargements for the extensive formulation of complex data connections, the actual transportation of the XML data, however, was only very rudimentary treated.

Traditional RPC protocols break down in the internet scenario, because they deal with defined net connections and unambiguously coordinated configurations of all participants. With HTTP, SOAP employs a transport protocol, onto which the entire internet infrastructure is coordinated. On closer reflection DCOM and IIOP are high propriety protocols of a few suppliers, though server components as a rule are not portable - although partly publicly documented from the OMG. SOAP on the other hand is simple and requires a small implementation expenditure.

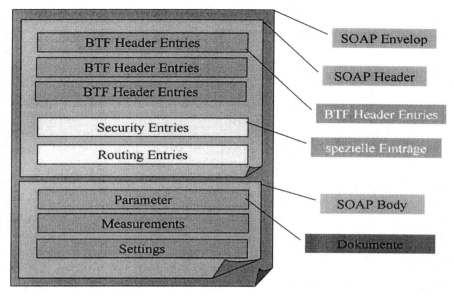

Figure 4: Construction of a SOAP message

7.   WHY XML AND NOT BINARY CODING?

The application of XML, in contrast to binary data, produces comparatively larger data sequences of tags, name space declarations and additional attributes. Consequently the processing speed does not seem to be suitable for employment in automation assignments. This disadvantage with reference to bandwidth requirements and resource consumption for the "packaging" of data in XML in contrast to binary protocols turns out, however, to be an advantage in heterogeneous system landscapes. Homogeneous middleware that is available on any platforms is currently only available on the basis of "virtual machines" and thus has a system inherent speed disadvantage. If SOAP is implemented on the basis of XML in a system-related way, then a decrease in the processing speed does not necessarily occur.

The same situation is also detectable when looking at the necessary bandwidths for data transfer. In addition to the bandwidth uses, the "work-loads" are also to be considered via the actual protocol. An URI (Uniform Resource Identifier) for the clear description of a SOAP end point with HTTP is mostly only a few bytes long. Binary protocols require the size of a standard frame to establish the connection and to identify the end point, even with the smallest data packages, and as a result load the network with 512 bytes or rather distinctly more; with that, a simple SOAP-Envelope is already completely described.

8.   SECURITY

If one views the SOAP specification 1.1 in an isolated way , one is initially attempted to reject the use of SOAP in every security-sensitive environment and thus, decline connection of the automation systems to the internet, because of missing security mechanisms. SOAP is not, however, responsible for the implementation of the security aspects. Several security mechanisms can be implemented in connection with SOAP, for example a password supported authentication or with HTTPS (SSL) a client-sided certification for the secure identification combined with the encoding of the SOAP packages [http-00].

A solution that is independent of the transport protocol, for encoding, authentication and signature of SOAP contents, can occur through the integration of additional header entries for security and routing into the SOAP Envelopes. These can either be directly inserted through the application, evaluated and tested at the opposite end, or these header entries can be implemented by suitable Switches/Firewalls that function as SOAP distributors (see also figure 4). Such implementations are legal in accordance with the enlargements planned in the SOAP specifications and very effective for specific enlargements of SOAP for the automation technology.

9.   RESULT

Once the net concepts and structures correspond to the modern requirements, then future operating models with suitable monitoring functions can be supported. It is necessary, that:
- The system, not the control system, is considered as the logical unit.
- A high flexibility through consequent use of TCP/IP and HTTP as a transportation and transfer protocol on the basis of Ethernet as net technology can be guaranteed.
- The communication of individual control units, sensors and actuators in a network based on switches with a quasi deterministic behaviour is allowed, whereby
- SOAP objects on the basis of XML guarantee the interoperability and
- Continuity from the sensor to internet and as a result enable a data continuity from the production to the EDP office without any overheads, this means that

- a sufficient compatibility and interoperability between the control IT and the office IT is achieved.
- Through internet-similar structures in the system, no additional programming costs for data communication and data synchronisation arise and secure, switched network hierarchies merge to a logical net layer, as a result the bottlenecks SPS-CPUs are cancelled.

10.   SUMMARY

With new system standards, instruments for control concepts are available that fulfil the complex requirements of an operating model:

Independent networking of sensors and actuators for a monitoring of the loads and for a deduction of maintenance and prognosis models for modular designed production systems in an independent way from the control-function.

11.   REFERENCES:

[KOE-00]   König, C.; Hirschbach, O.: Frühzeitig den Vorsprung sichern. In: Automobil-Produktion, Februar 2000, S. 48 – 50

[GAB-94]   Gabler, Th.: Gabler Wirtschaftslexikon; Gabler Verlag, 1994

[GER-99]   Gerhardt, A., Kurth, J.: Zum Engineering der Systemkosten im Maschinen-und Anlagenbau. In: ZWF 93 (1998) 9, Carl Hanser Verlag, München

[LUC-97]   Luczak, H.; Eversheim, W.: Entwicklung einer Vorgehensweise zur Planung eines potentialorientierten Dienstleistungsprogramms für kleine und mittelständische Unternehmen des Maschinen- und Anlagenbaus; Abschlussbericht des AiF-Forschungsvorhabens Nr. 10329, Aachen, 1997

[STE-87]   Steinhilper, R.: Produktrecycling im Maschinenbau
Springer Verlag Berlin Heidelberg, 1987; Dissertation Universität Stuttgart

[LUN-96]   Lund, R.T.: The Remanufacturing Industry: Hidden Giant; Boston University, Boston, MA, 1996

[SCH-00]   Schmälzle, A.: Bewertungssystem für die Generalüberholung von Montageanlagen - Ein Beitrag zur wirtschaftlichen Gestaltung geschlossener Facility-Management-Systeme im Anlagenbau; Dissertation, Universität Karlsruhe, 2000

[http-00]   http Authentification: Basic and Digest Authentification – RFC 2617 HTTP 1.1 (siehe http://sunsite.auc.dk/RFC/rfc/rfc2617.html)

[SOAP-00]   SOAP Microsoft Corp.; http://msdn.microsoft.com/soap/default.asp

[W3C-00]   W3C WEB Seite SOAP http://www.w3.org/TR/SOAP/

[KOR-99]   Koren, Y.; et al: Reconfigurable Manufacturing Systems. In: Annals of the CIRP Vol. 48/2/1999

[GMD-00]   Natur und Robotik GMD Spiegel ¾ 2000, S. 31 ff.

Solution of FMS Scheduling Problems Using the Hybrid Dynamical Approach

J. Somló, A. Savkin, V. Lukanyin

Abstract: Recently, one of the most promising approaches for the solution of FMS scheduling problems is the use of switching servers policies, or in other name the use of hybrid dynamical approach. In the literature the stability problem of the processes was formulated, different switching policies were analysed. Single-machine and multi-machine processing problems were drowning up.

In the present paper, new proposal for the determination of the <u>demand rates</u>, necessary for new approach, is given. Also, a novel proposal to use of the single-machine results for multi-machine case is given. This method is named: <u>The controlled buffer technique</u>.

Key words: Hybrid Dynamical Approach, FMS Scheduling, Demand Rate, Overlapping Production, Controlled Buffers.

1.   INTRODUCTION

By the paper Perkins, Kumar [3], " Stable Distributed, Real Time Scheduling of Flexible Manufacturing Assembly, Disassembly Systems", published in 1989, a new direction of manufacturing scheduling, the hybrid dynamical approach has started. In 1993 Chase, Serrano and Ramage [4] formulated the Switched Arrival and Switched Server Problems. Showing, that periodic trajectories are characteristic for the behaviour of the second and chaos for the first.

In the book Matveev, Savkin [5], " Qualitative Theory of Hybrid Dynamical Systems" research result for hybrid systems, that is, for system where network of digital and analogue devices figure, or a digital device interact with continuous environment are presented. The theory of differential automatons is used.

The given topic is very popular in literature and several hundred of papers, dealing with different aspects, were published.

In the present paper the demand rate determination and the multiple machine processing will be analysed and new proposals will be given.

2. HYBRID DYNAMICAL APPROACH

2.1 Production of High Numbers of parts (HNP)

Flexible Manufacturing Systems are dedicated to process one of a kind or small series in automated way. Nevertheless, some FMS-es are suitable to manufacture high number of parts (HNP). On high number we understand several hundred, several thousands or more parts (sometime less). Of course, the character of manufacturing depends not only on the number but on "size" of parts, too (the meaning of "size" here is the volume, the complication, etc.).

2.2 The Task and Data Base

The task of scheduling for an FMS is usually given by the <u>MRP subsystem of PPS</u>. (Formally, we use partly, the notation of French [1]):

One has:

n <u>jobs</u> ($J_1, J_2,..., J_n$), to be processed on m <u>machines</u> ($M_1, M_2,..., M_m$). On <u>machines</u> we understand an equivalent group of machines, which can be in practice one machine, too. Sometimes instead of word machine we use <u>server</u>. For all of the jobs the number of parts n_i the due date dd_i , the release date r_i (r_i- is the time at which J_i becomes available for processing) is given. T_{sch} is the scheduling period.

The <u>CAPP</u> subsystem determines for every job J_i (a job J_i is n_i parts of type i) the operation sequence o_{ij} ,where $j=1,2,..., jl$ (jl- is the number of operation sequences), o_{ij} –is an integer showing in which machine the given operation is performed. Very important information are the <u>processing times</u> q_{ij} . The processing times are determined by the operation planning subsystem of CAPP including the manufacturing data determination section. We note that the engineering database for scheduling is generated from CAPP results. But, it needs proper modification. In engineering database for

scheduling it is convenient to apply $\tau_{ij} = \dfrac{q_{ij}}{n_h}$, where n_h is the number of machines in the equivalent group. Sometimes it is convenient to use: $\tau_{ih} = \tau_{ij}$, or $\tau_{ij}^h = \tau_{ij}$. Where h is the index of the machine, on which the j-th operation is realized for the i-th part type. In the following always the most convenient form will be used hoping that it will not cause misunderstanding.

In Perkins, Kumar [3] (unlike French [1]) revisiting of the same machine is also considered. In this case $\tau_{ih}^{(k)} = \tau_{ih}$ is used at the k-th visit to the same machine.

2.3   Hybrid Dynamical Approach to the Solution of FMS Scheduling Problems

Perkins, Kumar [3] proposed to use the hybrid dynamical approach to solve FMS Scheduling Problems covering assembling and disassembling tasks as well. In the present paragraph, only the job-shop type scheduling problem will be discussed. The development to rather complicated task (assembling disassembling) is possible, but not discussed here.

When using the hybrid dynamical approach the parts are delivered to the machine groups with some constant demand rate d_i , i=1,2,.....I. Let $u_{ij}(t)$, $y_{ij}(t)$, $z_{ij}(t)$ the input, output and buffer content, respectively.

The hybrid dynamical approach, usually, is based on some switching policy. Let us consider one of the simplest (nevertheless effective) switching policy the so-called Clear the Largest Buffer Level Policy (CLB). According to that, always that buffer will be cleared (served) which has the largest buffer content when some other buffer has been emptied (see: Fig.1.). Let us consider a process beginning with point A where the buffer i was emptied ($z_{ij}(t_A) = u_{ij}(t_A) - y_{ij}(t_A) = 0$).

After t_A the server was occupied with some other (not i) process. Let us suppose that at t_{s1} one of the other buffers was emptied (for example: the one of the belonging to k-th flow). Let us also suppose that i had the largest buffer level. Then, after some inactivity period (due to set-up, transportation, and manipulation etc. times) the server begins to remove fluid from the i-th tank. As it can be recognized from Fig.1. in points A, B, C... the number of produced parts exactly coincide with the required number.

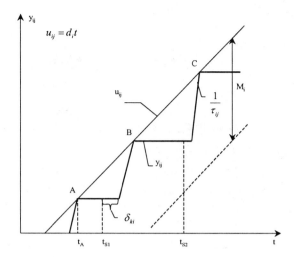

Figure 1. Clear the Largest Buffer Level Policy

Let us remark, that for manufacturing application only the time instants where z_{ij} and y_{ij} has integer value are meaningful.

2.3.1   Stability of Processes

In Perkins, Kumar [3] the definition of the stability of switching policies for a single machine (machine working in isolation) were given. According to that a switching policy is stable if

$$\sup_{t} z_i(t) \le M_i < \infty, \tag{1}$$

for $i = 1,2,...n$, and for any t

That is, the buffers levels at the machine are bounded.

Considering Fig.1. it means that a line parallel to u_{ij} can be drawn, so that y_{ij} never goes below it. The smaller M_i is the better the output characteristics of the system are. The better the output follows the required input.

In Perkins, Kumar [3] results for the estimation of the buffer bounds were given.

2.3.2   Necessary Condition for Stability

The necessary condition of the stability for a single machine is

$$\rho_j = \sum_{i=1}^{n} \tau_{ij} d_i < 1, \tag{2}$$

j - is the index of the machine, ρ_j - is the so-called machine load.

Indeed, let during the period t_j on the machine n_1, n_2..., n_i,... n_n part should be processed. The machine capacity should be higher, than the capacities needed for manufacturing the parts. That is

$$\sum_{i=1}^{n} \tau_{ij} n_i < t_j \tag{3}$$

$$\rho_j = \sum_{i=1}^{n} \tau_{ij} \frac{n_i}{t_j} = \sum_{i=1}^{n} \tau_{ij} d_i < 1 \tag{4}$$

If one chooses $d_i = \dfrac{n_i}{t_i}$ condition (2) is obtained, ρ_j -is the time needed

for processing the part. It can be recognized that the time $(1 - \rho_j)$ is the reserve during which set-up, transportation, manipulation, etc., tasks can be solved.

When a switching policy is applied to clear a buffer (a tank) it is easy to compute the switching time instants. Indeed, let us suppose that at the time instant t_{sk} a switch to serving part type i occurs, then

$$t_{s(k+1)} = t_{sk} + \delta_{ki} + z(t_{sk})\tau_{ij} + (t_{s(k+1)} - t_{sk})d_{ij}\tau_{ij} \tag{5}$$

From equation (5) one gets

$$t_{s(k+1)} = t_{sk} + [1 - d_{ij}\tau_{ij}]^{-1}[\delta_{ki} + z_i(t_{sk})\tau_{ij}] \tag{6}$$

It is also clear from Fig.1., that the following condition is valid

$$d_{ij} < 1/\tau_{ij} \tag{7}$$

In Perkins, Kumar [3] a rather general switching policy then the Clear the Largest Buffer Level was analyzed. Results for Clear the Fraction (CAF) and results for Clear the Largest Work (CLW) switching policies were given, too.

In Matveev, Savkin [5], cyclic switching policies and special kind of Clear the Largest Buffer Level were analyzed.

Until now we considered the processes of the system in isolation. That is we did not consider that the inputs (e.g. d_{ij}) of one level can be the output of the other levels. When the levels are in isolation the single machine approach can be used. But, when these are not in isolation it cannot. Indeed, if we consider Fig.1., and suppose $u_{i(j-1)}$ and $y_{i(j-1)}$, values as are given on the graphics, even if there will be an ideal input ($u_{i(j-1)} = d_{i(j-1)}t$) the output $y_{i(j-1)}$ will highly different from the ideal input u_{ij} for the next machine. If $y_{i(j-1)}$ will form the input u_{ij} dynamic instability (see: [3]) of the system can occur. Perkins, Kumar [3] could construct example for that. In the above paper the authors proposed a technique, the so-called CAF policy with Backoff, which can result stable performance. Nevertheless, because this leaves machines idle at some conditions the effectiveness of this policy is doubtful.

In the 4-th paragraph of the present paper the so-called controlled buffer technique is proposed which can help to overcome this difficulty.

3.   HYBRID DYNAMICAL APPROACH TO THE SOLUTION OF FMS SCHEDULING PROBLEMS

3.1   Demand Rate for Single Machine Processing

Now, let us try to determine the demand rates d_i; $i = 1,2...,n$, for single machine processing. It is clear that the upper limit value of demand rate is

$$\overline{\lim} d_i = 1/\tau_{ij}, \; i = 1,2,\ldots n, \tag{8}$$

j –is the machine group identification number. This is due to the machining intensity constraint. As it was mentioned τ_{ij} is determined on CAPP subsystem level. In practice

$$d_i \le (1/\tau_{ij}) - (\Delta d_{up})_i \tag{9}$$

should be used, where $(\Delta d_{up})_i$ is the upper demand rate reserve value necessary for the realization. The lower limit of the demand rate is

$$\underline{\lim} d_i = n_i /(dd_i - r_i). \; \text{Or} \tag{10}$$

$$\underline{\lim} d_i = n_i / T_{sch}, \tag{11}$$

which represent the production task requirement. These relation give the

$$d_i \ge n_i /(dd_i - r_i) + (\Delta d_{lo})_i \; \text{or} \tag{12}$$

$$d_i \ge n_i / T_{sch} + (\Delta d_{lo})_i, \tag{13}$$

relation, where $(\Delta d_{lo})_i$ is the lower demand rate reserve value. In the followings, for the sake of simplicity, only (11) and (13) will be considered. To generalize the results for the other case is a single task.

Another upper limit is given by the machine capacity constraint. It can be derived from the necessary condition of stability formulated above. Let us consider this relation in form

$$\sum_p n_p \tau_{pj} \le T_{sch}. \tag{14}$$

From that (see also relation (2))

$$\sum_p d_p \tau_{pj} \le 1 \tag{15}$$

Now, let us determine the coefficient η which make (14) and (15) to turn into equality.

$$\eta = T_{sch} / (\sum_p n_p \tau_{pj}) = T_{sch} / t_{\Sigma j}, \tag{16}$$

where $t_{\Sigma j} = \sum_p n_p \tau_{pj}$ is the summary machining time requirement on the machine j.

From the above the other upper limit value is

$$\overline{\lim} d_i = n_i / t_{\Sigma j}. \tag{17}$$

Taking into attention (17) the upper limit should be

$$\overline{\lim} d_i = Min\{(1/\tau_{ij}),(n_i / t_{\Sigma j})\}. \tag{18}$$

So, one gets the domain for demand rate

$$\frac{n_i}{T_{sch}} + (\Delta d_{lo})_i \le d_i \le Min\{(1/\tau_{ij}),(n_i / t_{\Sigma j})\} - (\Delta d_{lo})_i. \tag{19}$$

Let us notice that the increase of d_i is desirable. But, the necessary set-up, etc., times requirements constraint this increase. (This should be reflected in the value of- $(\Delta d_{up})_i$, too.) It seems to us that the best aids for final decisions are the simulation studies.

Now, let us turn to the case of multi-machine job-shop type manufacturing.

3.2   Demand Rate for Multi-Machine Processing. Overlapping Production

In the 1-st paragraph of the present paper, formulation of the classical problems of scheduling was given. Let us outline the basic features of the solution. The beginning of the first operation is at $t_{i1} \geq r_i$. The end of the last operation is at $t_{il} \leq dd_i$. At the input and output of the machines, buffers of size n_i are needed. At flexible manufacturing, overlapping production of parts is possible. This is due to the fact that buffers, transport and manipulation devices are automated. Set-up, transport, manipulation times are low. The overlapping production makes possible to decrease significantly the production times. In more details see: Somlo [7].

Now, let us deal with the determination of the demand rates for multi-machine processing. It is clear that the demand rate should satisfy the following upper limit

$$(d_i)_j < 1 / \tau_{ij}, \text{ for all } j, \tag{20}$$

where j indicates the machines part-type i is processed on, and $(d_i)_j$ is the demand rate constrained by the given machine conditions.

It is supposed that the rate is constant for some part-type, independently of the machine it is processed on. So, from (20) one gets

$$d_i < 1 / (\underset{j}{Max}\, \tau_{ij}). \tag{21}$$

Similarly as in the single-machine case the lower limit of demand rate is determined by

$$\underline{\lim} d_i = n_i / T_{sch}. \tag{22}$$

Let us deal with the determination of the upper limit value given by the total time necessary for the production of parts.

The upper limit value of the demand rates can be determined based on the overall time necessary for overlapped production (see: Somlo [7]). It can be proved that this time

$$t^a_{\Sigma l} \approx n_i \underset{j}{Max}\,\tau_{ij}. \tag{23}$$

So,

$$\overline{\lim}d_i = (n_i\,/\,\tau^a_{\Sigma l}) \tag{24}$$

Let us simplify the above result. It is easy to recognize that

$$n_i \underset{j}{Max}\,\tau_{ij} < t^a_{\Sigma l} \le (n_i - 1)\underset{j}{Max}\,\tau_{ij} + \sum_{j=1}^{l}\tau_{ij} \tag{25}$$

and so, for the demand rate one gets the upper limit given by relation (21). That is

$$d_i < 1\,/\,\underset{j}{Max}\ \tau_{ij} \tag{26}$$

The important relation (26) reflects the fact, that the overall machining time for overlapping production gives the same upper limit value for demand rate as the machining intensity constraint.

Until now, the set-up, etc. times effects were neglected. For taking into attention these, reserves should be introduced. For those, decrease of demand rates from the upper limit value is necessary. To produce n_i parts of type J_i MRP allows to use $(dd_i - r_i)$ or in analyzed case $T_{s\,ch}$ time period. It is usually much higher than $t^a_{\Sigma l}$ at overlapped production. Of course, it should be kept in mind that in these cases several HNP series are processed in parallel.

For these cases the necessary condition for stability (2) should also be satisfied. To provide it let us determine the upper limit by

$$\overline{\lim}\,d_i = (1\,/\,\underset{j}{Max}\ \tau_{ij})(t_{\Sigma}\,/\,T_{s\,ch})\eta \tag{27}$$

Here η is properly chosen coefficient. In relation (27) instead of $t_{\Sigma l}^{a}$ simply t_{Σ} is used.

According to the necessary condition of stability (see relation (2)),

$$\delta_h = \sum_{i=1}^{n} d_i \tau_{ih} \leq 1, \ h = 1,2,..,m.$$ (28)

Because (see (23)),

$$t_{\Sigma} \approx n_i \underset{j}{Max} \tau_{ij}$$ (29)

$$d_i = (n_i / T_{sch})\eta$$ (30)

Can be taken, and from (28)

$$\sum_{i=1}^{n} (n_i / T_{sch}) \tau_{ih} \eta \leq 1, \ h=1,2...,m$$ (31)

is obtained. In (31)

$$\sum_{i=1}^{n} n_i \tau_{ih} = t_h, \ h = 1,2...,m$$ (32)

is the summary working time to perform the production tasks on the machine with index h.

From relation (31) one gets the efficiency coefficient η, which turns the relation into equality for every machine group

$$(\eta)_h = T_{sch}/t_h, \ h=1,2...,m$$ (33)

Accordingly,

$$\eta \leq T_{sch} (1/ \underset{h}{Max} t_h), \ h=1,2...,m.$$ (34)

From (27), (29) and (33)

$$\overline{\lim}d_{\underset{h}{i}} = n_i / Maxt_h \tag{35}$$

Having (21) and (34) the upper limit value of demand rate is

$$\overline{\lim}d_i = Min\left\{ (1/Max\tau_{ij}), (n_i / Maxt_h) \atop j \qquad\qquad h \right\} \tag{36}$$

The lower limit value is given by (22). So

$$\frac{n_i}{T_{sch}} + (\Delta d_{lo})_i \le d_i \le Min\left\{ \frac{1}{Max\tau_{ij}}, \frac{n_i}{Maxt_h} \atop j \qquad\quad h \right\} - (\Delta d_{up})_i \tag{37}$$

Now, let us consider the demand rates for the consecutive machines. The first machine gets a d_i ramp-input signal. The consecutive machines get the same with τ_{ij} time delay as it is shown on Fig.2.

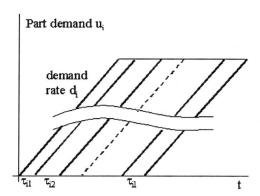

Figure 2. Demand rates at consecutive machine group

It is an interesting opportunity to use the secondary optimization method (see: Somló J., Nagy J. [2])) to get time gain in FMS scheduling (in more details see: Somló J. [7])

4.   THE CONTROLLED BUFFER TECHNIQUE

As it was mentioned, Perkins, Kumar [3] proposed to use distributed CAF (in particular Clear- the- Largest Buffer Level) policy with Backoff to provide stability of processes at general job shop type FMS scheduling. Matveev, Savkin [5] analyzed multiple server flow networks which also cover this field. They proposed to construct regularizable switched multiple server flow networks which exhibit a regular behavior.

Perkins, Humes, Kumar [6] proposed to use the so-called regulated buffer stabilizing techniques to reach stable performance for FMS scheduling problems.

Here we propose a different technique to be able <u>to use the results of hybrid dynamical approach for single machines in the case of multiple-machine, job shop type FMS scheduling.</u>

The following, so called controlled buffer technique is proposed here. Let us introduce into the system structure the, so called auxiliary buffers, as it is shown on Figure 3. These, together with the input buffers are controlled.<u>The controlled buffers act as follows.</u> At the input buffer of j-th machine always the ideal input (d_i demand rate) and the corresponding number of parts already machined on $(j-1)$ machine should be introduced. If the output buffer of machine $(j-1)$ is unable to cover demands then the auxiliary buffer comes to its place and delivers the proper number of parts (not finished production). The proper content of the auxiliary buffer is provided before the regular system actions. (It is also possible to include service sections during regular actions to fulfill this task.) There is always surplus which goes to fill back the auxiliary buffer. When the part type i production on machine $(j-1)$ begins the d_i demand rate on machine j is covered by this production. It can be recognized that the same number of parts that was taken of the auxiliary buffer, later, is returned from the output of the $(j-1)$-th machine. Indeed, because in an "activity" period the $(j-1)$ machine produces exactly the number of parts between two empty buffer position it will return the number of parts which were taken from the auxiliary buffer. In more details see: Somlo [7].

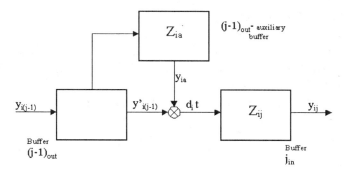

Figure 3. Controlled buffer technique

5.   CONCLUSIONS

The idea of hybrid dynamical approach to the solution of FMS scheduling problem give an opportunity to increase significantly the effectiveness of utilization of the equipment of these high value systems. So, the economic effect of the use of this may be very high.

The theoretical results obtained, recently concerning hybrid dynamical systems give strong basis for the solution of problems. Perkins, Kumar [3], Matveev, Savkin [5] can be considered as basic works.

In the present paper an attempt is made to clear the relation of classical scheduling problems formulation and hybrid dynamical approach. It has been shown that the essence of the effect of the last is the opportunity of overlapping production. The use of hybrid dynamical approach to a single-machine problem is well established with nice results. Not so clear is the situation when dealing with job-shop type multi-machine problems in FMS-es (with possible assembling and disassembling processes included). In the present paper the buffer control is proposed as a solution to reduce the problems to the use of the single-machine case.

The modern simulation methods (Taylor, Simple ++, etc.: and as well the continuos system simulation languages Matlab, Matrix-x, etc.) give excellent opportunity to analyze the application problems details concerning the use of hybrid- dynamical approach. These studies can lead to new results and better understanding of the problems.

ACKNOWLEDGEMENT

The authors express thanks to OTKA of the Hungarian Academy of Sciences (OTKA T 026407 and 0229072) and to the Australian Research Counsil for the support of the research.

Special thanks for the advices and exceptionally high help to Dr. F. Erdélyi.

REFERENCES

[1] French S. "Sequencing and Scheduling: An Introduction too the Mathematics of Job-Shop," Wiley. 1982, pp.245.

[2] Somló J., Nagy J. "A New Approach to Cutting Data Optimization," PROLAMAT'76 Stirling, Scotland (See also: North Holland 1976)

[3] Perkins J.R., Kumar P.R. "Stable, Distributed, Real-Time Scheduling of Flexible Manufacturing (Assembly) Disassembly Systems," IEEE Transactions on Automatic Control. Vol. 34, No:2. February 1989, pp. 139-148.

[4] Chase C., Serano J., Ramadge P.J. "Periodicity and Chaos from Switched Flow Systems: Contrasting Examples of Discretely Controlled Continuous Systems. IEE Transactions on Automatic Control, Vol. 38, No 1, January 1993, pp. 70-83.

[5] Matveev A.S., Savkin A.V., "Qualitative Theory of Hybrid Dynamical Systems," Birkhauser 2000, pp. 348.

[6] Perkins J.R., Humes C., Kumar P.R., "Distributed Scheduling of Flexible Manufacturing Systems: Stability and Performance," IEEE Transactions on Robotics and Automation, Vol.10, No: 2, April 1994.

[7] Watanabe T., "Intelligent Scheduling of FMS: The Optimization Approach," Japan/USA Symp. On Flexible Automation, Vol. 1, pp. 447-454, San Francisco 1992.

[8] Watanabe T., Sakamoto M., "On-line Scheduling for Adaptive Control Machine Tools in FMS," IFAC 9th World Congress, Budapest 1984.

[9] Somló J., "Hybrid Dynamical Approach Makes FMS Scheduling More Effective", to be published Periodica Politechnika Budapest University of Technology and Economics.

Simulation of Robot Arm Actions Realized by Using Discrete Event Simulation Method in LabVIEW

A. Anoufriev and Gy. Lipovszki
avanv007@hotmail.com; lipovszki@rit.bme.hu
Budapest University of Technology and Economics
Budapest 1111 Goldman Gy. tér 3.

Abstract: In the present paper a model for Robot Arm simulation is proposed. This model was developed using a new graphically programmable discreet event simulation system. This system was written in LabVIEW language and consists of several basic elements. Using these elements it is possible to construct new elements such as, for example, robot arm.

Key words: discrete event simulation, scheduling, robotics

1.   INTRODUCTION

Choosing a modeling approach instead of conducting real experiments does not automatically mean a choice for simulation on the contrary. When one is able to build a mathematical model, then this must be given preference. It is possible to realize only if the relationships that compose the model are simple enough. It may use mathematical methods (such as algebra, calculus, or probability theory) to obtain exact information on questions of interests; this is called *analytic solution*.

Most of real world systems are too complex to allow realistic models to be evaluated analytically [7]. So these models must be studied by simulation.

In simulation, one uses computers to evaluate the model performance numerically, and data are gathered in order to estimate the desired behavior of the model.

Reasons for choosing to work with models rather than experimenting are usually based on cost aspects. Other reasons also come to mind, for example safety or the timespan of an experiment. Simulation should be used in those cases also when an experiment in reality is impossible or undesirable to perform.

Numerous discreet event simulation systems are existing in the world. These have advantages and disadvantages. In our frame system we tried to reduce the disadvantages of simulation environment.

1) The advantage of the present simulation system is a very high level of man-computer interactivity. Programming with graphical picture assigned icons and studying the result with the best suitable diagram type here is basic request. This programming style gives possibility to test very complex model situations that in text oriented programming is very hard or complicate.

2) The most of existing discrete event simulation systems use their own programming language for constructing models. It is increasing the time of study period but later on drastically reduce the development time of new models.

3) Finally, the main aim of this development was to construct a "relatively cheap" discrete event simulation system, for education purposes.

The Robot Arm model proposed in the present paper is connected with PUMA 560 robot. Detailed information about this robot (among other publications) may be found in J.Somlo, B.Lantos, P.T.Cat [2]. Real-time realization of robot control laws is possible using the OSA (Open System Architecture) robot control described in [5]. But all the mentioned may be used for other robots, too.

2. Description of the system

Let us formulate the task as follows. Develop the model of the PUMA 560 robot arm, which service three production lines. The robot takes a work piece from the Buffer on input and load machine tool on the output. The time for empty and charge motions is different and programmable.

Let us to describe a modeling environment of the Robot Arm model. A simple graphical interpretation is given on Fig.1. The model of the production system contains a number of model objects. Inputs of the production system are Source objects and outputs are Sink objects. Any number of objects can be between Source and Buffer as well as between Machine and Sink objects. Let us to describe some of the objects which are used in the system.

The **Source object** produces new entities with distribution functions given in interval. This subroutine is shown on Fig. 2. We can see the input parameters on left side of the picture and output parameters on the right side. Let us describe some of input and output parameters.

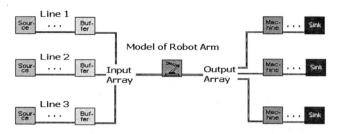

Fig. 1. Production System.

The "Container Object Name" and "Source Object Name" define given object relatively another elements of model. The "Source Object Name" is Name, which we give to this object and the "Container Object Name" is the so called, parent name.

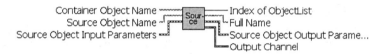

Fig. 2. Source object.

For example, if a factory has several workshops and every workshop has several production lines the "Full Name" of an element will be " 'Container Object Name' & 'Source Object Name'". The "Index of ObjectList" is an identification number of this object in a list of objects, which are used in the system. The "Output Channel" is output that entities go out through. The "Source Object Output Parameters" is output, which provides the user of the model with full information about processes inside subroutine. The "Source Object Input Parameters" are tuning up in work of the object. It is possible to program interval time between appearances of entities, the first creation time, etc.

The work of the **Sink object** is opposite to the work of the Source object. This element takes out entities from the system. This object is shown on Fig. 3.

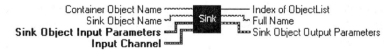

Fig. 3. Sink object.

The functions of the "Container Object Name", "Sink Object Name", "Index of ObjectList", "Full Name" and "Sink Object Output Parameters" are similar to functions with same name in Source object. These inputs and outputs have the same functionality in every object.

The "Input Channel" is the entry point of this object where the entities are going in. The "Sink Object Input Parameters" are used for switch on/off this object in simulation.

The **Buffer object** collects entities as if a real buffer in production system collects work pieces. It is possible to program a Buffer "Capacity" as property of the "Buffer Object Input Parameters". The Buffer object is shown on Fig. 4.

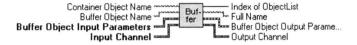

Fig. 4. Buffer object.

The **Machine object** is shown on Fig. 5. The main task of this object is to delay the motion of entity with calculated processing time. The calculation is performed by using stochastic distributions described above. The kind and parameters of distribution are input values of this object. The object calculates the next time of launching as an output parameter. More about probability distribution method are given bellow.

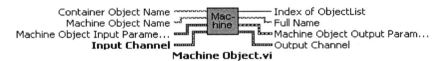

Fig. 5. Machine object.

In simulation, we use probability distribution functions to describe the stochastic character of system elements. Examples of stochastic quantities are: arrival of work piece in a system, length of the processes, breakdown behavior of a machine etc. .

There are two kinds of distributions existing: empirical and theoretical. If it is possible we use theoretical distributions. That is because there are usually not available historic data, and the theoretical distribution gives better results and easier to use. In this frame system five types of distribution functions were used: ***constant, exponential, normal, triangle and uniform distribution.***

The program of object calculates processing time or time of appearing new entity accordingly with one of the probability distributions used in the system. The advancing of the time of the process is not a trivial question in modeling (discrete event simulation).

Discrete simulator programs are using the "time slicing" or the "next event" technique for advancing the simulation clock.

Time slicing means that time passes in constant steps with other words this is a sampled time system. At each point of time, it is checked whether an event was overtaken or not. If the event was overtaken the system state is recalculated. A disadvantage of this method is that it is often slow; because it is evaluating every time step for the full system whether the system change or not. The next event technique does not have this disadvantage. The clock jumps from event to event.

Let us to explain advantages and disadvantages of these methods in an example [see 4]. The next event technique makes use of the fact that, in discrete processes, the state of the system changes erratically. If a product is processed at time 10 seconds, and the process takes 7.3 seconds, then we know the process will be finished at time 17.3 seconds. The next event simulator will determine at time 10 seconds (of simulation time) that the next event will take place at time 17.3 and (assuming that no events occur before that time) the clock will then jump from 10 to 17.3.

With the time slicing technique, time will pass according to fixed intervals and the simulation algorithm will continually ask: process ready?

For example:

t=11: Process is ready? Answer: No.

t=12: Process is ready? Answer: No.

....

t=17: Process is ready? Answer: No.

t=18: Process is ready? Answer: Yes.

One can see that this method requests more calculation, and we can also see the size of the time steps determines the accuracy of the simulation. If one wants to simulate the processing time more accurately, than he should have taken steps of 0.1 seconds: i.e. even more calculation covers the period of 7.3 seconds. The advantage of the time slicing technique is the easier program organization.

In our frame system (simulator) the "next event" technique was used.

3. Robot Arm model

A new object for solving the given task was developed. This new object is the **Robot Arm**. It does not exist as a basic object, so we developed one with given conditions. The Robot Arm (in future RA) model can open/close individually all its inputs/outputs. Having an input and an output strategy RA can get objects from the chosen input channel and is able to send it to a selected output channel. This task could not be solved by using the existing basic objects. So, we had to build their network and develop some new elements. This **Robot Arm** object can be used like another basic objects as base for constructing new objects. This program is shown on Fig. 6 .

Two inputs "Status of Next Line" (machine states connected to Robot Arm output) and "Contents of Previous Line" (content of Buffers connected to Robot Arm input) are used for choosing of input and output indexes.

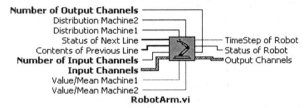

Fig. 6. Robot object.

The structure shown on Fig. 7 was used in **Robot Arm** object. This model contains a **Join**, a **Selector** and two **Machine** subroutines. Short descriptions of these objects is given bellow.

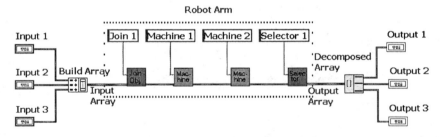

Fig. 7. Model of the robot.

The Machine1 calculate time for moving without work piece and the Machine2 for moving with work piece in a gripper. These calculations fulfilled by using stochastic distribution methods.

Let us describe the main objects of the model. The aim of **Join** is choosing from entities of inputs according with control policy described below. Join as all another basic objects is subprogram with input and output parameters (see Fig. 8).

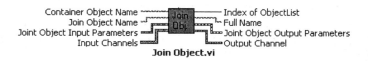

Fig. 8. Join object.

One can see the input parameters on the left side of the picture and output parameters on the right side of it. It is possible to switch off /on this subroutine from simulation.

Let us describe some of the input and output parameters. The main parameter is the "Input Channels". The entities go into Join object through this input. Every entity brings information about its number in object list and about its current channel. The "Output Channel" contains the entity from selected input channel. The input parameters of Join are "Capacity" (number of entities inside the Join object), the "Number of Input Channels" (dimension of input array) and the "Index of Input Channel". The Join object output parameters indicate the content, work and stage of this object during a simulation. Every object has its individual name. It gives us the possibility to use the same subroutine many times in the simulation model. The "Container Object Name" is a so-called parent name of object. The "Full Name" includes name of container and name of object. The "Index of Object List" is the identification number of object in list of objects used in the system. The work of **Selector object** is opposite to work of Joint. Selector takes the input and inserts it in output array accordingly with output control policy described bellow. Selector subroutine is shown on Fig. 9 .

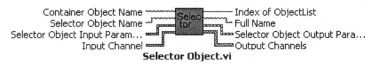

Fig. 9. Selector subroutine.

The input and output parameters are similar to a Join's parameters but it has an array of channels on exit and single channel on the input.

By using of these objects we started to develop the Robot Arm object.

Let us assume that the control policy for choosing of an input is based on *Work from the Buffer with the Largest Queue.*

The largest queue policy work in the model is the next. Let $T_0=0$. At time $t=T_n$ let the robot choose that buffer to work which has the largest current level. If there is more than one buffer content which are not zero we find the maximum content buffer index. If there are more buffer with equal content we choose from sequence of "not empty buffers". After this process the robot picks up the waiting entity and move it to an idle state machine tool. The choosing of machine tool fulfilled accordingly with one output choosing policy described bellow. This process is repeated before the production is stopped.

The following matrix determines the output choosing policy:

$$\begin{pmatrix} a_{11} & \cdots & a_{1j} \\ \vdots & \ddots & \vdots \\ a_{n1} & \cdots & a_{nj} \end{pmatrix}.$$

The matrix has dimension $N \times J$ where N is a number of rows and J is number of columns. In our case $N=J=3$ and n corresponds to input and j corresponds to output. The robot chooses the output channel with non-zero element in the row. The output matrix, for example is hard determined as in a "ProductSelectingTable":

$$\begin{pmatrix} 1 & 0 & 1 \\ 0 & 1 & 0 \\ 0 & 1 & 1 \end{pmatrix}.$$

The program gives possibility to fill up control matrix accordingly with one of the rules specified below:

1. Specific channel: always send to channel j.
2. By percentage: for example 90 % of products go to channel j, the remaining percentage go to channel $j+1$ (Bernoulli distribution).
3. By entity name: if the entity name of the 1st object in the queue matches "EntityName" then send to channel j else $j+1$.
4. By "User Attribute" value (direct): the channel number is written directly on the label named "OutputName" of the 1st entity in the queue.
5. By label value (conditional): if the value on the "User Attribute" of the 1st entity in the queue is less than a given value then send to channel 1, equal then channel 2 else 3.
6. Round robin: all output channels are used in rotation. If channel is closed, then wait till open.
7. Largest queue: Send to the channel connected to the entity with the largest queue.
8. By lookup table: Send to the channel specified in row 1 column 2 of global table named "ProductSelectingTable".

Let us shortly describe one of the twelve output selecting methods enumerated above. Let the method that choose output would be based on Bernoulli distribution (5th in the list of the methods). The essence of the method is that the system send X percent of all entities which arrive on three inputs of the robot to the first output Y percent to the second and 100-(X+Y) to third output. X and Y are programmable values. Graphical interpretation of this method is shown on Fig. 10. The program generates values from 1 to 100 after arriving new entity to the robot's inputs. The probability of appearance every value from 0 to 100 is equivalent.

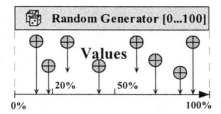

Fig. 10. Bernoulli distribution.

For example, let us send 20% of all entities to the output 1, 30% to the output 2 and leftover 50% to the output 3. It means that if generated value is less or equal 20 the model sends it to output 1, if one is more 20 and less or equal 50 the model sends it to output 2 and if generated value is more 50 the model sends it to output 3.

This strategy can be easily realized on LabView program-ming language (see Fig. 11).

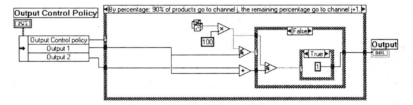

Fig. 11. Realization of Output Choosing Strategy.

The system gives information about the results in the form of Gantt diagram. The Gantt diagram is a XY graph. Time is shown along X-axis and on Y-axis are placed the states of objects in the system. The Gantt diagram gives us real information about stage of object in every time slice. An example of Gantt diagram is given on Fig. 12. Usually, Gantt charts are used in the production management for visualization the solution of scheduling problems.

The numbers along X-axis are time units and colored areas on the chart show periods when objects are busy. The program builds Gantt diagram for the simulated production period. The objects of system placed along Y-axis by the following way:

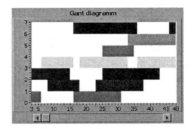

Fig. 12. Example of Gantt chart.

Nominations of Fig 12. can be seen in Table 1.

Table 1. Nomination of GANTT chat.

Position on Y axis	1.1.1.1.1.1.1.1 **Name of Object**
between 0-1	Machine1 (at the input of RobotArm)
between 1-2	Machine2 (at the input of RobotArm)
between 2-3	Machine3 (at the input of RobotArm)
between 3-4	RobotArm
between 4-5	Machine4 (at the output of RobotArm)
between 5-6	Machine5 (at the output of RobotArm)
between 6-7	Machine6 (at the output of RobotArm)

The Machine1 and Machine4 belong to the production line number one. The Machine2 and Machine5 belong to the production line number two and Machine3 and Machine6 are elements of the production line number three.

The scheduling problem formulation and solution can be **static** and **deterministic** [see 6]. When the number of jobs and their ready times are known and fixed our system is static. When processing times and all other parameters are known and fixed the system is deterministic. Our system can solve **dynamic** and **stochastic scheduling** problems. It means that jobs can arrive randomly over a period of time (dynamic) and the processing times, times of appearances of new entities etc. can be uncertain (stochastic).

All these reasons can lead to situations when static and deterministic scheduling solutions which were produced by a manager are quite different from scheduling composed automatically by the program.

4. Planed Future Developments

The constructing of Robot Arm model is the first step of development. We are planning to develop all elements, which are needed for simulation production systems and flexible manufacturing systems. The model of transportation system and Robocar will be developed in the near future.

5. SUMMARY

In this article we introduced a new discrete event development system and its application in developing of new element named Robot Arm. We can use any number of objects as reference of the basic class objects. It is possible to develop new object by using the base objects or other (previously) developed objects. The first tests of the presented system have shown that the development system is working properly and development of a new object is not too hard. The system gives the possibility to build up Gantt diagrams and can be applied for verification of the solution of scheduling problems.

6. References

[1]   Gy. Lipovszki "Diszkrét Esemény Szimulátor LabVIEW programnyelven" Budapesti Műszaki és Gazdaságtudományi Egyetem Rendszer- és Irányítástechnika Tanszék, Budapest, 2000 szeptember.

[2]   J.Somlo, B.Lantos, P.T.Cat "Advanced Robot Control" Akadémiai Kiadó. Budapest 1997.

[3]   Averill M. Law, W. David Kelton "Simulation Modeling and Analysis", Second edition, McGraw-Hill, Inc

[4]   "Simulation. A pragmatic guide to discrete event simulation in business and engineering." F&H Ltd. The Nederland's. 1998

[5]   Somlo Janos, Loginov Alexander, Sokolov Alexei "Optimal Robot Motion Planning
        and Realization Using LabView" INES'99 IEEE International Conference on
        Intelligent Engineering Systems, 1999 November 1-3, Stara Lesna, Slovak Republic.

[6]   S. French "Sequencing and scheduling: An introduction to the Mathematics of the
        Job-Shop" Ellis Horwood Limeted, 1982.

[7]   Ivan L. Ermolov, Philip R. Moore, Jury V. Poduraev "Modeling and visualization for
        mobile robots working in severe environment" IFAC Symposium on Manufacturing,
        Modeling, Management and Control. Patras Greece, July 2000.

Alexandre ANOUFRIEV was born in Moscow, Russia in 1976. He received M.Sc. degree in mechanical engineering from Budapest University
of Technology and Economics (major in Robotics) and Bauman State Technical University (major in MECHATRONICS and control theory), in 1999 and 2000 respectively.

At present moment he is Ph.D. candidate at Budapest University of Technology and Economics. His current research interests are in Scheduling theory, Simulation and Optimization of Manufacturing processes, Production Planning.

György LIPOVSZKI was born in Miskolc, Hungary in 1950. He received M.Sc. degree in electrical engineering from Budapest University of Technology and Economics, at 1975.

He has received his Ph.D. degree at Budapest University of Technology and Economics in 1997. He is associate professor of Department Systems and Control Engineering at Faculty of Mechanical Engineering. His main research and education topic is continuous and discrete event simulation and control

Robotic equipment for deep-sea operation: Digital mock-up and assessment

E. Cavallo, R.C. Michelini, R.M. Molfino and R.P. Razzoli
Industrial Robot Design Research Group - University of Genova

Abstract: The paper presents a robotic arm, carried by a deep-see vehicle and purposely designed to perform precision handling for specialised maintenance tasks. The duties, up now performed by trained divers, are suitably defined, but remains rather compelling, due to the surroundings and with rising risk, as the depth increases. Resort to general-purpose, high-versatility robots happens to be unfair, as for sophistication requirements, and ineffective, as for duty bent; specialisation, the other hand, needs careful operation acknowledgement and proper task training. The considered problem deals with routine operations, and a purposely adapted contrivance is developed: the functional bent suggests to resort to a parallel kinematics manipulator. Design difficulties are the main drawbacks; the proposed solution, then, largely deals with digital mock-up and virtual testing assessment, for initially choosing the duty tailored fixture, thereafter for the life-cycle checks up to extreme deep-sea struggles.

Key Words: Integrated design, Digital mock-up, Instrumental robotics

1. INTRODUCTION

Underwater robotics has a long history, with, possibly, military-driven applications and, in many cases, looking after high-tech implementations, as particularly demanding tasks are sought. The situation is fast evolving, with the broader expansion of routine-applications, required to exploit the natural resources for mankind benefit everywhere they are placed. The preliminary undersea-surveying has generally been accomplished by the direct actions of frogmen or by remote-manipulation supervised from special vehicles by front-end operators. The unfriendly surroundings dramatically bounds the reachable work-areas and the duties that can be put in executions, without a raising in costs, not easily achieving balanced return on investments. This is suggesting a completely new approach in underwater robotics [1], [9], [3], [4] to look after duty-driven contrivances, properly goal-oriented.

The solution effectiveness depends on the ability of transferring well established technologies, so that low-cost devices could be developed to face the sub-set of adverse tasks as these pop out each time new surrounding are brought to work. At the moment, the off-shore structures for oil drawing out represent widely spread plants, requiring life-long inspection, regulation and maintaining operations, most of the time, strictly specialised, so that duty-driven fixtures could be devised to reach reliable efficiency. At dismissal, the wrecked plants become a non-bearable nuisance for the sea-life and the existing national bylaws require the safe and efficient restitution of previous conditions by dismantling and removing the old crocks.

The oil and gas industry is thus becoming an important purchaser of these diving-tools, and robotics founds out a promising application domain, on condition that transfer of the appropriate technology is suitably modulated to face actual tasks. In this paper, example topics on the transfer of parallel kinematics manipulators are recalled. The investigation has already dealt with the main conditioning facts [10], [11] hereafter, the procedure of linking structural architectures and functional properties is presented and appropriate digital mock-ups are developed, to supply virtual testing means, according to integrated design rules.

2.   PARALLEL KINEMATICS MANIPULATION

Parallel kinematics manipulators are very little exploited, even when they might represent the most effective answer for a given requirement, due to the difficulties in trajectory tracking and path planning brought back by solving the backward kinematics. The hindrance is factually removed if proper CAD codes are available, so that the designer directly addresses the conditioning requirements, without a beforehand immersion in the peculiarities of closed-loop mechanisms. In the following, only typical features of this integrated design process are summarised, sending back to the specialised literature for the pertinent details. For explanatory purposes, the study considers a typical six-degrees-of-freedom lay-out, Fig. 1, since it can be taken to represent a standard reference for the development process. The sequence of the topics is organised as follows: first, the qualitative description of the equipment properties is given; the design framework to be used for achieving proper functional performance is, then, presented, with the example mapping of the duty behavioural characteristics. Concluding comments provide further hints worked out from the example development.

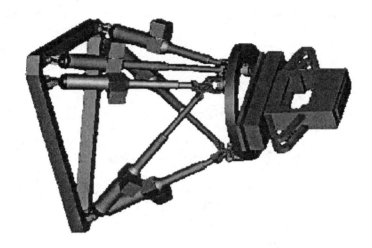

Figure 1. The 6 d.o.f. parallel platform lay-out

2.1 Functional characterisation

High stiffness and low inertial coupling are peculiar features of parallel mechanisms; work-space bounds and path singularities are main drawbacks. Thereafter, high force actuation along a properly bounded trajectory is considered typical work-cycle, to be performed with large acceleration and accuracy even if a simple open-loop position command is used. The example choice considers a six-legs Stewart platform, employed to accomplish the replacement of faulty electrovalves, placed in the control manifold of the X-mas trees laying on the sea bottom of a gas drawing site. The task requests comparatively high forces and very accurate positioning; the correct analysis of the relationship between, on one side, the joint-space velocities and actuators forces and, the other side, the work-space velocities and effector forces (applied by the moving platform) needs reach properly balanced issues.

The investigation is carried on, referring to the forward kinematics of the mechanism; virtual assessments play a relevant role to work out quantitative figures of the performance for the different required engagements. The topic has been investigated in the past, and several authors [7], [6], have described the velocities and the force transmission characteristics by means of varying gain couplings, represented by the local Jacobian matrices, shown as hyper-ellipsoids around every task location of the work-space.

For design purposes, the six-dimensional input-output transformation of the chosen six-legs Stewart platform is split into two couples of sub-spaces, binding linear and angular displacements and, respectively, force and torque transmission. The manipulator behaviour, as a all, is highly non-linear, but the analysis makes possible the calculation of the locally linearised gains, so that the command can have before-hand setting and trimming, once the duty is selected.

2.2   Characterising design framework

The design framework, as previously stated, separately deals, splitting the manipulator Jacobian J in a translation and in a rotation part.

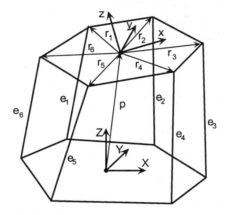

Figure 2. Geometry reference vectors

According to the notation explained in Fig. 2, the 6x6 Jacobian matrix of a Stewart like platform can be expressed as:

$$\mathbf{J} = \begin{bmatrix} \mathbf{r}_1 \times \mathbf{e}_1 & \mathbf{r}_2 \times \mathbf{e}_2 & \cdots & \mathbf{r}_6 \times \mathbf{e}_6 \\ \mathbf{e}_1 & \mathbf{e}_2 & \cdots & \mathbf{e}_6 \end{bmatrix}^T \tag{1}$$

It is well known that the relation between the input actuation and the output actions, force and torque, of the moving platform is locally expressed by the transposed Jacobian:

$$\mathbf{F} = \mathbf{J}^T \mathbf{f} \tag{2}$$

where: the vector $\mathbf{f} = (f_1,...,f_6)^T$ represents the actuation force and the vector $\mathbf{F}=(f_o^T,m_0^T)^T$ represents output force/torque exerted by the moving platform.

In order to simplify the mechanism analysis, is convenient to consider separately the output force and torque; such a formulation is very useful for the mechanism design steps, when the attention is focused on the net output *force* either *torque* components, as, depending on task requests, one has to evaluate the force the robot can exert without generating any torque, or vice versa.

On these premises, the generic expression (2) is decomposed into:

$$\begin{pmatrix} f_0 \\ m_0 \end{pmatrix} = \begin{bmatrix} J_t^T \\ J_r^T \end{bmatrix} f \qquad (3)$$

where: the two 3x6 matrices J_t^T and J_r^T express the relations between the actuation input and the effector *net force* either respectively *net torque*.

Applying the definition of the velocity and force /torque ellipsoids to the two simplified Jacobian matrices [8], we obtain that the output force exerted by the moving platform form an ellipsoid, whose axes lie on the direction of the eigenvectors of the matrix $[J_tJ_t^T]$ and their magnitudes are the square root of the matrix eigenvalues. Similarly, it is possible to express the net torque ellipsoid making use of the $[J_rJ_r^T]$ matrix.

The analysis has to be repeated for every point of the work-space. The robot behaviour is, thereafter, is fully specified evaluating, for all authorised manipulator configurations corresponding to chosen tip position and attitude, the exerted:

force, applied without generating moment (net force ellipsoid); or:

torque, transmitted without generating effort (net torque ellipsoid).

Remembering that the evaluation of the amplification between internal and external velocity is in the same way linked to the inverse of the two above matrices, the inquiry is easily extended to the linear and rotational velocity ellipsoids; moving accordingly, the velocity ellipsoids are evaluated distinguishing linear from rotational displacement missions:

linear local *path* of the platform centroid (net displacement ellipsoid);

angular local *attitude* around the platform centroid (net angular ellipsoid).

In this case, it is useful to remember that, due to the definition, the principal axes of the velocity and of the force ellipsoids coincide, and the length of the axes are in inverse proportions. In fact, the force ellipsoid, defined by $[JJ^T]$, and the velocity ellipsoid, given by $[JJ^T]^{-1}$, have the same eigenvectors and the eigenvalues are in reciprocal proportion.

The reciprocity between velocity and force ellipsoids has an important consequence: the configurations in which the force transmission reaches the maximum value is the worst for the control of the manipulator, when related to the velocity transmission. The final setting of the manipulator, thereafter, can actually be achieved once the work-space analysis is completed and the appropriate parameter (size, geometry, actuation gains ...) balancing ensures the task execution.

This kind of investigation has been accomplished. For example purposes, a customisable Stewart-like 6 d.o.f. parallel platform is considered and the repeated solutions for every configurations are numerically obtained. The set of net force ellipsoids, Fig. 3, of net torque ellipsoids, Fig. 4, of net displacement ellipsoids, Fig. 5, net angular ellipsoids, Fig. 6, show the local manipulator behaviours and allow to foresee the functional characteristics of the mechanism, when characterised on merely input-output actuation gains, either, velocity ratios.

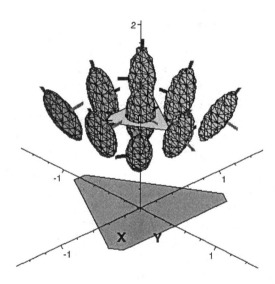

Figure 3. Net force ellipsoids for 9 different workspace locations

The parallel kinematics architectures, as previously recalled, are mainly used for comparatively large (generalised) force transmission, along high-accuracy trajectories; in these situations, they grant built-in propensity and can avoid the addition of sensing elements on the effector and feedback loop, always critical for under-sea applications. The sketched transform matrices description provides an effective aid for tuning the driving force actuation in dependence of the desired work-cycles and with balanced concern with the tracked trajectory accuracy. The actually selected robot comes out removing the fear of addressing *unconventional* mechanical structures, as the useful design information are made available as virtual benchmark .

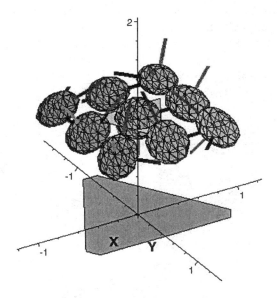

Figure 4. Net torque ellipsoids for 9 different workspace locations

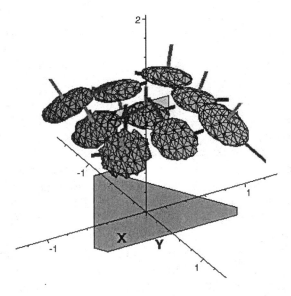

Figure 5. Net linear velocity ellipsoids at the same work-space locations

The sample issues given in the paper aims at providing explanatory hints of the design process in view of recurrently trimming force and displacement transmission. The overall project carries on the investigation by kinetic and dynamic simulations on multi-body models and by FEM analysis, to enrich the knowledge about the machine behaviour. The preliminary investigation ends by setting the robot structural lay-out and by selecting the set of sub-assemblies assuring the sought performance (in this case, for instance, the functional and structural optimisation of the top double spherical and of the bottom universal joints is sought).

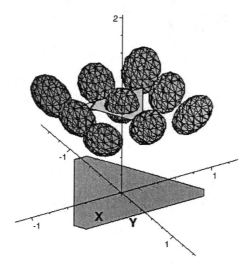

Figure 6. Net rotational velocity ellipsoids at the same work-space locations

The complete set of data on the mechanism structure permits the designer to define, since the first steps of the process, the better choices to achieve the desired performance. For deep-sea applications, effective virtual prototyping [5] and testing bear particular relevance, as the experimentation on real settings is dangerous and expensive. Happily enough, nowadays, the utilisation of software packages available for geometric, structural, cinematic and control simulations gives not only an added value to the quality of products, but also an indispensable contribution to achieve significant life-cycle results, with friendly graphic rendering. This is the way, we intend "virtual reality", that is the description of every peculiar feature of the object by means of the appropriate model and the analysis to extract characteristics and information about the system behaviour under actual running conditions. The opportunity was largely exploited along with the project development, and the principal results where discussed in previous papers; here details on the joint- to the work-space transforms are recalled due to their effects on the diffusion of parallel kinematics mechanisms, even when these happen to be best choice.

3.   CONCLUDING COMMENTS

The instrumental robots are properly selected due to their functional-bent. To that purport, increasing attention has recently been focussed on parallel kinematics manipulator, due to their ability of high accuracy path tracking, with open loop command. The spreading out of the technology founds great hindrance, because of difficulties in solving the forward kinematics. In the paper, a procedure leading to an effective choice of a typical 6 d.o.f. parallel manipulator is recalled, giving the basic reference, based on standard CAD tools, for assessing the work-space cinematic and dynamic performance.

The analysis is grounded on numerical computations, where the Jacobian matrices are properly used; these relate a six dimensional input vector to a six dimensional output vector, making possible the full work-space analysis in terms of generalised force, either, generalised displacement transmission matrices. For practical purposes, the investigation is better tackled, splitting the analysis as previously recalled; in fact, actual duties suitably distinguish force from torque tasks (and linear from angular displacement tasks). This makes possible very effective restitution into tree-dimensions, suggesting the setting of the open-loop gains according to functional-driven requests.

The deep-sea development resulted in a very effective setting; the special purpose robot, based on the chosen Stewart-like 6 d.o.f. parallel platform, makes possible to move heavy objects with high accelerations and accuracy, achieving the expected performance with a simple and rugged device, not requiring sophisticated sensors and feed-backs. According to this study, the peculiar problem of plugging-in electrovalves on submarine control panels is efficiently solved using a task oriented parallel architecture, able to exert due force to execute the insertion and extraction operations.

Summing up, the proposal did resort to a comparatively unconventional architecture; to assess the benefits, virtual prototyping and testing [2] are used to look after the operation reliability of the device, by duty-driven assignments, to lower development investment and rise life-cycle effectiveness. In the future, the replacing of divers will become appropriate technology, whether the anthropocentric view is dropped and rigs suitably designed following engineering rules are procured. This means to accomplish feasibility studies, to compare competing solutions, to acknowledge on-process performance, to test recovery sequences, etc. before starting the manufacturing of any special artefacts; and "virtual reality" testing is basic tool to this scope.

REFERENCES

[1]   Acaccia, G.M., Callegari, M., Michelini, R.C., Molfino, R.M., Razzoli, R.P.: "Underwater robotics: example survey and suggestions for effective devices", 4th. ECPD Intl. Conf. Advanced Robotics, Intelligent Automation & Active Systems, Moscow, Aug. 24-26, 1998, pp. 409-416.

[2]   Acaccia, G.M., Cavallo, E., Garofalo, E., Michelini, R.C., Molfino, R.M., Callegari, M.: "Remote manipulator for deep-see operations: animation and virtual reality assessment", in Proc. 2nd Workshop on Harbour, Maritime & Logistics Modelling and Simulation (HMS99), 16-18 September 1999, Genova, Italy, pp.57-62, ISBN 1 56555 175 3.

[3]   Batlle, J., Fuertes, J.M., Martì, J., Pacheco, Ll. and Meléndez, J.: "Telemanipulated arms for underwater applications", Proceedings, Seventh Intl. Conf. on Advanced Robotics ICAR '95, Sant Feliu de Guixols, Catalonia, Spain, Vol. 1, pp. 267-272, 1995.

[4]   Bruzzone, L.E., Cavallo, E.M., Michelini, R.C., Molfino, R.M,. Razzoli, R.P.: "The design of a robotic equipment for deep sea maintenance operation", Proc. ASME Engineering Systems Design and Analysis Conference 2000, Montreaux, Switzerland, July 10-13, 2000.

[5]   Callegari, M., Cavallo, E., Garofalo, E., Michelini, R.C., Molfino, R.M., Razzoli, R.P.: "The design of diving robots: set-up assessment by virtual mock-ups", Proc. of the XI ADM Intl. Conf. on Design Tools and Methods in Industrial Engineering, Palermo 8-12 Dic. 1999, vol. C - pp. 147-154.

[6]   Chiu, S.: "Kinematic characterization of manipulators: an approach to defining optimality", IEEE Int. Conf. Robotics and Automation (1988), pp. 828-833

[7]   Kosuge, K., Okuda, M., Fukida, T., Koduka, T., Mizuno, T.: "Input/output force analysis of Stewart platform type of manipulators", Proc. Of 1993 IEEE/RSJ Intl. Conf. On Intelligent Robots and Systems, Yokohama, Japan, July 26-30, 1993, pp. 1666-1673.

[8]   Ma, O., Angeles, J.: "Architecture singularities of platform manipulators", Proceedings of the 1991 IEEE Intl. Conf. On Robotics and Automation, Sacramento, California, pp. 1542-1547.

[9]   Maddalena, D., Prendin, W., and Terribile, A.: "Underwater Telerobotics reaches the field", Proceedings, 27th ISIR Robotics towards 2000, Milan, Italy, pp. 167-172, 1996.

[10]   Michelini, R.C., Molfino, R.M., Razzoli, R.P., Rio, A., Truffelli, F.: "A parallel kinematics robotic arm for deep sea operations", Proc. Intl. Workshop on Parallel Kinematics Machines (PKM99), Milano, Italy, Nov. 30th, 1999, pp. 243-250.

[11]   Sayers, C.P., Yoerger, D.R., Paul, R.P., Lisievich, J.S.: "A manipulator work package for teleoperation from unmanned untethered vehicles: current feasibility and future applications", Proceedings, 6th IARP Workshop on Underwater Robotics, Toulon-La Seyne, France, 1996.

Assisted mails sorting and forwarding stands: performance analysis and ergonomic assessment

G.M. Acaccia*, M. Corvi°, W. Ferebauer*, R.C. Michelini*
*) *Industrial Robot Design Research Group - University of Genova - Italy*
°) *ELSAG SpA - Divisione Postale - Genova - Italy*

Abstract: Industrial automation to-day covers broader areas, as robotic equipment provides technically sound solutions to remove on-process workers once task-cycle are properly assessed. Return on investment, however, remains sometimes open point and this is central question when competition is particularly strong. The postal services face all over the world drastic changes due to new means (information and communication technology) and business options (removal of governmental restriction on utilities), with enhanced attention on effectiveness by lean work-organisation, namely, mixed settings, where human operators are primary resource within sophisticated automatic processing lines. This is not without drawbacks, with implications on the plant ergonomics, and throughout productivity checks need be ran for actual operation lay-outs. The paper tackles such problems, exploiting simulation tools to accomplish beforehand analyses by virtual testing, in terms of resource allocation, process planning and labour protection.

Key words: Integrated design, Economics & social impact and justification

1. INTRODUCTION

The exponential growth of information and communication technologies has drastically modified traditional activities, such as the mailing services, that happened to be consolidated as governmental branches or protected utilities. Now, the world-wide web supports an open space for information transmission and share, making unsound sheltered postal services, disjoined from duty-driven competitiveness, flexible specialisation and business integration. The full re-thinking of conventional mailing organisations goes through the identification of the main mission, by setting out the core business, up to the development of processing machines and related work-cycle plans consistent with scope-oriented enterprises. These new horizons led to the special branch of *postal automation*, with relevant issues on facilities and techniques, to enhance the productivity with careful check of the economical return.

This branch quite soon represented a challenging domain, aiming, on one side, at *intelligent* automation set-ups (handwriting automatic reading units, etc.), the other side, at *men-machine* sustainable accountability (mixed human- and robot-attended work-lines, etc.). Technology-driven goals, thereafter, drastically modify postal organisations, with special requests about efficiency, as National borders do not turn down any more outsiders and world-wide-web alternatives can cover a large extent of communication needs; struggle to survive would follow on condition to carefully balance technology innovation against investment leanness. The choice of effective set-ups requires the accurate foresight of the life-cycle performance for actual work frames; a goal efficiently achieved by proper resort to CAD-based digital mock-up and virtual reality testing. In the paper, these aspects of the design and development process are summarised, looking after an example case.

Figure 1. Typical in-feed magazine

2. AN EXAMPLE MAIL PROCESSING STAND

Universal postal services face challenging tasks to compete within sectional classes (as for handled objects, conveying means, receiving points, etc.), but benefit from good advance with the processing of "flat" items of varying size and weigh, *Fig. 1*, presenting widely scattered addressees. Thus, restricting the present analysis to the generic "flats" class, an effective set-up would typically include automatic and manually operated units, to join high flexibility and productivity. The outfit should be tailored to the identified duty requirements and suitably built as a modular set-up, easily down- or up-graded as the case arises. Basically, we might distinguish:

- an input section, with items supplying stations and a set of automatic either manual in-feed modules;
- a processing section, with items recognition and coding modules, followed by dispatching and sorting equipment;
- an output section, with items routing and piling up into labelled assemblies.

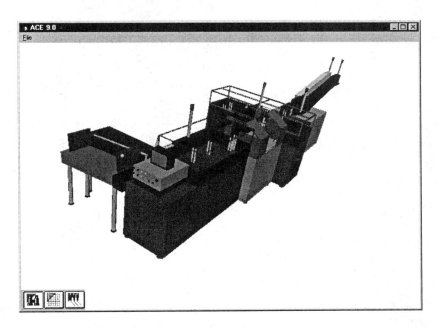

Figure 2. Automatic in-feed station (provided by the simulator)

In this example: - the automatic in-feed is done by a high-speed flat-feeder, HSFF, *Fig. 2*, via the in-feed magazine, the pick-off device and the double control-unit; - the processing is performed by an image lifter module, ILM, via the size measuring-unit, the bar code reader and the optical character recognition unit, and a the sorting and dispatching equipment, via the flat delay line, FDL, the label application module, LAM, the indexing module, IM and the sorter interface module, SIM; - the output delivery, finally, leads to the items boxing, via the diverting/forwarding devices, to the chosen tray/bin/bag out-feed units, or to the local reject module. The generic facility builds up as *compact flat sorter module*, CFSM, with main duties, *Fig. 3*; played by the in-feed buffer/sorter, IBS, the optical character recognition, OCR, device and the transport/singling system.

Figure 3. Example output delivery station

The operators are critically engaged at the sorting stands, where manual/automatic in-feed is scheduled, at the equipment up-keeping, for jam removal, and at the output section, to help forwarding the selected piled-up items. For the present study, focus is on the feeding stands, where, basically, three operator types distinguish: - *ot*, assigned to transfer the input trays; - *op*, involved at superintending the automatic portage; - *om*, employed for the manual feeding. The individual work-cycles are properly characterised as for timing and required energy, *Fig. 4*, with staffing depending on work-loads. The inclusion of on-process men provides simple supplying means for the automatic tracks, with the enhanced versatility of enabling manual by-passes when out of standard items shall be processed.

2.1   The Operation Model and Task Scheduling

To assess the proper outfits, alternative lay-outs have to be compared for actual working conditions. Resort to virtual prototyping and simulation runs makes these checks possible for wide operation spectra; to that goal, the AutoMod environment (by AutoSimulations Inc.) is used, due to the related graphic editor, to sketch animated plant units (machines, operators, transfer units, etc.) and the governing blocks, to enable the tasks evolution with recording of actual achievements. The 3D models assure a realistic view of the work-places and make easy to acknowledge the ergonomic figures of different duty sequences (as for operators' postures or efforts) and surroundings settings (as for noise, temperature, light, etc.), according to (off-line specified) job-allotment paradigms.

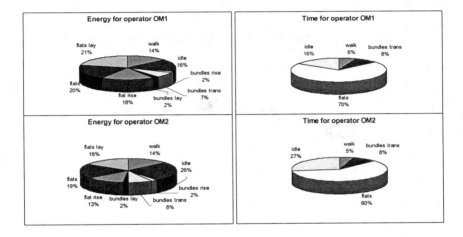

Figure 4. Example duty assignment (of the **om** operators)

Looking at the functional specification of a sorting and forwarding stand, the working out of the *model* implies to specify, and to encode as 'objects' (with properties and methods), all the physical and logic resources, relevant for describing the work-flow. Three object classes distinguish:

— machines: automatic in-feed module, manual in-feed module, sorter module, etc.;

— operators: *ot* (red), *op* (green), *om* (yellow);

— processed material: mix of flat items and related portage out-fits;

then a lay-out establishes, *Fig. 5*, with, e.g., two manual and three automatic in-feed units, the sorter module, the proper distribution of operators and the set of auxiliary facilities (trolleys with full/empty containers, tables with temporary mails stacks, etc.).

The generic feeding duty needs consider:

— the set of shuttles, carts or trolleys, each one carrying several containers (trays), with inside loose or bundled *flats*;

— the containers, with various mixes of loose *flats* or *bundles* of known amount of flats assorted by size;

- the bundles, with series of journals or magazines, large wrappers, standard envelopes, etc.;
- the (out-of-standard) flats, loose or different thickness/length/height items.

The *ot* operator moves one tray at a time from the cart to a table, leaves the contents there and brings the empty container back; from their buffer table, the *op* operators supply an automatic in-feed unit and look after jams removal, as case arises; similarly, from their buffer table, the *om* operators recognise the flats, encode the address and manually in-feed the items toward the sorting line.

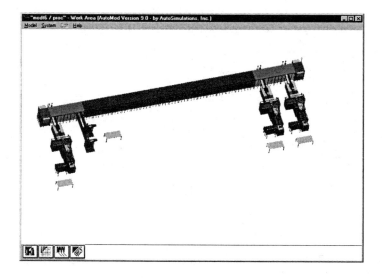

Figure 5. Virtual lay-out of the simulated plant

2.2   The Reference Ergonomic Requirements

The ergonomic checks deals with the relations: - humans vs. scheduled activity; - humans vs. machines; - humans vs. processed materials; - humans vs. surroundings. Proper choices are fixed by standards, such as, e.g., the ones, in the USA, of NIOSH, *National Institute for Occupational Safety and Health*. The physical activity is weighed by a *labour index*, ratio of the handled load and a 'recommended weight limit'; this is analysed into a series of factors, say: the actual weight, the vertical and horizontal location of the operator's hands by respect of his feet, the load displacement, the path off-set out of front manoeuvres, the operation frequency and the type of grasping device. The factors are ranged into classes and modified by figures, as for duty severity; this labour index should not exceed a given threshold; when an activity splits into several subtasks, each needs be weighed and shall not exceed the threshold.

The analysis leads to evaluate the '*ergonomic stress index*', ESI, once fulfilled the *scientific* job-allotment description. Thereafter, the overall assigned duty-cycle is assessed in terms of metabolic energy expenditure, distinguishing five levels, from rest to very severe conditions and computing final estimates, with account of different operators and surroundings. During the design steps, the estimates combine the subtasks, to obtain the duty-cycle average figure. Experimental tests shall, generally, be ran on real set-ups, measuring the operators' oxygen consumption into actual work conditions, with: *partial* checks, covering the activity periods only, for light duties; or *integral* checks, covering as well the rest periods, for heavy duties.

The man/machine relations ought be verified as for the passive and active protection rules and as for the psycho-physic stress levels. The re-design of the work set-ups makes possible to rise the effectiveness, suggested by the habits of trained operators. The preparation of the material to be handled has relevant impacts: too huge or too heavy trays shell be avoided; for assorted items, the pre-processing of bundles mix enhances efficiency. Finally, the work-place benefits from properly fit lay-outs: for the in/out path, the light direction, the attention preservation, etc.; further requirements deals with health and safety prescriptions.

3. MODELLING AND SIMULATION CHECKS

The choice of the mail processing fixtures requires checks on the technical soundness and on the economical effectiveness; the former looks at ergonomic settings, the latter aims at fully exploiting the enabled resources. These checks start at the early ideation phases and need be accomplished for actual running conditions; resort to digital mock-ups makes easy to verify the achieved performance for the pertinent operation environments. By virtual testing, different lay-outs, obtained combining the basic modules and alternative auxiliary teams, are compared: during the design steps, to find out the technically optimised solution; during the operation steps, to balance the subsidiary man-made jobs within the work-flow.

The simulation snapshots aim at the realistic restitution of the surroundings, at least, so forth as items fetch, handling and in-feed are considered. The AutoMod (by AutoSimulations Inc.) supplies comparatively simple 3D displays of the basic modules, such as, for instance, the in-feed station, *Fig. 6*, the sorter, *Fig. 7*, etc., so that the finally assembled work space can be tested with actual running operation conditions. The description of a typical arrangement, *Fig. 5*, leads to the selected modules, with the related input queues (generated by the arrivals of items to be processed) and the assigned operators, with the subsidiary carts and tables; at the interface,

physical and logical entities distinguish:
- the former details the current *loads*, e.g.: carrying-trays (with related attributes: transfer tasks, bundles identifiers, flats sizes and mixes, etc.); jam-occurrences (an *initialiser* randomly affects the individual items, to generate the drawback); jam-localisers (red/yellow lights show where jams occur); reset-flags (to identify actual lay-outs); cart-flags (to specify the status); *ot-*, *op-*, *om-* operators-labels (with attached on-going actions, say: transfer tasks from/to carts, tables, etc.; ancillary tasks for in-feed, jam removal, etc.; manual tasks for fetching and feeding, etc.); process-flags (to rule items dispatch and sort flow, included jam generation occurrences);
- the latter deals with the task inventory and data transfer, e.g.: *loads* lists (for trays: cart/table fetch/delivery queues; for bundles: automatic/manual in-feed queues; for flats: hierarchy for automatic/manual entry and wait queues; for jams: delay to removal actions; for operators: delays before subsequent tasks); state variables (the occurrence-driven processes use *loads* to look into the scheduling list and reach the *end-of-task* when all actions are fulfilled).

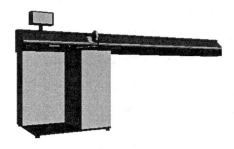

Figure 6. In-feed station for the virtual testing

3.1   The Govern Logic and Simulation Setting

The AutoMod code, further to realistic restitution capabilities, provides effective means to govern the facility time behaviour, combining queues and branches rules with functions and routines algorithms. The acknowledged *loads* have, actually, embedded attributes, so that task progression is directed by the chosen schedules; the simulation, however, profits by referring to *loads* managers, supplying explicit specification of the process govern. Few hints follow.

Several operators are concurrently engaged; their work-cycle is analysed to single out run-chains and cross-links and *loads* managers are encoded to synchronise the actions by message-passing queues: every operator has own jobs agenda, *loads* queue and wait lists. The govern unit of the *ot*-operator, for instance, deals with the work-cycle: - to pick up a full tray from a cart; - to bring and place the tray on a table; - to empty the tray; - to pick up the empty tray; - to bring back the tray to a cart. The cycle repeats until all trays are processed. The task starts by means of two routines: cart assignment and table search; during the execution, the *loads* queue can modify the mail run-chain or establish new cross-links; the cart search routine ends the cycle and is re-ran once a cart is completed. Similar analyses establish for every operator, looking at proper routines (for instance: for the *op-* , jam overseeing and removal; for the *om-* , flats singling out, reading and coding) and enabling the check and branch assignments.

Figure 7. Sorter station for the virtual testing

The *loads* managers associated with *ot-*, *op-*, *om-* operators are re-set depending on resource allocation (number of men vs. carts, tables and stations) and efficiency patterns (superposed waviness and unexpected occurrences). Two further *loads* managers specify the govern rules of the automatic in-feed fixtures and, respectively, of the cart dispatching service, by means of the appropriate queues and wait lists and the related run-chains and cross-links.

3.2 Example Simulation Studies and Assessments Issues

Example developments are shortly summarised in the following for explanatory purpose. By simulation, the designer aims at verifying the technical appropriateness of the chosen sets of fixtures, the suitableness of given lay-outs and subsidiary resources, the plant effectiveness for the expected work-loads and the actually needed personnel depending on the throughput and on ergonomic figures. The study shall repeat in function of the facility characteristics, the work conditions, the operators' number and duties, etc. assessing, for given schedules, plant performance, resources' exploitation ratios, personnel stress and fatigue levels or other data on the reached issues. The digital mock-up is simply required to acknowledge the ergonomic features, to detect critical situations and to provide hints for better work conditions and plant efficiency, and AutoMod looks to be an adequate aid to that purpose, with effective restitution of the main functional and structural properties of the included resources, *Fig. 8.*

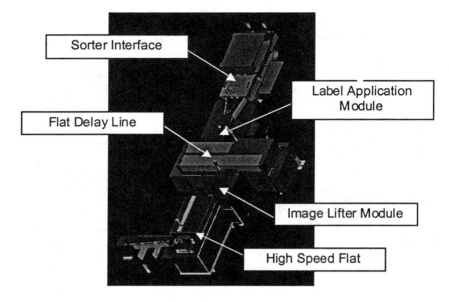

Figure 8. Functional and structural details of the simulated set-up

The generic simulation shall cover different work-shifts, each time with the resources (men and facilities) allocation depending on the *loads* agenda (e.g.: shifts, with high volume *standard* items; shifts, with low volumes *special*, as for size or weigh, items; etc.). The selected software does not offer a graphic page for re-defining each time the current lay-out; the drawback is removed by specifying a maximal configuration and declaring the sub-set of resources timely not included. Thereafter, at the input section two areas distinguish, for the automatic, either the manual in-feed; each area receives a set of carts (bearing five shelves with up to four trays each) and has a series of tables where the mail is temporarily placed. The *ot*-operators work between carts and tables; the *op*-operators feed the automatic stations (at the average rate: 0.5-0.6 s/flat); and the *om*-operators, the manual ones (at the average rate: 1.7-2 s/flat, including the reading/coding actions). The maximal configuration includes: up to three automatic and up to two manual in-feed stations; up to four *ot-* , three *op-* and two *om*-operators; the sorter unit has, therefore, to deal with that maximum throughput during the peak shifts.

To identify the operation characteristics, let refer to the main features, say: work-schedules, items yields and men commitment. For the scheduling, input data (single-index array) are the sets of the arrived full trays (and returned empty ones). The processed *loads* are described by double-index arrays: the first index, defining the affected operator; the second, the handled item (standard mail, packed bundle, special flat, etc.); each array is up-dated, as a new item is withdrawn from a tray and the pointer acknowledges the incumbency. This comes out from the duty analysis, namely, by assessing the time for trays handling, for items transportation, for mails manipulation, for flats in-feed, for jams removal, etc. including auxiliary actions (e. g., walking, taking position, and the likes).

The simulation data setting has to deal with:

– the agendas: lists of full trolleys (one for each of the enabled processing areas) and of related full trays, up-dated every time a new arrival is prompted;
– the operators' duties: amount of full/empty trays (in each area) on the trolleys/tables, up-dated every time a cycle is ended; amount of bundles/items ready for in-feed operations;
– the throughput: overall flats batches and mixes, given by two (one for the automatic, one for the manual in-feed) double-index arrays, to specify processing unit and items' type;
– the disturbances: jam occurrences, double-index array, to specify the line and the location;
– the labour indices: lists of actual actions (to stand by, walk, pick-up/carry/lay-down a tray, take-out/lift/handle/place an item, acknowledge/encode a flat, recognise/remove/re-set a jam) and accounts of weighed figures.

Along with the daily work-shifts, the requests change, from early large amounts (e.g., due to the night mail piling-up), to variable steady conditions (e.g., depending on yearly seasons). The suitable mail sorter should grant the largest throughput and have sufficient buffering capacity in-between the in-feed units so to avoid slowdown at bundles batch processing. The analysis aims at verifying the actual distribution of the operators during peak shifts and at assessing the manpower balance for typical steady runs, in terms of (average) duty-cycle times and of (standard) energy consumption and in front of the sought throughputs. The checks provide detailed views of the requested times and energies, showing whether critical situations arise or when proper balancing would improve the productivity; the results help the planning out of work-shifts on a day- or on a week-base, once the mails arrival is accordingly forecast. A set of example results will be orally presented and discussed, for explanatory purposes. Different processing strategies and resource allocation policies are compared to give evidence of situations that might approach critical issues.

4.   CONCLUSIONS

The life-cycle design of customer-driven artefacts or user-oriented fixtures is becoming winning skill, and resort to digital mock-up and computer simulation are effective means to immediately give factual assessments of the achieved performances. The paper is directly concerned by the ergonomic constraints of processing stands where operators and automatic machines are each other interfaced. The over-all effectiveness is highly affected by the agendas organisation and by the lay-out reflected conditions; the AutoMod models help the reaching of straightforward pictures for the on-progress plant behaviour and the simulation campaign readily supply the data to evaluate benefits and drawbacks of alternative set-ups, with direct measurement of the expected return on investment for any given mails sorting and forwarding stand. The tool deserves increasing relevance in the domain of assisted mailing machines, as competitiveness spreads on world-wide contexts and specialised services arise segmenting a market already deeply modified by the emerging information and communication technologies.

The paper first gives a bird's eye view of the assisted mails sorting and forwarding stands, based on integrated set-ups with operators, in-feed modules and mail sorting machines. The attention is turned to the mass-handling of 'flat items' mixes with multiple-routing request, this being typical duty for universal postal utilities. The lay-out is obtained combining basic modules; the CAD tool is exploited, at the ideation stage, for the choice of the modules and the built up of the appropriate fit-outs, translating the all the all into an AutoMod code. The reference work-patterns requirements and related ergonomic prescriptions are, furthermore, considered to verify the effectiveness among competing set-ups, in terms of productivity and, both, to give evidence that the actions, actually selected for the duties fulfilment, lay within the labour protection acts, issued by the different national rules. These kind of tests is, on one side, required by the trade unions with due account of the social impact ranges; it is, the other side, critical reference to evaluate the return on investments.

The study, finally, refers to an example explanatory case, to compare competing lay-outs, accomplishing "flats" handling, sorting and dispatching operations, once the reference input mixes and output forwarding are given. The set-up includes manual and automatic stations and the simulation covers different options as for schedules, task engagement and work cycles. The issues give evidence of possible inconsistencies and hints on how to modify the lay-out to improve the productivity, while keeping optimal work-conditions.

INDEX OF CONTRIBUTORS